BEYOND PERTURBATION
INTRODUCTION TO THE HOMOTOPY ANALYSIS METHOD

CRC Series: Modern Mechanics and Mathematics

Series Editors: David Gao and Ray W. Ogden

PUBLISHED TITLES

Beyond Perturbation: Introduction to The Homotopy Analysis Method
by Shijun Liao

Mechanics of Elastic Composites
by Nicolaie Dan Cristescu, Eduard-Marius Craciun, and Eugen Soós

FORTHCOMING TITLES

Continuum Mechanics and Plasticity
by Han-Chin Wu

Hybrid Incompatible Finite Element Methods
by Theodore H.H. Pian, Chang-Chun Wu

Microstructural Randomness in Mechanics of Materials
by Martin Ostroja Starzewski

BEYOND PERTURBATION
INTRODUCTION TO THE HOMOTOPY ANALYSIS METHOD

Shijun Liao

CHAPMAN & HALL/CRC

A CRC Press Company
Boca Raton   London   New York   Washington, D.C.

Library of Congress Cataloging-in-Publication Data

Liao, Shijun.
  Beyond perturbation : introduction to homotopy analysis method / Shijun Liao.
    p. cm. — (Modern mechanics and mathematics ; 2)
  Includes bibliographical references and index.
  ISBN 1-58488-407-X (alk. paper)
  1. Homotopy theory. 2. Mathematical analysis. I. Title. II. Series.

  QA612.7.L57 203
  514'.24—dc22   2003055776

This book contains information obtained from authentic and highly regarded sources. Reprinted material is quoted with permission, and sources are indicated. A wide variety of references are listed. Reasonable efforts have been made to publish reliable data and information, but the author and the publisher cannot assume responsibility for the validity of all materials or for the consequences of their use.

Neither this book nor any part may be reproduced or transmitted in any form or by any means, electronic or mechanical, including photocopying, microfilming, and recording, or by any information storage or retrieval system, without prior permission in writing from the publisher.

The consent of CRC Press LLC does not extend to copying for general distribution, for promotion, for creating new works, or for resale. Specific permission must be obtained in writing from CRC Press LLC for such copying.

Direct all inquiries to CRC Press LLC, 2000 N.W. Corporate Blvd., Boca Raton, Florida 33431.

Trademark Notice: Product or corporate names may be trademarks or registered trademarks, and are used only for identification and explanation, without intent to infringe.

Visit the CRC Press Web site at www.crcpress.com

© 2004 by CRC Press LLC

No claim to original U.S. Government works
International Standard Book Number 1-58488-407-X
Library of Congress Card Number 2003055776
Printed in the United States of America 1 2 3 4 5 6 7 8 9 0
Printed on acid-free paper

To my wife, Shi Liu

Preface

In general it is difficult to obtain analytic approximations of nonlinear problems with strong nonlinearity. Traditionally, solution expressions of a nonlinear problem are mainly determined by the type of nonlinear equations and the employed analytic techniques, and the convergence regions of solution series are strongly dependent of physical parameters. It is well known that analytic approximations of nonlinear problems often break down as nonlinearity becomes strong and perturbation approximations are valid only for nonlinear problems with weak nonlinearity.

In this book we introduce an analytic method for nonlinear problems in general, namely the homotopy analysis method. We show that, even if a nonlinear problem has a unique solution, there may exist an *infinite* number of different solution expressions whose convergence region and rate are dependent on an auxiliary parameter. Unlike all previous analytic techniques, the homotopy analysis method provides us with a simple way to *control* and *adjust* the convergence region and rate of solution series of nonlinear problems. Thus, this method is valid for nonlinear problems with strong nonlinearity. Moreover, unlike all previous analytic techniques, the homotopy analysis method provides great freedom to use different base functions to express solutions of a nonlinear problem so that one can approximate a nonlinear problem more efficiently by means of better base functions. Furthermore, the homotopy analysis method logically contains some previous techniques such as Adomian's decomposition method, Lyapunov's artificial small parameter method, and the δ-expansion method. Thus, it can be regarded as a unified or generalized theory of these previous methods.

The book consists of two parts. Part I (Chapter 1 to Chapter 5) deals with the basic ideas of the homotopy analysis method. In Chapter 2, the homotopy analysis method is introduced by means of a rather simple nonlinear problem. The reader is strongly advised to read this chapter first. In Chapter 3, a systematic description is given and a convergence theorem is described for general cases. In Chapter 4 we show that Lyapunov's artificial small parameter method, the δ-expansion method, and Adomian's decomposition method are simply special cases of the homotopy analysis method. In Chapter 5 the advantages and limitations of the homotopy analysis method are briefly discussed and some open questions are pointed out. In Part II (Chapter 6 to Chapter 18), the homotopy analysis method is applied to solve some nonlinear problems, such as simple bifurcations of a nonlinear boundary-value problem (Chapter 6), multiple solutions of a nonlinear boundary-value problem (Chapter 7), eigenvalue and eigenfunction of a nonlinear boundary-value problem (Chapter 8), the Thomas-Fermi atom model (Chapter 9), Volterra's population model (Chapter 10), free oscillations of conservative systems with odd nonlinearity (Chapter 11), free oscillations of conservative systems with quadratic nonlinearity (Chapter 12), limit cycle in a multidimensional system (Chapter 13), Blasius' viscous flow (Chapter 14), boundary-layer flows with exponential property (Chapter 15), boundary-layer flows with algebraic property (Chapter 16), Von Kármán swirling viscous flow (Chapter 17), and nonlinear progressive waves in deep water (Chapter 18). In Part II, only Chapters 14, 15, and 18 are adapted from published articles of the author.

I would like to express my sincere thanks to Professor P. Hagedorn (Darmstadt University of Technology, Germany) and Professor Y.Z. Liu (Shanghai Jiao Tong University, China) for reading Part I of the manuscript and giving their valuable comments. Thanks to Robert B. Stern, Jamie B. Sigal, and Amy Rodriguez (CRC Press) for their editorial help as well as Nishith Arora for assistance on LaTeX. I would like to express my sincere acknowledgement to Professor J.M. Zhu and Professor Y.S. He (Shanghai Jiao Tong University, China), Professor Chiang C. Mei (Department of Civil and Environmental Engineering, Massachusetts Institute of Technology, Cambridge, MA) and Professor D.Y. Hsieh (Division of Applied Mathematics, Brown University, Providence, RI) for their continuous encouragement over the years. Thanks to my co-authors of some articles, Professor Antonio Campo (College of Engineering, Idaho State University); Professor Kwok F. Cheung (Department of Ocean and Resources Engineering, University of Hawaii at Monoa); Professor Allen T. Chwang (Department of Mechanical Engineering, Hong Kong University, Hong Kong, China); and Professor Ioan Pop (Faculty of Mathematics, University of Cluj, Romania), for their cooperation and valuable discussions. This work is partly supported by National Natural Science Fund for Distinguished Young Scholars of China (Approval No. 50125923), Li Ka Shing Foundation (Cheung Kong Scholars Programme), Ministry of Education of China, Shanghai Jiao Tong University, and German Academic Exchange Service (DAAD, Sandwich Programme).

Finally, I would like to express my pure-hearted thanks to my wife for her love, understanding, and encouragement.

Contents

PART I Basic Ideas		**1**
1	**Introduction**	**3**
2	**Illustrative description**	**9**
	2.1 An illustrative example	9
	2.2 Solution given by some previous analytic techniques	10
	2.2.1 Perturbation method	10
	2.2.2 Lyapunov's artificial small parameter method	11
	2.2.3 Adomian's decomposition method	11
	2.2.4 The δ-expansion method	12
	2.3 Homotopy analysis solution	14
	2.3.1 Zero-order deformation equation	14
	2.3.2 High-order deformation equation	16
	2.3.3 Convergence theorem	18
	2.3.4 Some fundamental rules	19
	2.3.5 Solution expressions	21
	2.3.6 The role of the auxiliary parameter $\hbar$	31
	2.3.7 Homotopy-Padé method	38
3	**Systematic description**	**53**
	3.1 Zero-order deformation equation	53
	3.2 High-order deformation equation	55
	3.3 Convergence theorem	57
	3.4 Fundamental rules	60
	3.5 Control of convergence region and rate	62
	3.5.1 The $\hbar$-curve and the valid region of $\hbar$	63
	3.5.2 Homotopy-Padé technique	64
	3.6 Further generalization	66
4	**Relations to some previous analytic methods**	**69**
	4.1 Relation to Adomian's decomposition method	69
	4.2 Relation to artificial small parameter method	73
	4.3 Relation to δ-expansion method	75
	4.4 Unification of nonperturbation methods	78

5	**Advantages, limitations, and open questions**	**79**
	5.1 Advantages	79
	5.2 Limitations	80
	5.3 Open questions	81

PART II Applications   83

6	**Simple bifurcation of a nonlinear problem**	**85**
	6.1 Homotopy analysis solution	86
	6.1.1 Zero-order deformation equation	86
	6.1.2 High-order deformation equation	88
	6.1.3 Convergence theorem	89
	6.2 Result analysis	90
7	**Multiple solutions of a nonlinear problem**	**99**
	7.1 Homotopy analysis solution	100
	7.1.1 Zero-order deformation equation	100
	7.1.2 High-order deformation equation	101
	7.1.3 Convergence theorem	103
	7.2 Result analysis	104
8	**Nonlinear eigenvalue problem**	**115**
	8.1 Homotopy analysis solution	116
	8.1.1 Zero-order deformation equation	116
	8.1.2 High-order deformation equation	117
	8.1.3 Convergence theorem	120
	8.2 Result analysis	121
9	**Thomas-Fermi atom model**	**133**
	9.1 Homotopy analysis solution	133
	9.1.1 Asymptotic property	133
	9.1.2 Zero-order deformation equation	134
	9.1.3 High-order deformation equations	136
	9.1.4 Recursive expressions	137
	9.1.5 Convergence theorem	138
	9.2 Result analysis	140
10	**Volterra's population model**	**149**
	10.1 Homotopy analysis solution	149
	10.1.1 Zero-order deformation equation	149
	10.1.2 High-order deformation equation	152
	10.1.3 Recursive expression	154
	10.1.4 Convergence theorem	155
	10.2 Result analysis	156
	10.2.1 Choosing a plain initial approximation	156

Table of Contents xi

    10.2.2 Choosing the best initial approximation 157

11 Free oscillation systems with odd nonlinearity   165
  11.1 Homotopy analysis solution 165
      11.1.1 Zero-order deformation equation 165
      11.1.2 High-order deformation equation 167
  11.2 Illustrative examples 170
      11.2.1 Example 11.2.1 170
      11.2.2 Example 11.2.2 171
      11.2.3 Example 11.2.3 172
  11.3 The control of convergence region 173

12 Free oscillation systems with quadratic nonlinearity   179
  12.1 Homotopy analysis solution 179
      12.1.1 Zero-order deformation equation 179
      12.1.2 High-order deformation equation 182
  12.2 Illustrative examples 185
      12.2.1 Example 12.2.1 185
      12.2.2 Example 12.2.2 187

13 Limit cycle in a multidimensional system   197
  13.1 Homotopy analysis solution 198
      13.1.1 Zero-order deformation equation 198
      13.1.2 High-order deformation equation 201
      13.1.3 Convergence theorem 204
  13.2 Result analysis 205

14 Blasius' viscous flow   217
  14.1 Solution expressed by power functions 218
      14.1.1 Zero-order deformation equation 218
      14.1.2 High-order deformation equation 219
      14.1.3 Convergence theorem 220
      14.1.4 Result analysis 221
  14.2 Solution expressed by exponentials and polynomials 223
      14.2.1 Asymptotic property 223
      14.2.2 Zero-order deformation equation 224
      14.2.3 High-order deformation equation 225
      14.2.4 Recursive expressions 225
      14.2.5 Convergence theorem 227
      14.2.6 Result analysis 228

15 Boundary-layer flows with exponential property   239
  15.1 Homotopy analysis solution 241
      15.1.1 Zero-order deformation equation 241
      15.1.2 High-order deformation equation 242

15.1.3 Recursive formulae 243
15.1.4 Convergence theorem 246
15.2 Result analysis 247

16 Boundary-layer flows with algebraic property   257
16.1 Homotopy analysis solution 257
    16.1.1 Asymptotic property 257
    16.1.2 Zero-order deformation equation 258
    16.1.3 High-order deformation equation 260
    16.1.4 Recursive formulations 262
    16.1.5 Convergence theorem 263
16.2 Result analysis 265

17 Von Kármán swirling viscous flow   275
17.1 Homotopy analysis solution 276
    17.1.1 Zero-order deformation equation 277
    17.1.2 High-order deformation equation 280
    17.1.3 Convergence theorem 283
17.2 Result analysis 285

18 Nonlinear progressive waves in deep water   295
18.1 Homotopy analysis solution 296
    18.1.1 Zero-order deformation equation 296
    18.1.2 High-order deformation equation 299
18.2 Result analysis 304

Bibliography   311

Index   321

PART I

BASIC IDEAS

The way that can be spoken of is not the constant way;
The name that can be named is not the constant name.

Lao Tzu, an ancient Chinese philosopher

1

Introduction

Most phenomena in our world are essentially nonlinear and are described by nonlinear equations. Since the appearance of high-performance digit computers, it becomes easier and easier to solve a linear problem. However, generally speaking, it is still difficult to obtain accurate solutions of nonlinear problems. In particular, it is often more difficult to get an analytic approximation than a numerical one of a given nonlinear problem, although we now have high-performance supercomputers and some high-quality symbolic computation software such as Mathematica, Maple, and so on. The numerical techniques generally can be applied to nonlinear problems in complicated computation domain; this is an obvious advantage of numerical methods over analytic ones that often handle nonlinear problems in simple domains. However, numerical methods give discontinuous points of a curve and thus it is often costly and time consuming to get a complete curve of results. Besides, from numerical results, it is hard to have a whole and essential understanding of a nonlinear problem. Numerical difficulties additionally appear if a nonlinear problem contains singularities or has multiple solutions. The numerical and analytic methods of nonlinear problems have their own advantages and limitations, and thus it is unnecessary for us to do one thing and neglect another. Generally, one delights in giving analytic solutions of a nonlinear problem.

There are some analytic techniques for nonlinear problems, such as perturbation techniques [1, 2, 3, 4, 5, 6, 7, 8, 9, 10, 11, 12] that are well known and widely applied. By means of perturbation techniques, a lot of important properties and interesting phenomena of nonlinear problems have been revealed. One of the astonishing successes of perturbation techniques is the discovery of the ninth planet in the solar system, found in the vast sky at a predicted point. Recently, the singular perturbation techniques are considered to be one of the top 10 progresses of theoretical and applied mechanics in the 20th century [13]. It is therefore out of question that perturbation techniques play important roles in the development of science and engineering. For further details, the reader is referred to the foregoing textbooks of perturbation methods.

Perturbation techniques are essentially based on the existence of small or large parameters or variables called perturbation quantity. Briefly speaking, perturbation techniques use perturbation quantities to transfer a nonlinear problem into an infinite number of linear sub-problems and then approximate it by the sum of solutions of the first several sub-problems. The existence of perturbation quantities is obviously a cornerstone of perturbation techniques, however, it is the perturbation quantity that brings perturbation techniques some serious restrictions. Firstly, it is impossible that every nonlinear problem contains such a perturbation quantity. This is an obvious restriction of perturbation techniques. Secondly, analytic approximations of nonlinear problems often break down as nonlinearity becomes strong, and thus perturbation approximations are valid only for nonlinear problems with weak nonlinearity. Consider the drag of a sphere in a uniform stream, a classical nonlinear problem in fluid mechanics governed by the famous Navier-Stokes equation, for example. Since 1851 when Stokes [14] first considered this problem, many scientists have attacked it by means of linear theories [15, 16], straightforward perturbation technique [17], and matching perturbation method [18, 19]. However, all these previous theoretical drag formulae agree with experimental data only for small Reynolds number, as shown in Figure 1.1. Thus, as pointed out by White [20], "the idea of using creeping flow to expand into the high Reynolds number region has not been successful". This might be partly due to the fact that perturbation techniques do not provide us with any ways to adjust convergence region and rate of perturbation approximations.

There are a few nonperturbation techniques. The dependence of perturbation techniques on small/large parameters can be avoided by introducing a so-called artificial small parameter. In 1892 Lyapunov [21] considered the equation

$$\frac{dx}{dt} = A(t)\, x,$$

where $A(t)$ is a time periodic matrix. Lyapunov [21] introduced an artificial parameter ϵ to replace this equation with the equation

$$\frac{dx}{dt} = \epsilon\, A(t)\, x$$

and then calculated power series expansions over ϵ for the solutions. In many cases Lyapunov proved that series converge for $\epsilon = 1$, and therefore we can put in the final expression by setting $\epsilon = 1$. The above approach is called Lyapunov's artificial small parameter method [21]. This idea was further employed by Karmishin et al. [22] to propose the so-called δ-expansion method. Karmishin et al. [22] introduced an artificial parameter δ to replace the equation

$$x^5 + x = 1 \tag{1.1}$$

with the equation

$$x^{1+\delta} + x = 1 \tag{1.2}$$

and then calculated power series expansions over δ and finally gained the approximations by converting the series to [3,3] Padé approximants and setting $\delta = 4$. In essence, the δ-expansion method is equivalent to the Lyapunov's artificial small parameter method. Note that both methods introduce an artificial parameter, although it appears in a different place and is denoted by

Introduction

different symbol in a given nonlinear equation. We additionally have great freedom to replace Equation (1.1) by many different equations such as

$$\delta\, x^5 + x = 1. \tag{1.3}$$

As pointed out by Karmishin et al. [22], the approximation given by the above equation is much worse than that given by Equation (1.2). Both the artificial small parameter method and the δ-expansion method obviously need some fundamental rules to determine the place where the artificial parameter ϵ or δ should appear. Like perturbation techniques, both the artificial small parameter method and the δ-expansion method themselves do not provide us with a convenient way to adjust convergence region and rate of approximation series.

Adomian's decomposition method [23, 24, 25] is a powerful analytic technique for strongly nonlinear problems. The basic ideas of Adomian's decomposition method is simply described in §4.1. Adomian's decomposition method is valid for ordinary and partial differential equations, no matter whether they contain small/large parameters, and thus is rather general. Moreover, the Adomian approximation series converge quickly. However, Adomian's decomposition method has some restrictions. Approximates solutions given by Adomian's decomposition method often contain polynomials. In general, convergence regions of power series are small, thus acceleration techniques are often needed to enlarge convergence regions. This is mainly due to the fact that power series is often not an efficient set of base functions to approximate a nonlinear problem, but unfortunately Adomain's decomposition method does not provide us with freedom to use different base functions. Like the artificial small parameter method and the δ-expansion method, Adomian's decomposition method itself also does not provide us with a convenient way to adjust convergence region and rate of approximation solutions.

In summary, neither perturbation techniques nor nonperturbation methods such as the artificial small parameter methods, the δ-expansion method, and Adomian's decomposition method can provide us with a convenient way to *adjust* and *control* convergence region and rate of approximation series. The efficiency to approximate a nonlinear problem has not been taken into enough account, therefore it is necessary to develop some new analytic methods such that they

1. Are valid for strongly nonlinear problems even if a given nonlinear problem does not contain any small/large parameters

2. Provide us with a convenient way to adjust the convergence region and rate of approximation series

3. Provide us with freedom to use different base functions to approximate a nonlinear problem.

A kind of analytic technique, namely the homotopy analysis method [26, 27, 28, 29, 30], was proposed by means of homotopy [31], a fundamental concept of topology [32]. The idea of the homotopy is very simple and straightforward. For example, consider a differential equation

$$\mathcal{A}[u(t)] = 0, \tag{1.4}$$

where $\mathcal{A}$ is a nonlinear operator, t denotes the time, and $u(t)$ is an unknown variable. Let $u_0(t)$ denote an initial approximation of $u(t)$ and $\mathcal{L}$ denote an auxiliary linear operator with the property

$$\mathcal{L}f = 0 \qquad \text{when } f = 0. \tag{1.5}$$

We then construct the so-called homotopy

$$\mathcal{H}[\phi(t;q);q] = (1-q)\,\mathcal{L}\,[\phi(t;q) - u_0(t)] + q\,\mathcal{A}[\phi(t;q)], \tag{1.6}$$

where $q \in [0,1]$ is an embedding parameter and $\phi(t;q)$ is a function of t and q. When $q = 0$ and $q = 1$, we have

$$\mathcal{H}[\phi(t;q);q]|_{q=0} = \mathcal{L}[\phi(t;0) - u_0(t)]$$

and

$$\mathcal{H}[\phi(t;q);q]|_{q=1} = \mathcal{A}[\phi(t;1)],$$

respectively. Using (1.5), it is clear that

$$\phi(t;0) = u_0(t)$$

is the solution of the equation

$$\mathcal{H}[\phi(t;q);q]|_{q=0} = 0.$$

And

$$\phi(t;1) = u(t)$$

is therefore obviously the solution of the equation

$$\mathcal{H}[\phi(t;q);q]|_{q=1} = 0.$$

As the embedding parameter q increases from 0 to 1, the solution $\phi(t;q)$ of the equation

$$\mathcal{H}[\phi(t;q);q] = 0$$

depends upon the embedding parameter q and varies from the initial approximation $u_0(t)$ to the solution $u(t)$ of Equation (1.4). In topology, such a kind of continuous variation is called deformation.

Based on the idea of homotopy, some numerical techniques such as the continuation method [33] and the homotopy continuation method [34] were

Introduction

developed. In fact, the artificial small parameter method and the δ-expansion method can be described by the homotopy if we replace the artificial parameter ϵ or δ by the embedding parameter q, as shown in Chapter 4. However, although the above-mentioned traditional way to construct the homotopy (1.6) might be enough from viewpoints of numerical techniques, it is not good enough from viewpoints of analytic ones. This is manly because we have great freedom to choose the so-called auxiliary operator $\mathcal{L}$ and the initial approximations but lack any rules to direct their choice. More importantly, the traditional way to construct a homotopy cannot provide a convenient way to adjust convergence region and rate of approximation series.

In this book the basic ideas of the homotopy analysis method are described in details and some typical nonlinear problems in science and engineering are employed to illustrate its validity and flexibility. To simply show the validity of the homotopy analysis method, we point out that the 10th-order drag formula of a sphere in a uniform stream given by the homotopy analysis method agrees well with experimental data in a considerably larger region than all previous theoretical drag formulae published in the past 150 years, as shown in Figure 1.1. In short, the homotopy analysis method is based on the concept of homotopy. However, instead of using the traditional homotopy (1.6), we introduce a nonzero auxiliary parameter $\hbar$ and a nonzero auxiliary function $H(t)$ to construct such a new kind of homotopy

$$\tilde{\mathcal{H}}(\Phi; q, \hbar, H) = (1-q)\, \mathcal{L}[\Phi(t; q, \hbar, H) - u_0(t)] - q\, \hbar\, H(t)\, \mathcal{A}[\Phi(t; q, \hbar, H)], \quad (1.7)$$

which is more general than (1.6) because (1.6) is only a special case of (1.7) when $\hbar = -1$ and $H(t) = 1$, i.e.,

$$\mathcal{H}(\phi; q) = \tilde{\mathcal{H}}(\Phi; q, -1, 1). \tag{1.8}$$

Similarly, as q increases from 0 to 1, $\Phi(t; q, \hbar, H)$ varies from the initial approximation $u_0(t)$ to the exact solution $u(t)$ of the original nonlinear problem. However, the solution $\Phi(t; q, \hbar, H)$ of the equation

$$\tilde{\mathcal{H}}[\Phi(t; q, \hbar, H)] = 0 \tag{1.9}$$

depends not only on the embedding parameter q but also on the auxiliary parameter $\hbar$ and the auxiliary function $H(t)$. So, at $q = 1$, the solution still depends upon the auxiliary parameter $\hbar$ and the auxiliary function $H(t)$. Thus, different from the traditional homotopy (1.6), the generalized homotopy (1.7) can provide us with a *family* of approximation series whose convergence region depends upon the auxiliary parameter $\hbar$ and the auxiliary function $H(t)$, as illustrated later in this book. More importantly, this provides us with a simple way to *adjust* and *control* the convergence regions and rates of approximation series.

The homotopy analysis method is rather general and valid for nonlinear ordinary and partial differential equations in many different types. It has been successfully applied to many nonlinear problems such as nonlinear oscillations [35, 36, 37, 38, 39], boundary layer flows [28, 29, 40, 41, 42, 43], heat transfer [44, 45], viscous flows in porous medium [46], viscous flows of Oldroyd 6-constant fluids [47], magnetohydrodynamic flows of non-Newtonian fluids [48], nonlinear water waves [49, 50], Thomas-Fermi equation [51], Lane-Emden equation [30], and so on. To show its validity and flexibility, we give many new applications of the homotopy analysis method in this book.

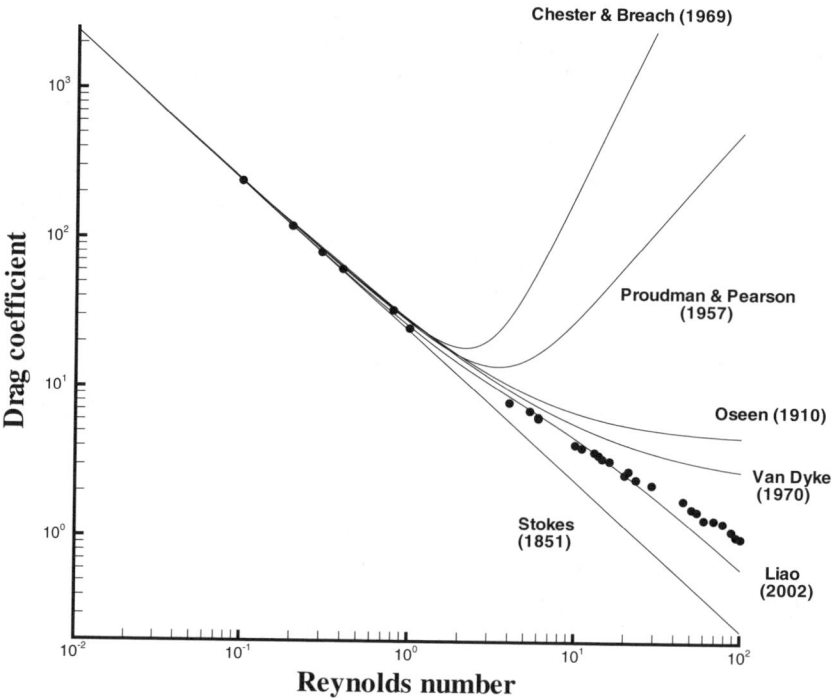

FIGURE 1.1
Comparison of experimental data of drag coefficient of a sphere in a uniform stream with theoretical results. Symbols: experimental data; solid line: theoretical results. (Modified from *International Journal of Non-Linear Mechanics*, 37, Shi-Jun Liao, "An analytic approximation of the drag coefficient for the viscous flow past a sphere", 1-18, Copyright (2002), with permission from Elsevier)

2
Illustrative description

In this chapter we use a simple nonlinear ordinary differential equation as an example to introduce the basic ideas of the homotopy analysis method.

2.1 An illustrative example

Consider a free sphere dropping in the air from a static state. Let $\tilde{t}$ denote the time, $U(\tilde{t})$ the velocity of the sphere, m the mass, and g the acceleration of gravity. Assume that the air resistance on the sphere is $a\,U^2(\tilde{t})$, where a is a constant. Then, due to Newton's second law, it holds

$$m\frac{dU(\tilde{t})}{d\tilde{t}} = mg - aU^2(\tilde{t}), \qquad (2.1)$$

subject to the initial condition

$$U(0) = 0. \qquad (2.2)$$

Physically speaking, the speed of a freely dropping sphere is increased due to the gravity until a steady velocity U_∞ is reached. So, even not knowing the solution $U(\tilde{t})$ in detail, we can gain the limit velocity U_∞ directly from (2.1), i.e.,

$$U_\infty = \sqrt{\frac{mg}{a}}. \qquad (2.3)$$

Using U_∞ and U_∞/g as the characteristic velocity and time, respectively, and writing

$$\tilde{t} = \left(\frac{U_\infty}{g}\right) t, \qquad U(\tilde{t}) = U_\infty V(t), \qquad (2.4)$$

we have the dimensionless equation

$$\dot{V}(t) + V^2(t) = 1, \qquad t \geq 0, \qquad (2.5)$$

subject to the initial condition

$$V(0) = 0, \qquad (2.6)$$

where t denotes the dimensionless time and the dot denotes the derivative with respect to t. Obviously, as $t \to +\infty$, i.e., $\tilde{t} \to \infty$ and $U(\tilde{t}) \to U_\infty$, we have from (2.4) that
$$\lim_{t \to +\infty} V(t) = 1, \tag{2.7}$$
even without solving Equations (2.5) and (2.6).

The exact solution of Equations (2.5) and (2.6) is
$$V(t) = \tanh(t), \tag{2.8}$$
useful for the comparisons of different approximations.

2.2 Solution given by some previous analytic techniques

For the sake of comparison, we first apply some well-known previous analytic techniques to solve the illustrative nonlinear problem.

2.2.1 Perturbation method

To give perturbation approximation, we assume that the dimensionless time t is a small variable (called perturbation quantity) and then express $V(t)$ in a power series
$$V(t) = \alpha_0 + \alpha_1 t + \alpha_2 t^2 + \alpha_3 t^3 + \cdots. \tag{2.9}$$
Using the initial condition (2.6) we gain $\alpha_0 = 0$. Then, substituting the above expression into Equation (2.5), we obtain
$$\sum_{k=0}^{+\infty} \left[(k+1)\alpha_{k+1} + \sum_{j=0}^{k} \alpha_j \alpha_{k-j} \right] t^k = 1,$$
which holds for any $t \geq 0$, provided
$$\alpha_1 = 1, \tag{2.10}$$
$$\alpha_{k+1} = -\frac{1}{k+1} \sum_{j=0}^{k} \alpha_j \alpha_{k-j}, \quad k \geq 1. \tag{2.11}$$
We therefore have the perturbation solution
$$V_{pert}(t) = t - \frac{1}{3}t^3 + \frac{2}{15}t^5 - \frac{17}{315}t^7 + \cdots = \sum_{n=0}^{+\infty} \alpha_{2n+1} t^{2n+1}, \tag{2.12}$$
which converges in a rather small region $0 \leq t < \rho_0$, where $\rho_0 \approx 3/2$, as shown in Figure 2.1. Note that the convergence region and rate of the perturbation solution (2.12) are uniquely determined.

2.2.2 Lyapunov's artificial small parameter method

By Lyapunov's artificial small parameter method we first replace Equation (2.5) by the equation
$$\dot{V}(t) + \varepsilon\, V^2(t) = 1 \tag{2.13}$$
and then write
$$V(t) = V_0(t) + \varepsilon\, V_1(t) + \varepsilon^2\, V_2(t) + \cdots, \tag{2.14}$$
where ε is an artificial small parameter. Substituting Equation (2.14) into Equations (2.13) and (2.6) and then balancing the coefficients of power series of ε, we obtain the equations
$$\dot{V}_0(t) = 1, \quad V_0(0) = 0,$$
$$\dot{V}_1(t) + V_0^2(t) = 0, \quad V_1(0) = 0,$$
$$\vdots$$
which give
$$V_0(t) = t, \quad V_1(t) = -\frac{t^3}{3}, \quad V_2(t) = \frac{2t^5}{15}, \cdots,$$
successively. Finally, setting $\varepsilon = 1$ in Equation (2.14), we have
$$V(t) = t - \frac{1}{3}t^3 + \frac{2}{15}t^5 - \frac{17}{315}t^7 + \cdots = \sum_{n=0}^{+\infty} \alpha_{2n+1}\, t^{2n+1}, \tag{2.15}$$
which is exactly the same as the perturbation solution (2.12) and thus is valid in a rather restricted region of t, as shown in Figure 2.1. Note that the convergence region and rate of the solution (2.15) given by Lyapunov's artificial small parameter method are also uniquely determined.

2.2.3 Adomian's decomposition method

By Adomian's decomposition method we first replace Equations (2.5) and (2.6) by
$$V(t) = t - \int_0^t V^2(t)dt. \tag{2.16}$$
The solution is given by
$$V(t) = V_0(t) + \sum_{k=1}^{+\infty} V_k(t),$$
where
$$V_0(t) = t,$$
$$V_k(t) = -\int_0^t A_{k-1}(t)\, dt, \quad k \geq 1$$

and
$$A_k(t) = \sum_{n=0}^{k} V_n(t)\, V_{k-n}(t)$$
is the so-called Adomian polynomial. We gain successively
$$V_1(t) = -\frac{t^3}{3}, \quad V_2(t) = \frac{2t^5}{15}, \quad V_3(t) = -\frac{17}{315}t^7, \quad \ldots$$
such that
$$V(t) = t - \frac{1}{3}t^3 + \frac{2}{15}t^5 - \frac{17}{315}t^7 + \cdots = \sum_{n=0}^{+\infty} \alpha_{2n+1}\, t^{2n+1}, \qquad (2.17)$$
which is exactly the same as the perturbation result (2.12), and thus is also valid in a rather small region, as shown in Figure 2.1. Note that the convergence region and rate of the solution (2.17) given by Adomian's decomposition method are also uniquely determined.

2.2.4 The δ-expansion method

By the δ-expansion method we first replace Equation (2.5) by
$$\dot{V}(t) + V^{1+\delta}(t) = 1, \qquad (2.18)$$
where δ is a real number. Write
$$V(t) = V_0(t) + \sum_{n=1}^{+\infty} V_n(t)\, \delta^n. \qquad (2.19)$$
Then we expand $V^{1+\delta}(t)$ in the power series of δ as follows:
$$V^{1+\delta} = V_0 + [V_1 + V_0 \ln V_0]\, \delta$$
$$+ \left[V_1(1 + \ln V_0) + \frac{1}{2} V_0 \ln^2 V_0 + V_2 \right] \delta^2 + \cdots. \qquad (2.20)$$
Substituting Equations (2.19) and (2.20) into Equation (2.18) and balancing the power series of δ, we have the following equations:
$$\dot{V}_0 + V_0 = 1, \quad V_0(0) = 0,$$
$$\dot{V}_1 + V_1 = -V_0 \ln V_0, \quad V_1(0) = 0,$$
$$\dot{V}_2 + V_2 = -V_1(1 + \ln V_0) - \frac{1}{2} V_0 \ln^2 V_0, \quad V_2(0) = 0,$$
$$\dot{V}_3 + V_3 = -V_2(1 + \ln V_0) - V_1\left(1 + \frac{1}{2} \ln V_0\right) \ln V_0$$
$$- \frac{1}{6} V_0 \ln^3 V_0 - \frac{V_1^2}{2 V_0}, \quad V_3(0) = 0,$$
$$\vdots$$

Illustrative Description

Solving the above linear equations successively, we obtain

$$V_0(t) = 1 - \exp(-t),$$
$$V_1(t) = \exp(-t)\left[t - \frac{\pi^2}{6} + P_2^L(e^{-t})\right] - (1 - e^{-t})\ln(1 - e^{-t}),$$
$$\vdots$$

where

$$P_n^L(z) = \sum_{k=1}^{+\infty} \frac{z^k}{k^n}$$

is the nth polylogarithm function of z. The first-order approximation is

$$V(t) \approx 1 + \exp(-t)\left[t - \frac{\pi^2}{6} - 1 + P_2^L(e^{-t})\right] - (1 - e^{-t})\ln(1 - e^{-t}). \quad (2.21)$$

The first several approximations of this kind of solution seem to be valid in the whole region $0 \leq t < +\infty$. However, due to the appearance of the special function $P_n^L(z)$, it becomes more and more difficult to get higher order approximations.

It is interesting that the solutions given by the perturbation method, Lyapunov's artificial small parameter method, and Adomian's decomposition method are the *same* for the illustrative problem. However, this solution is valid in a rather small region $0 \leq t < 3/2$. This illustrates that, like perturbation approximations, solutions given by previous nonperturbation techniques such as Lyapunov's artificial small parameter method and Adomian's decomposition method might break down as physical parameters or variables increase and nonlinearity goes stronger. This fact also implies that there might exist some relationships between them. It should be emphasized that the convergence region and rate of solutions given by *all* of these previous analytic methods are uniquely determined, and neither perturbation techniques nor previous nonperturbation methods such as Lyapunov's artificial small parameter method, Adomian's decomposition method, and the δ-expansion method can provide us with a convenient way to *control* and *adjust* convergence region and rate of solution series. Furthermore, the important property (2.7) of $V(t)$ at infinity, which we obtain without solving the problem, seems useless for all of these previous analytic techniques. This example illustrates that, in essence, neither perturbation techniques nor the previous nonperturbation methods can provide us with any ways to utilize such kinds of valuable information to approximate a given nonlinear problem more efficiently. Finally, note that the artificial parameters ϵ and δ appear in the different places in Equations (2.13) and (2.18), respectively, but the solution (2.21) given by the δ-expansion method is valid in the whole region $0 \leq t < +\infty$ and thus is much better than the solution (2.15) given by Lyapunov's artificial small parameter method. As mentioned in Chapter 1, Equations (2.13) and (2.18) can be regarded as a kind of homotopy, if we define ϵ and δ as the embedding parameter. This example illustrates that, in order to approximate a given nonlinear problem more efficiently, it seems important to properly construct a homotopy. However, up to now, we have not had any fundamental rules to direct us.

2.3 Homotopy analysis solution

In this section the basic ideas of the homotopy analysis method are introduced by the same illustrative problem mentioned above.

2.3.1 Zero-order deformation equation

Let $V_0(t)$ denote an initial guess of $V(t)$, which satisfies the initial condition (2.6), i.e.,
$$V_0(0) = 0. \qquad (2.22)$$
Let $q \in [0, 1]$ denote the so-called embedding parameter. The homotopy analysis method is based on a kind of continuous mapping $V(t) \to \Phi(t; q)$ such that, as the embedding parameter q increases from 0 to 1, $\Phi(t; q)$ varies from the initial guess $V_0(t)$ to the exact solution $V(t)$. To ensure this, choose such an auxiliary linear operator as
$$\mathcal{L}[\Phi(t; q)] = \gamma_1(t) \frac{\partial \Phi(t; q)}{\partial t} + \gamma_2(t) \, \Phi(t; q), \qquad (2.23)$$
where $\gamma_1(t) \neq 0$ and $\gamma_2(t)$ are real functions to be determined later. From Equation (2.5), we define the nonlinear operator
$$\mathcal{N}[\Phi(t; q)] = \frac{\partial \Phi(t; q)}{\partial t} + \Phi^2(t; q) - 1. \qquad (2.24)$$
Let $\hbar \neq 0$ and $H(t) \neq 0$ denote the so-called auxiliary parameter and auxiliary function, respectively. Using the embedding parameter $q \in [0, 1]$, we construct a family of equations
$$(1 - q) \, \mathcal{L}\left[\Phi(t; q) - V_0(t)\right] = \hbar \, q \, H(t) \, \mathcal{N}[\Phi(t; q)], \qquad (2.25)$$
subject to the initial condition
$$\Phi(0; q) = 0. \qquad (2.26)$$
It should be emphasized that we have great freedom to choose the auxiliary parameter $\hbar$, the auxiliary function $H(t)$, the initial approximation $V_0(t)$, and

Illustrative Description

the auxiliary linear operator $\mathcal{L}$. It is such freedom that plays important roles and establishes the cornerstone of the validity and flexibility of the homotopy analysis method, as shown later in this book.

When $q = 0$, Equation (2.25) becomes

$$\mathcal{L}\left[\Phi(t;0) - V_0(t)\right] = 0, \quad t \geq 0, \tag{2.27}$$

subject to the initial condition

$$\Phi(0;0) = 0. \tag{2.28}$$

According to Equations (2.22) and (2.23), the solution of Equations (2.27) and (2.28) is simply

$$\Phi(t;0) = V_0(t). \tag{2.29}$$

When $q = 1$, Equation (2.25) becomes

$$\hbar\, H(t)\, \mathcal{N}[\Phi(t;1)] = 0, \quad t \geq 0, \tag{2.30}$$

subject to the initial condition

$$\Phi(0;1) = 0. \tag{2.31}$$

Since $\hbar \neq 0$, $H(t) \neq 0$ and by means of the definition (2.24), Equations (2.30) and (2.31) are equivalent to the original equations (2.5) and (2.6), provided

$$\Phi(t;1) = V(t). \tag{2.32}$$

Therefore, according to Equations (2.29) and (2.32), $\Phi(t;q)$ varies from the initial guess $V_0(t)$ to the exact solution $V(t)$ as the embedding parameter q increases from 0 to 1. In topology, this kind of variation is called *deformation*, and Equations (2.25) and (2.26) construct the homotopy $\Phi(t;q)$. For brevity, Equations (2.25) and (2.26) are called *the zero-order deformation equations*.

Having the freedom to choose the auxiliary parameter $\hbar$, the auxiliary function $H(t)$, the initial approximation $V_0(t)$, and the auxiliary linear operator $\mathcal{L}$, we can assume that all of them are properly chosen so that the solution $\Phi(t;q)$ of the zero-order deformation equations (2.25) and (2.26) exists for $0 \leq q \leq 1$, and besides its mth-order derivative with respect to the embedding parameter q, i.e.,

$$V_0^{[m]}(t) = \left.\frac{\partial^m \Phi(t;q)}{\partial q^m}\right|_{q=0} \tag{2.33}$$

exists, where $m = 1, 2, 3, \cdots$. For brevity, $V_0^{[m]}(t)$ is called *the mth-order deformation derivative*. Define

$$V_m(t) = \frac{V_0^{[m]}(t)}{m!} = \frac{1}{m!}\left.\frac{\partial^m \Phi(t;q)}{\partial q^m}\right|_{q=0}. \tag{2.34}$$

By Taylor's theorem, we expand $\Phi(t;q)$ in a power series of the embedding parameter q as follows:

$$\Phi(t;q) = \Phi(t;0) + \sum_{m=1}^{+\infty} \frac{1}{m!} \left.\frac{\partial^m \Phi(t;q)}{\partial q^m}\right|_{q=0} q^m. \tag{2.35}$$

From Equations (2.29) and (2.34), the above power series becomes

$$\Phi(t;q) = V_0(t) + \sum_{m=1}^{+\infty} V_m(t)\, q^m. \tag{2.36}$$

Assume that the auxiliary parameter $\hbar$, the auxiliary function $H(t)$, the initial approximation $V_0(t)$, and the auxiliary linear operator $\mathcal{L}$ are so properly chosen that the series (2.36) converges at $q = 1$. Then, at $q = 1$, the series (2.36) becomes

$$\Phi(t;1) = V_0(t) + \sum_{m=1}^{+\infty} V_m(t). \tag{2.37}$$

Therefore, using Equation (2.32), we have

$$V(t) = V_0(t) + \sum_{m=1}^{+\infty} V_m(t). \tag{2.38}$$

The above expression provides us with a relationship between the initial guess $V_0(t)$ and the exact solution $V(t)$ by means of the terms $V_m(t)$ ($m = 1, 2, 3, \cdots$), which are unknown up to now.

2.3.2 High-order deformation equation

Define the vector

$$\vec{V}_n = \{V_0(t), V_1(t), V_2(t), \cdots, V_n(t)\}.$$

According to the definition (2.34), the governing equation and corresponding initial condition of $V_m(t)$ can be deduced from the zero-order deformation equations (2.25) and (2.26). Differentiating Equations (2.25) and (2.26) m times with respect to the embedding parameter q and then setting $q = 0$ and finally dividing them by $m!$, we have the so-called mth-order deformation equation

$$\mathcal{L}[V_m(t) - \chi_m V_{m-1}(t)] = \hbar\, H(t)\, R_m(\vec{V}_{m-1}), \tag{2.39}$$

subject to the initial condition

$$V_m(0) = 0, \tag{2.40}$$

where

$$R_m(\vec{V}_{m-1}) = \frac{1}{(m-1)!} \frac{\partial^{m-1}\mathcal{N}[\Phi(t;q)]}{\partial q^{m-1}}\bigg|_{q=0} \qquad (2.41)$$

and

$$\chi_m = \begin{cases} 0 \text{ when } m \leq 1, \\ 1 \text{ otherwise.} \end{cases} \qquad (2.42)$$

From Equations (2.24) and (2.41), we have

$$R_m(\vec{V}_{m-1}) = \dot{V}_{m-1}(t) + \sum_{j=0}^{m-1} V_j(t) V_{m-1-j}(t) - (1 - \chi_m). \qquad (2.43)$$

Notice that $R_m(\vec{V}_{m-1})$ given by the above expression is only dependent upon

$$V_0(t), V_1(t), V_2(t), \cdots, V_{m-1}(t),$$

which are known when solving the mth-order deformation equations (2.39) and (2.40). Thus, according to the definition (2.23) of the auxiliary operator $\mathcal{L}$, Equation (2.39) is a linear first-order differential equation, subject to the linear initial condition (2.40). Therefore, the solution $V_m(t)$ of high-order deformation equations (2.39) and (2.40) can be easily gained, especially by means of computation software such as Mathematica, Maple, MathLab, and so on. According to (2.38), we in essence transfer the original nonlinear problem, governed by Equations (2.5) and (2.6), into an infinite number of linear sub-problems governed by high-order deformation equations (2.39) and (2.40), and then use the sum of the solutions $V_m(t)$ of its first several sub-problems to approximate the exact solution. Note that such a kind of transformation needs not the existence of any small or large parameters in governing equation and initial/boundary conditions.

The mth-order approximation of $V(t)$ is given by

$$V(t) \approx \sum_{n=0}^{m} V_n(t). \qquad (2.44)$$

It should be emphasized that the zero-order deformation equation (2.25) is determined by the auxiliary linear operator $\mathcal{L}$, the initial approximation $V_0(t)$, the auxiliary parameter $\hbar$, and the auxiliary function $H(t)$. Theoretically speaking, the solution $V(t)$ given by the above approach is dependent of the auxiliary linear operator $\mathcal{L}$, the initial approximation $V_0(t)$, the auxiliary parameter $\hbar$, and the auxiliary function $H(t)$. Thus, unlike all previous analytic techniques, the convergence region and rate of solution series given by the above approach might not be uniquely determined. This is indeed true, as shown later in this chapter.

2.3.3 Convergence theorem

THEOREM 2.1
As long as the series (2.38) converges, where $V_m(t)$ is governed by the high-order deformation equations (2.39) and (2.40) under the definitions (2.42) and (2.43), it must be the exact solution of Equations (2.5) and (2.6).

Proof: If the series
$$\sum_{m=0}^{+\infty} V_m(t)$$
converges, we can write
$$S(t) = \sum_{m=0}^{+\infty} V_m(t)$$
and it holds
$$\lim_{m \to +\infty} V_m(t) = 0. \tag{2.45}$$

Using the definition (2.42) of χ_m, we have
$$\sum_{m=1}^{n} [V_m(t) - \chi_m V_{m-1}(t)]$$
$$= V_1 + (V_2 - V_1) + (V_3 - V_2) + \cdots + (V_n - V_{n-1})$$
$$= V_n(t),$$

which gives us, according to (2.45),
$$\sum_{m=1}^{+\infty} [V_m(t) - \chi_m V_{m-1}(t)] = \lim_{n \to +\infty} V_n(t) = 0.$$

Furthermore, using the above expression and the definition (2.23) of $\mathcal{L}$, we have
$$\sum_{m=1}^{+\infty} \mathcal{L}[V_m(t) - \chi_m V_{m-1}(t)] = \mathcal{L} \sum_{m=1}^{+\infty} [V_m(t) - \chi_m V_{m-1}(t)] = 0.$$

From the above expression and Equation (2.39), we obtain
$$\sum_{m=1}^{+\infty} \mathcal{L}[V_m(t) - \chi_m V_{m-1}(t)] = \hbar\, H(t) \sum_{m=1}^{+\infty} R_m(\vec{V}_{m-1}) = 0$$

which gives, since $\hbar \neq 0$ and $H(t) \neq 0$, that
$$\sum_{m=1}^{+\infty} R_m(\vec{V}_{m-1}) = 0. \tag{2.46}$$

Illustrative Description

From (2.43), it holds

$$\sum_{m=1}^{+\infty} R_m(\vec{V}_{m-1}) = \sum_{m=1}^{+\infty}\left[\dot{V}_{m-1}(t) + \sum_{j=0}^{m-1} V_j(t)V_{m-1-j}(t) - (1-\chi_m)\right]$$

$$= \sum_{m=0}^{+\infty}\dot{V}_m(t) - 1 + \sum_{m=1}^{+\infty}\sum_{j=0}^{m-1} V_j(t)V_{m-1-j}(t)$$

$$= \sum_{m=0}^{+\infty}\dot{V}_m(t) - 1 + \sum_{j=0}^{+\infty}\sum_{m=j+1}^{+\infty} V_j(t)V_{m-1-j}(t)$$

$$= \sum_{m=0}^{+\infty}\dot{V}_m(t) - 1 + \sum_{j=0}^{+\infty}V_j(t)\sum_{i=0}^{+\infty}V_i(t)$$

$$= \dot{S}(t) + S^2(t) - 1. \tag{2.47}$$

From Equations (2.46) and (2.47), we have

$$\dot{S}(t) + S^2(t) - 1 = 0, \qquad t \geq 0.$$

From Equations (2.22) and (2.40), it holds

$$S(0) = \sum_{m=0}^{+\infty} V_m(0) = V_0(0) + \sum_{m=1}^{+\infty} V_m(0) = V_0(0) = 0.$$

Therefore, according to the above two expressions, $S(t)$ must be the exact solution of Equations (2.5) and (2.6). This ends the proof.

Note that the above theorem is valid for the auxiliary linear operator $\mathcal{L}$ defined by (2.23) in a rather general form, where $\gamma_1(t) \neq 0$ and $\gamma_2(t)$ can be different functions, as illustrated later. This convergence theorem is important. It is because of this theorem that we can focus on ensuring that the approximation series converge. It is clear that the convergence of the series (2.38) depends upon the auxiliary parameter $\hbar$, the auxiliary function $H(t)$, the initial guess $V_0(t)$, and the auxiliary linear operator $\mathcal{L}$. Fortunately, the homotopy analysis method provides us with great freedom to choose all of them. Thus, as long as $\hbar$, $H(t)$, $V_0(t)$, and $\mathcal{L}$ are so properly chosen that the series (2.38) converges in a region $0 \leq t \leq t_0$, it *must* converge to the exact solution in this region. Therefore, the combination of the convergence theorem and the freedom of the choice of the auxiliary parameter $\hbar$, the auxiliary function $H(t)$, the initial guess $V_0(t)$, and the auxiliary linear operator $\mathcal{L}$ establishes the cornerstone of the validity and flexibility of the homotopy analysis method.

2.3.4 Some fundamental rules

As mentioned above, we have great freedom to choose the auxiliary linear operator $\mathcal{L}$, the initial approximation $V_0(t)$, and the auxiliary function $H(t)$

to construct the zero-order deformation equation. Theoretically, the foregoing freedom is so great that we can choose a lot of different auxiliary functions $H(t)$, initial approximations $V_0(t)$, and auxiliary linear operators $\mathcal{L}$. However, from the practical point of view, the freedom seems too great and it is necessary for us to have some fundamental rules to direct us.

Given a nonlinear problem, the essence of analytic approximation is to express its solution by a proper set of base functions. It is well known that a real function $f(x)$ can be approximated by many different base functions and thus can be more efficiently approximated by a relatively better set of base functions. The type of base functions is therefore rather important for the efficiency of approximating a nonlinear problem. The key of efficiently approximating a given nonlinear problem is to choose a relatively better set of base functions. Fortunately, using the freedom in the choice of the auxiliary linear operator $\mathcal{L}$, the initial approximation $V_0(t)$, and the auxiliary function $H(t)$, we can obtain many solution expressions of $V(t)$ presented by different base functions from which we might choose a better one to approximate a given nonlinear problem more efficiently.

In many cases, by mean of analyzing its physical background and/or its initial/boundary conditions and/or its type of nonlinearity, we might know what kinds of base functions are proper to represent the solution, even without solving a given nonlinear problem. For example, let

$$\{e_k(t) \mid k = 0, 1, 2, \cdots\} \tag{2.48}$$

denote such a set of base functions for the illustrative problem considered in this chapter. We can represent the solution in a series

$$V(t) = \sum_{n=0}^{+\infty} c_n \, e_k(t), \tag{2.49}$$

where c_n is a coefficient. As long as such a set of base functions is determined, the auxiliary function $H(t)$, the initial approximation $V_0(t)$, and the auxiliary linear operator $\mathcal{L}$ must be chosen in such a way that all solutions of the corresponding high-order deformation equations exist and can be expressed by this set of base functions. This provides us with a fundamental rule to direct the choice of the auxiliary function $H(t)$, the initial approximation $V_0(t)$, and the auxiliary linear operator $\mathcal{L}$, called the *rule of solution expression*. This rule plays an important role in the frame of the homotopy analysis method, as shown in this chapter.

As mentioned above, a real function $f(x)$ might be expressed by many different base functions. Thus, there might exist some different kinds of *rule of solution expressions* and all of them might give accurate approximations for a given nonlinear problem. In this case we might gain the best one by choosing the best set of base functions.

To further restrict the choice of the auxiliary function $H(t)$, it seems necessary to propose the so-called *rule of coefficient ergodicity*, i.e., all coefficients

Illustrative Description

in the solution expression, such as c_n in (2.49), can be modified to ensure the completeness of the set of base functions. In many cases, by means of the *rule of solution expression* and the *rule of coefficient ergodicity*, auxiliary functions can be uniquely determined. It is clear that the high-order deformation equations should be closed and have solutions. This provides us with the so-called *rule of solution existence*.

The so-called *rule of solution expression*, *rule of coefficient ergodicity*, and *rule of solution existence* play important roles and greatly simplify the application of the homotopy analysis method.

2.3.5 Solution expressions

Different from the foregoing solution expressions given by the perturbation and nonperturbation methods mentioned above, the solution given by the homotopy analysis method can be represented by many different base functions, as shown in this subsection.

2.3.5.1 Solution expressed by polynomial functions

Note that the perturbation solution (2.12) is a power series of t. So, it is straightforward to use the set of base functions

$$\{t^{2m+1} \mid m = 0, 1, 2, 3, \cdots\} \tag{2.50}$$

to represent $V(t)$, i.e.,

$$V(t) = \sum_{m=0}^{+\infty} a_m \, t^{2m+1}, \tag{2.51}$$

where a_m is a coefficient. This provides us with the first *rule of solution expression* of the illustrative problem.

Under the first *rule of solution expression* and according to the initial condition (2.22), it is straightforward to choose

$$V_0(t) = t \tag{2.52}$$

as the initial approximation of $V(t)$, and to choose an auxiliary linear operator

$$\mathcal{L}[\Phi(t;q)] = \frac{\partial \Phi(t;q)}{\partial t} \tag{2.53}$$

with the property

$$\mathcal{L}(C_1) = 0, \tag{2.54}$$

where C_1 is an integral constant. Under the first *rule of solution expression* denoted by (2.51) and from Equation (2.39), the auxiliary function $H(t)$ can be chosen in the form

$$H(t) = t^{2\kappa}. \tag{2.55}$$

According to (2.54), the solution of Equation (2.39) becomes

$$V_m(t) = \chi_m V_{m-1}(t) + \hbar \int_0^t \tau^{2\kappa} R_m(\vec{V}_{m-1}) \, d\tau + C_1,$$

where the integral constant C_1 is determined by the initial condition (2.40). It is found that when $\kappa \leq -1$ the term t^{-1} appears in the solution expression of $V_m(t)$, which disobeys the first *rule of solution expression* denoted by (2.51). In addition, when $\kappa \geq 1$, the base t^3 always disappears in the solution expression of $V_m(t)$ so that the coefficient of the term t^3 is always zero and thus cannot be modified even if the order of approximation tends to infinity. This, however, disobeys the so-called *rule of coefficient ergodicity*. In order to obey both of the first *rule of solution expression* denoted by (2.51) and the *rule of coefficient ergodicity*, we had to set $\kappa = 0$. This uniquely determines the corresponding auxiliary function

$$H(t) = 1. \qquad (2.56)$$

We now successively obtain

$$V_1(t) = \frac{1}{3}\hbar t^3,$$

$$V_2(t) = \frac{1}{3}\hbar(1+\hbar)t^3 + \frac{2}{15}\hbar^2 t^5,$$

$$V_3(t) = \frac{1}{3}\hbar(1+\hbar)^2 t^3 + \frac{2}{15}\hbar^2(1+\hbar)t^5 + \frac{17}{315}\hbar^3 t^7,$$

$$\vdots$$

It is found that the corresponding mth-order approximation can be expressed by

$$V(t) \approx \sum_{k=0}^{m} V_k(t) = \sum_{n=0}^{m} \mu_0^{m,n}(\hbar) \left[\alpha_{2n+1} \, t^{2n+1} \right], \qquad (2.57)$$

where α_{2n+1} is the same coefficient as that which appeared in the perturbation solution (2.12), and the function $\mu_0^{m,n}(\hbar)$ is defined by

$$\mu_0^{m,n}(\hbar) = (-\hbar)^n \sum_{j=0}^{m-n} \binom{n-1+j}{j} (1+\hbar)^j. \qquad (2.58)$$

Although the initial approximation $V_0(t)$, the auxiliary linear operator $\mathcal{L}$, and the auxiliary function $H(t)$ have been determined, we still have freedom to choose a proper value of the auxiliary parameter $\hbar$. Equation (2.57) denotes a *family* of solution expressions in auxiliary parameter $\hbar$. It is easy to prove that the function $\mu_0^{m,n}(\hbar)$ mentioned above has the property

$$\mu_0^{m,n}(-1) = 1, \quad \text{when } n \leq m. \qquad (2.59)$$

Illustrative Description

For any a finite positive integer n, it holds

$$\lim_{m \to +\infty} \mu_0^{m,n}(\hbar) = \begin{cases} 1, & \text{when } |1 + \hbar| < 1, \\ \infty, & \text{when } |1 + \hbar| > 1. \end{cases} \quad (2.60)$$

These two properties will be proved later in this chapter. So, when $\hbar = -1$, we have from (2.59), (2.57), and (2.12) that

$$V(t) = V_{pert}(t). \quad (2.61)$$

Therefore, the perturbation solution (2.12) is only a special case of the solution expression (2.57) when $\hbar = -1$, as are the solution (2.15) given by Lyapunov's artificial small parameter method and the solution (2.17) given by Adomian's decomposition method. Equation (2.57) logically contains the solution expression given by perturbation method, Lyapunov's artificial small parameter method, and Adomian's decomposition method and thus is more general.

Note that the coefficients of the solution expression (2.57) depend upon the auxiliary parameter $\hbar$. According to (2.60), the necessary condition for the series (2.57) to be convergent is $|1 + \hbar| < 1$, i.e.,

$$-2 < \hbar < 0.$$

It is interesting that the convergence region of the solution series (2.57) depends upon the value of $\hbar$. The closer the value of $\hbar$ ($-2 < \hbar < 0$) is to zero, the larger the convergence region of the series (2.57), as shown in Figure 2.1. It is found that the solution series (2.57) converges in the region

$$0 \le t < \rho_0 \sqrt{\frac{2}{|\hbar|} - 1},$$

where $\rho_0 \approx 3/2$ is the convergence radius of the perturbation solution (2.12). So, as $\hbar$ ($-2 < \hbar < 0$) tends to zero from below, the solution series (2.57) converges to the exact solution $V(t) = \tanh(t)$ in the whole region

$$0 \le t < +\infty.$$

Unlike all previous analytic techniques, we can adjust and control the convergence region of the solution series (2.57) by assigning $\hbar$ a proper value. The auxiliary parameter $\hbar$ therefore provides us with a convenient way to adjust and control convergence regions of solution series.

2.3.5.2 Solution expressed by fractional functions

Although the solution expression (2.57) represented by the base functions (2.50) can be valid in the whole region

$$0 \le t < +\infty$$

as $\hbar$ ($-2 < \hbar < 0$) tends to 0, the order of approximation must be very high to give an accurate enough result when the absolute value of $\hbar$ ($-2 < \hbar < 0$) is small. This kind of approximation is barely efficient, although theoretically it is better and more general than the solution (2.12) given by the perturbation method, Lyapunov's artificial small parameter method, and Adomian's decomposition method. It is therefore necessary to choose a better set of base functions to approximate $V(t)$ more efficiently.

As mentioned before, even without solving Equations (2.5) and (2.6), it is easy to know the limit velocity

$$V(+\infty) = 1.$$

The initial approximation (2.52) obviously does not satisfy this property. Generally, a power series converges in a finite region, therefore, the set (2.50) of polynomial functions is not proper to efficiently approximate $V(t)$ in the whole region $0 \le t < +\infty$.

Notice that it holds

$$\lim_{t \to +\infty} \frac{1}{(1+t)^m} = 0, \quad m \ge 1.$$

Thus, a function expressed by the set of base functions

$$\{(1+t)^{-m} \mid m = 0, 1, 2, 3, \cdots\} \tag{2.62}$$

has a finite value as $t \to +\infty$. We can assume that the solution $V(t)$ can be expressed by

$$V(t) = \sum_{m=0}^{+\infty} \frac{b_m}{(1+t)^m}, \tag{2.63}$$

where b_m is a coefficient to be determined. This provides us with the second *rule of solution expression* of the illustrative problem.

Under the second *rule of solution expression* and using the initial condition (2.6) and the limit velocity (2.7), it is straightforward to choose

$$V_0(t) = 1 - \frac{1}{1+t} \tag{2.64}$$

as the initial approximation of $V(t)$, and to choose the corresponding auxiliary linear operator

$$\mathcal{L}[\Phi(t;q)] = (1+t) \frac{\partial \Phi(t;q)}{\partial t} + \Phi(t;q) \tag{2.65}$$

with the property

$$\mathcal{L}\left(\frac{C_2}{1+t}\right) = 0, \tag{2.66}$$

Illustrative Description

where C_2 is an integral constant. Under the definition (2.65) of $\mathcal{L}$, the solution of the high-order deformation equation (2.39) becomes

$$V_m(t) = \chi_m V_{m-1}(t) + \frac{\hbar}{1+t}\int_0^t H(\tau)\,R_m(\vec{V}_{m-1})\,d\tau + \frac{C_2}{1+t}, \quad m \geq 1,$$

where the integral constant C_2 is determined by the initial condition (2.40). Under the second *rule of solution expression* denoted by (2.63) and from Equation (2.39), the auxiliary function $H(t)$ should be in the form

$$H(t) = \frac{1}{(1+t)^\kappa}, \tag{2.67}$$

where κ is an integer. It is found that, when $\kappa \leq 0$, the solutions of the high-order deformation equations (2.39) contain the term

$$\frac{\ln(1+t)}{1+t}$$

which incidentally disobeys the second *rule of solution expression* denoted by (2.63). When $\kappa > 1$, the base $(1+t)^{-2}$ disappears in the solution expression of $V_m(t)$ so that the coefficient of the term $(1+t)^{-2}$ is always zero and thus cannot be modified even if the order of approximation tends to infinity. This, however, disobeys the so-called *rule of coefficient ergodicity*. Thus, to obey both the second *rule of solution expression* and *rule of coefficient ergodicity*, we had to choose $\kappa = 1$, which uniquely determines the corresponding auxiliary function

$$H(t) = \frac{1}{1+t}. \tag{2.68}$$

Thereafter, we successively obtain

$$V_1(t) = -\frac{\hbar}{1+t} + \frac{2\hbar}{(1+t)^2} - \frac{\hbar}{(1+t)^3},$$

$$V_2(t) = -\hbar\left(1 + \frac{7}{12}\hbar\right)\frac{1}{1+t} + \frac{2\hbar(1+\hbar)}{(1+t)^2}$$
$$-\hbar\left(1 + \frac{7}{2}\hbar\right)\frac{1}{(1+t)^3} + \frac{10\hbar^2}{3(1+t)^4} - \frac{5\hbar^2}{4(1+t)^5},$$

$$\vdots$$

It is found that the corresponding mth-order approximation of $V(t)$ can be expressed by

$$V(t) \approx \sum_{n=0}^{2m+1} \frac{\beta_{m,n}(\hbar)}{(1+t)^n}, \tag{2.69}$$

where $\beta_{m,n}(\hbar)$ is a coefficient dependent upon $\hbar$.

Note that we still have freedom to choose the auxiliary parameter $\hbar$. So, (2.69) is in fact a new family of solution expressions. To investigate the influence of $\hbar$ on the solution series (2.69), we can first consider the convergence of some related series such as $V'(0), V''(0), V'''(0)$, and so on. It is found that $V'(0) = 1$ holds for all results at any order of approximations, thus it cannot provide us with any useful information about the choice of $\hbar$. However, $V''(0)$ and $V'''(0)$ are dependent of $\hbar$. Let $\mathbf{R}_\hbar$ denote a set of all possible values of $\hbar$ by means of which the corresponding series of $V''(0)$ converges. For brevity, we call the set $\mathbf{R}_\hbar$ *the valid region* of $\hbar$ for $V''(0)$. According to Theorem 2.1, for each $\hbar \in \mathbf{R}_\hbar$, the corresponding series of $V''(0)$ converges to the same result. The curve $V''(0)$ versus $\hbar$ contains a horizontal line segment above the valid region $\mathbf{R}_\hbar$. We call such a kind of curve *the $\hbar$-curve*, which clearly indicates the valid region $\mathbf{R}_\hbar$ of a solution series. The so-called $\hbar$-curves of $V''(0)$ and $V'''(0)$ given by the solution expression (2.69) are as shown in Figure 2.2. From Figure 2.2 it is clear that the series of $V''(0)$ and $V'''(0)$ given by the solution series (2.69) are convergent when

$$-3/2 \leq \hbar \leq -1/2.$$

This is indeed true. For example, for five different values of $\hbar$ in the region $-3/2 \leq \hbar \leq -1/2$, the series of $V''(0)$ and $V'''(0)$ given by (2.69) converge to the corresponding exact value 0 and -2, respectively, as shown in Tables 2.1 and 2.2. It is interesting that the convergence rate of the approximation series depends upon the value of $\hbar$, and the series of $V''(0)$ and $V'''(0)$ given by (2.69) converge fastest when $\hbar = -1$, as shown in Tables 2.1 and 2.2. This indicates that we can adjust the convergence rate of the solution series (2.69) by means of the auxiliary parameter $\hbar$. It is also true that, as long as the series of $V''(0)$ and $V'''(0)$ are convergent, the series (2.69) converge in the whole region $0 \leq t < +\infty$. Thus, according to Theorem 2.1, all of these convergent series must be the exact solution of the original nonlinear problem. For example, when $\hbar = -1$, the series (2.69) converges to the exact solution in the whole region $0 \leq t < +\infty$, as shown in Table 2.3. In general, by means of the so-called $\hbar$-curves, it is straightforward to know the corresponding valid region of $\hbar$. Choosing a value of $\hbar$ in the valid region, we can ensure that the corresponding solution series is convergent. In this way, we can control and adjust the convergence region and rate of solution series. Thus, the auxiliary parameter $\hbar$ plays an important role within the frame of the homotopy analysis method.

Unlike the solution (2.12) given by the perturbation method, Lyapunov's artificial small parameter method, and Adomian's decomposition method, the solution series (2.69) converges to the exact solution in the whole region $0 \leq t < +\infty$ when $-3/2 \leq \hbar \leq -1/2$. The solution series (2.69) is therefore almost more efficient than (2.57), although, theoretically speaking, both of them may converge to the exact solution in the whole region $0 \leq t < +\infty$. This is mainly because, for the illustrative problem, the base functions (2.62) are better and thus more efficient than (2.50).

Illustrative Description

Finally, let us investigate the relationship between the solution expression (2.69) with the perturbation solution (2.12). Substituting $\hbar = -1$ and

$$\frac{1}{1+t} = 1 - t + t^2 - t^3 + \cdots$$

into the 10th-order approximation of $V(t)$ given by (2.69), we have

$$V(t) \sim t - \frac{1}{3}t^3 + \frac{2}{5}t^5 - \frac{17}{315}t^7 + \frac{62}{2835}t^9 + \cdots,$$

whose first several terms are exactly the same as the perturbation series (2.12).

2.3.5.3 Solution expressed by exponential functions

It is well known that

$$\lim_{t \to +\infty} \exp(-nt) = 0, \quad n \geq 1.$$

So, a function expressed by the set of base functions

$$\{ \exp(-nt) \mid n \geq 0 \} \tag{2.70}$$

is finite as t tends to infinity. Considering the limit velocity (2.7), the above base functions are better than (2.50). Assume that $V(t)$ can be expressed by

$$V(t) = \sum_{n=0}^{+\infty} c_n \exp(-nt), \tag{2.71}$$

where c_n is a coefficient. This provides us with the third *rule of solution expression* of the illustrative problem.

Under the third *rule of solution expression* and from (2.6) and (2.7), it is straightforward to choose

$$V_0(t) = 1 - \exp(-t) \tag{2.72}$$

as the initial approximation of $V(t)$, and to choose the auxiliary linear operator

$$\mathcal{L}[\Phi(t;q)] = \frac{\partial \Phi(t;q)}{\partial t} + \Phi(t;q) \tag{2.73}$$

with the property

$$\mathcal{L}\left[C_3 \exp(-t)\right] = 0, \tag{2.74}$$

where C_3 is an integral coefficient. In this case, the solution of the mth-order deformation equation (2.39) becomes

$$V_m(t) = \chi_m V_{m-1}(t) + \hbar \exp(-t) \int_0^t \exp(\tau) \, H(\tau) \, R_m(\vec{V}_{m-1}) \, d\tau$$
$$+ C_3 \exp(-t), \quad m \geq 1,$$

where the integral constant C_3 is determined by (2.40). According to the third *rule of solution expression* denoted by (2.71) and from Equation (2.39), the auxiliary function $H(t)$ should be in the form

$$H(t) = \exp(-\kappa\, t), \qquad (2.75)$$

where κ is an integer. It is found that, when $\kappa \leq 0$, the solutions of the high-order deformation equations (2.39) contain the term

$$t\, \exp(-t),$$

which incidentally disobeys the third *rule of solution expression* denoted by (2.71). When $\kappa \geq 2$, the base $\exp(-2t)$ always disappears in the solution expressions of the high-order deformation equation (2.39) so that the coefficient of the term $\exp(-2\,t)$ cannot be modified even if the order of approximation tends to infinity. This, however, disobeys the so-called *rule of coefficient ergodicity*. Thus, to obey both of the third *rule of solution expression* denoted by (2.71) and the *rule of coefficient ergodicity*, we had to set $\kappa = 1$, which uniquely determines the corresponding auxiliary function

$$H(t) = \exp(-t). \qquad (2.76)$$

Therefore, we have

$$V_1(t) = -\frac{\hbar}{2}\, e^{-t} + \hbar\, e^{-2t} - \frac{\hbar}{2}\, e^{-3t},$$

$$V_2(t) = -\frac{\hbar}{2}\left(1 + \frac{\hbar}{2}\right) e^{-t} + \hbar\left(1 + \frac{\hbar}{2}\right) e^{-2t} - \frac{\hbar}{2}(1+\hbar)\, e^{-3t}$$
$$+ \frac{\hbar^2}{2} e^{-4t} - \frac{\hbar^2}{4} e^{-5t},$$

$$\vdots$$

It is found that the corresponding mth-order approximation of $V(t)$ can be generally expressed by

$$V(t) \approx \sum_{n=0}^{2m+1} \gamma_{m,n}(\hbar)\, \exp(-n\, t), \qquad (2.77)$$

where $\gamma_{m,n}(\hbar)$ is a coefficient dependent of $\hbar$.

Equation (2.77) is also a family of solution expressions in the auxiliary parameter $\hbar$. To investigate the influence of $\hbar$ on the convergence of the solution series (2.77), we first plot the so-called $\hbar$-curves of $V''(0)$ and $V'''(0)$, as shown in Figure 2.3. According to these $\hbar$-curves, it is easy to discover the valid region of $\hbar$, which correspond to the line segments nearly parallel to the horizontal axis. The so-called valid regions of $\hbar$ are enlarged as the order of

Illustrative Description

approximations increases, as shown in Figure 2.3. So, it is clear that the series of $V''(0)$ and $V'''(0)$ given by (2.77) converge if $\hbar$ belongs to the corresponding valid regions of $\hbar$. According to Theorem 2.1, they must converge to the exact values of $V''(0)$ and $V'''(0)$, respectively. This is indeed true, as shown in Tables 2.4 and 2.5 when $\hbar = -3/2, -5/4, -1, -3/4$ and $-1/2$. Note that the series seems to converge fastest when $\hbar = -1$. Furthermore, as long as the series of $V''(0)$ and $V'''(0)$ converge, the corresponding solution series (2.77) of $V(t)$ also converges to the exact solution (2.8) in the whole region $0 \leq t < +\infty$. For example, the approximation result of $V(t)$ given by (2.77), when $\hbar = -1$, agrees well with the exact result (2.8), as shown in Table 2.6. Generally, it is convenient to investigate the influence of $\hbar$ on the convergence of solution series by means of such kinds of $\hbar$-curves.

Note that the solution expression (2.77) is valid in the whole region $0 \leq t < +\infty$. Comparing Tables 2.4 to 2.6 with Tables 2.1 to 2.3, respectively, we find that, by means of the same value of $\hbar$, the solution series (2.77) converges faster than (2.69), and even the 10th-order approximation of (2.77) when $\hbar = -1$ agrees very well with the exact solution. So, the solution expression (2.77) is better and thus more efficient than (2.69) and, as mentioned before, (2.69) is more efficient than (2.57). These therefore illustrate that we may approximate a given nonlinear problem more efficiently by choosing a better set of base functions within the frame of the homotopy analysis method.

It is found that the mth-order approximation (2.77) of $V(t)$ can be explicitly expressed by

$$V(t) \approx 1 + 2 \sum_{n=1}^{m} [(-1)^n \exp(-nt)] \mu_0^{m,n}\left(\frac{\hbar}{2}\right)$$
$$- \exp(-t) \left[\left(1 + \frac{\hbar}{2}\right) + \frac{\hbar}{2} \exp(-2t)\right]^m, \qquad (2.78)$$

where the function $\mu_0^{m,n}(x)$ is defined by (2.58). It is interesting that the function $\mu_0^{m,n}(\hbar)$ appears again. Due to the property (2.59), the above expression becomes, when $\hbar = -2$,

$$V(t) \approx 1 + 2 \sum_{n=1}^{m} (-1)^n \exp(-nt) + (-1)^{m+1} \exp[-(2m+1)t]. \qquad (2.79)$$

Note that the exact solution (2.8) can be expanded as a series

$$V(t) \approx 1 + 2 \sum_{n=1}^{+\infty} (-1)^n \exp(-nt), \qquad (2.80)$$

which converges to the exact solution in the region $0 < t < +\infty$ but diverges at the point $t = 0$ where it gives either 1 or -1. However, with an additional term

$$(-1)^{m+1} \exp[-(2m+1)],$$

the expression (2.79) converges to the exact solution in the whole region $0 \le t < +\infty$ including the point $t = 0$. In fact, even the third-order approximation of $V(t)$ given by (2.79), i.e.,

$$V(t) \approx 1 - 2\exp(-2t) + 2\exp(-4t) - 2\exp(-6t) + \exp(-7t), \qquad (2.81)$$

agrees very well with the exact solution, as shown in Figure 2.4.

The idea to avoid the appearance of the term such as

$$\ln(1+t)/(1+t), \quad t\exp(-t)$$

in approximate expansions is not new. To gain uniformly valid approximations, some perturbation techniques were developed to avoid the appearance of the so-called secular terms such as

$$t\sin t, \; t\cos t$$

in perturbation solutions. This kind of technique goes back to various scientists in the 19th century such as Lindstedt [52], Bohlin [53], Poincaré [54], Gyldén [55], and so on. The idea was further developed by many scientists such as Lighthill [56, 57], Malkin [58], Kuo [59, 60], and Tsien [61]. However, the terms $\ln(1+t)/(1+t)$ and $t\exp(-t)$ tend to zero as $t \to +\infty$. Therefore, these terms do not belong to the so-called secular term in perturbation techniques. Thus, the *rule of solution expression* can be seen as the generalization of this idea.

To show that the term $t\exp(-t)$ indeed does not belong to the so-called secular terms in perturbation methods, we point out that $V(t)$ can be expressed by the base functions

$$\{t^m \exp(-nt) \mid m \ge 0, n \ge 1\}. \qquad (2.82)$$

Using the same initial approximation as (2.72), the same auxiliary linear operator as (2.73) but the auxiliary function

$$H(t) = 1 \qquad (2.83)$$

different from (2.76), we can obtain in the similar way the corresponding mth-order approximation of $V(t)$, which can be explicitly expressed by

$$V(t) \approx 1 + 2 \sum_{n=1}^{m+1} \sum_{k=0}^{m+1-n} \sigma_0^{m,n,k}(\hbar) \left[(-1)^n \frac{(-nt)^k}{k!} \exp(-nt) \right], \qquad (2.84)$$

where

$$\sigma_0^{m,n,k}(\hbar) = \frac{1}{2}\left[\mu_0^{m,n+k}(\hbar) + \mu_0^{m,n+k-1}(\hbar)\right]. \qquad (2.85)$$

It is interesting that the function $\mu_0^{m,n}(\hbar)$ appears once again. By means of the so-called $\hbar$-curves of the corresponding $V''(0)$ and $V'''(0)$, it is found that,

Illustrative Description 31

when $-2 < \hbar < 0$, the solution series (2.84) converges to the exact solution (2.8) in the whole region $0 \le t < +\infty$, as shown in Table 2.7.

It should be emphasized that, in the frame of the homotopy analysis method, the solution $V(t)$ may be expressed by four different base functions (2.50), (2.62), (2.70), and (2.82), although the illustrative problem has only a unique solution. Correspondingly, we gain four families of solution expressions (2.57), (2.69), (2.78), and (2.84). Theoretically, all of them can converge to the same exact solution $V(t) = \tanh(t)$ in the whole region $0 \le t < +\infty$. However, the solution expression (2.57) is least efficient among the four solution expressions and is therefore the worst because it is convergent in a finite region for a given value of $-2 < \hbar < 0$. By means of comparing Tables 2.3, 2.6, and 2.7 with each other, the solution expression (2.78) based on the pure exponential functions is more efficient than the solution expression (2.69) based on fractional functions and the solution expression (2.84) based on combined polynomial and exponential functions, and thus is the best. The solution expression (2.84) is more efficient than the solution expression (2.69). This example clearly illustrates that, in the frame of the homotopy analysis method, the solution of a given nonlinear problem can be expressed by many different base functions and thus can be more efficiently approximated by a better set of base function, even if the solution is unique.

Indeed, this illustrative example is very simple and the exact solution is known. However, it clearly illustrates that, by means of the homotopy analysis method, convergence region and rate of solution series can be adjusted and controlled by means of plotting the so-called $\hbar$-curves and then choosing $\hbar$ in the corresponding valid regions of $\hbar$. Even when it is unnecessary to enlarge convergence regions, we can give a more efficient solution series by assigning $\hbar$ a proper value. This illustrative example also shows the important roles of the *rule of solution expression* and *rule of coefficient ergodicity* in choosing the initial approximation, the auxiliary linear operator, and the auxiliary function.

2.3.6 The role of the auxiliary parameter $\hbar$

As mentioned before, the homotopy analysis method is based on the homotopy, a basic concept of topology. The nonzero auxiliary parameter $\hbar$ is introduced to construct the so-called zero-order deformation equation, which gives a more general homotopy than the traditional one. Thus, unlike all previous analytic techniques, the homotopy analysis method provides us with a family of solution expressions in the auxiliary parameter $\hbar$. As a result, the convergence region and rate of solution series are dependent upon the auxiliary parameter $\hbar$ and thus can be greatly enlarged by means of choosing a proper value for $\hbar$. This provides us with a convenient way to adjust and control convergence region and rate of solution series given by the homotopy analysis method, as illustrated above.

In this subsection we prove in a completely different way that convergence regions of series can be indeed adjusted and controlled by introducing an auxiliary parameter. This proof can provide us with a rational base for the validity of the homotopy analysis method.

First, we emphasize that the definition (2.58) of $\mu_0^{m,n}(\hbar)$ is gained in the homotopy analysis method, which appears in the solution expressions (2.57), (2.78), and (2.84). It is interesting that the same definition can be deduced directly from the famous Newtonian binomial theorem. To show this, consider a series

$$\frac{1}{1+t} = 1 - t + t^2 - t^3 + \cdots = \lim_{m \to +\infty} \sum_{n=0}^{m}(-1)^n t^n, \quad |t| < 1. \qquad (2.86)$$

Define

$$x = 1 + \hbar + \hbar\, t,$$

which gives

$$\frac{1}{1+t} = -\frac{\hbar}{(1-x)}.$$

When $|x| = |1 + \hbar + \hbar\, t| < 1$ and $|1 + \hbar| < 1$, i.e.,

$$-1 < t < \frac{2}{|\hbar|} - 1, \quad -2 < \hbar < 0,$$

it holds

$$\frac{1}{1+t} = -\frac{\hbar}{1-x} = -\hbar\left(1 + x + x^2 + x^3 + \cdots\right) = -\hbar \sum_{n=0}^{+\infty}(1 + \hbar + \hbar\, t)^n.$$

Thus,

$$\frac{1}{1+t} = \lim_{m \to +\infty}\left[-\hbar \sum_{n=0}^{m}(1 + \hbar + \hbar\, t)^n\right]$$

is valid in the region

$$-1 < t < \frac{2}{|\hbar|} - 1 \quad (-2 < \hbar < 0).$$

We have

$$-\hbar \sum_{n=0}^{m}(1 + \hbar + \hbar\, t)^n$$

$$= -\hbar \sum_{n=0}^{m}\sum_{k=0}^{n}\binom{n}{k}(1+\hbar)^{n-k}(\hbar\, t)^k$$

$$= -\hbar \sum_{k=0}^{m}\sum_{n=k}^{m}\binom{n}{k}(1+\hbar)^{n-k}\hbar^k\, t^k$$

Illustrative Description

$$= \sum_{k=0}^{m} (-1)^k \, t^k (-\hbar)^{k+1} \sum_{i=0}^{m-k} \binom{k+i}{k} (1+\hbar)^i$$

$$= \sum_{k=0}^{m} (-1)^k \, t^k \left[(-\hbar)^{k+1} \sum_{i=0}^{m-k} \binom{k+i}{i} (1+\hbar)^i \right]$$

$$= \sum_{n=0}^{m} (-1)^n \, t^n \, \mu_{-1}^{m,n}(\hbar),$$

where

$$\mu_{-1}^{m,n}(\hbar) = (-\hbar)^{n+1} \sum_{j=0}^{m-n} \binom{n+j}{j} (1+\hbar)^j. \tag{2.87}$$

Comparing the above to the definition (2.58), we gain the relationship

$$\mu_{-1}^{m,n}(\hbar) = \mu_0^{m+1,n+1}(\hbar). \tag{2.88}$$

Thus, it holds

$$\frac{1}{1+t} = \lim_{m \to +\infty} \sum_{n=0}^{m} \mu_0^{m+1,n+1}(\hbar) \, [(-1)^n \, t^n] \tag{2.89}$$

in the region

$$-1 < t < \frac{2}{|\hbar|} - 1 \quad (-2 < \hbar < 0).$$

Obviously, the convergence region is $-1 < t < 1$ when $\hbar = -1$, $-1 < t < 3$ when $\hbar = -1/2$, and $-1 < t < 99$ when $\hbar = -1/50$, respectively. In particular, the convergence region becomes

$$-1 < t < +\infty$$

as $\hbar$ tends to zero from below. Thus, the convergence region of the series (2.89) can be indeed adjusted and controlled by the auxiliary parameter $\hbar$. What we should emphasize here is that the same definition $\mu_0^{m,n}(\hbar)$ is first obtained in the frame of the homotopy analysis method and then deduced from the famous Newtonian binomial theorem. This fact logically shows the validity and reasonability of the homotopy analysis method.

The foregoing ideas can be employed to give such a generalized theorem as the one below.

THEOREM 2.2
It holds

$$(1+t)^\alpha = \lim_{m \to +\infty} \sum_{n=0}^{m} \mu_\alpha^{m,n}(\hbar) \binom{\alpha}{n} t^n \tag{2.90}$$

for a real number α ($\alpha \neq 0, 1, 2, 3, \cdots$) in the region

$$-1 < t < \frac{2}{|\hbar|} - 1 \quad (-2 < \hbar < 0),$$

where

$$\binom{\alpha}{n} = \frac{\alpha(\alpha-1)(\alpha-2)\cdots(\alpha-n+1)}{n!}$$

and

$$\mu_\alpha^{m,n}(\hbar) = (-\hbar)^{n-\alpha} \sum_{j=0}^{m-n} (-1)^j \binom{\alpha-n}{j} (1+\hbar)^j. \tag{2.91}$$

Proof: Write $x = 1 + \hbar + \hbar\, t$. Let $|x| < 1$ and $|1 + \hbar| < 1$, i.e.,

$$-1 < t < \frac{2}{|\hbar|} - 1, \quad -2 < \hbar < 0.$$

By the traditional Newton binomial theorem [62], it holds when $|x| < 1$ and $|1 + \hbar| < 1$ that

$$(1+t)^\alpha = (-\hbar)^{-\alpha}(1-x)^\alpha = (-\hbar)^{-\alpha} \sum_{n=0}^{+\infty} (-1)^n \binom{\alpha}{n} x^n$$

$$= (-\hbar)^{-\alpha} \sum_{n=0}^{+\infty} (-1)^n \binom{\alpha}{n} (1+\hbar+\hbar\, t)^n$$

$$= \lim_{m \to +\infty} (-\hbar)^{-\alpha} \sum_{n=0}^{m} (-1)^n \binom{\alpha}{n} (1+\hbar+\hbar\, t)^n.$$

The sum of the first m terms of above series is given by

$$(-\hbar)^{-\alpha} \sum_{n=0}^{m} (-1)^n \binom{\alpha}{n} (1+\hbar+\hbar\, t)^n$$

$$= (-\hbar)^{-\alpha} \sum_{n=0}^{m} (-1)^n \binom{\alpha}{n} \sum_{j=0}^{n} \binom{n}{j} (1+\hbar)^{n-j} \hbar^j\, t^j$$

$$= (-\hbar)^{-\alpha} \sum_{j=0}^{m} t^j \sum_{n=j}^{m} (-1)^n \binom{\alpha}{n} \binom{n}{j} (1+\hbar)^{n-j} \hbar^j$$

$$= (-\hbar)^{-\alpha} \sum_{j=0}^{m} t^j \sum_{i=0}^{m-j} (-1)^{i+j} \binom{\alpha}{i+j} \binom{i+j}{j} (1+\hbar)^i \hbar^j$$

$$= (-\hbar)^{-\alpha} \sum_{j=0}^{m} t^j \sum_{i=0}^{m-j} (-1)^{i+j} \binom{\alpha}{j} \binom{\alpha-j}{i} (1+\hbar)^i \hbar^j$$

$$= \sum_{j=0}^{m}\left[\binom{\alpha}{j}t^{j}\right]\sum_{i=0}^{m-j}(-1)^{i}\binom{\alpha-j}{i}(1+\hbar)^{i}(-\hbar)^{j-\alpha}$$

$$= \sum_{n=0}^{m}\mu_{\alpha}^{m,n}(\hbar)\left[\binom{\alpha}{n}t^{n}\right],$$

where

$$\mu_{\alpha}^{m,n}(\hbar) = (-\hbar)^{n-\alpha}\sum_{j=0}^{m-n}(-1)^{j}\binom{\alpha-n}{j}(1+\hbar)^{j}.$$

This ends the proof.

Although the definition (2.91) is deduced for real numbers $-\infty < \alpha < +\infty$ except integers $\alpha = 0, 1, 2, 3, \cdots$, it is valid for all real numbers. For any integer k, we have using the definition (2.91) that

$$\mu_{k}^{m,n}(\hbar) = (-\hbar)^{n-k}\sum_{j=0}^{m-n}\binom{n-k-1+j}{j}(1+\hbar)^{j}, \qquad (2.92)$$

which contains the definition (2.58) of $\mu_{0}^{m,n}(\hbar)$ and the definition (2.87) of $\mu_{-1}^{m,n}(\hbar)$. It can be proved that for any real number $\alpha \in (-\infty, +\infty)$ it holds

$$\mu_{\alpha}^{m,n}(-1) = 1 \qquad (2.93)$$

and

$$\lim_{m \to +\infty}\mu_{\alpha}^{m,n}(\hbar) = 1, \quad \text{when } |1+\hbar| < 1 \qquad (2.94)$$

for any finite positive integer n. According to the definitions (2.58) and (2.91), it holds for the integer $l \geq 0$ that

$$\mu_{-l}^{m,n}(\hbar) = \mu_{0}^{m+l,n+l}(\hbar). \qquad (2.95)$$

The proof of (2.93) is straightforward. When $|1+\hbar| < 1$ it holds from the definition (2.91) that

$$\lim_{m \to +\infty}\mu_{\alpha}^{m,n}(\hbar)$$

$$= (-\hbar)^{n-\alpha}\sum_{k=0}^{+\infty}(-1)^{k}\binom{\alpha-n}{k}(1+\hbar)^{k}$$

$$= (-\hbar)^{n-\alpha}\sum_{k=0}^{+\infty}\binom{\alpha-n}{k}(-1-\hbar)^{k}$$

$$= (-\hbar)^{n-\alpha}[1+(-1-\hbar)]^{\alpha-n}$$

$$= 1.$$

This ends the proof of (2.94).

It should be emphasized that $\mu_0^{m,n}(\hbar)$ defined by (2.58) and $\mu_{-1}^{m,n}(\hbar)$ defined by (2.87) are only special cases of (2.91) when $\alpha = 0$ and $\alpha = -1$, respectively. All of these logically verify the reasonableness and validity of the solutions (2.57), (2.78) and (2.84) given by the homotopy analysis method.

The rationality of solutions (2.57), (2.78), and (2.84) can be explained another way. It is known that the Taylor series of any a given function is unique. According to the property (2.60), when $|1 + \hbar| < 1$, it holds for any a given finite positive integer N that

$$\lim_{m \to +\infty} \sum_{n=0}^{N} \left(\alpha_{2n+1} t^{2n+1} \right) \mu_0^{m,n}(\hbar) = \sum_{n=0}^{N} \alpha_{2n+1} t^{2n+1}.$$

Therefore, given any finite positive integer N, the sum of the first N terms of the solution (2.57) is the same as the sum of the first N terms of the perturbation solution (2.12), if the order of approximation tends to infinity. Thus, the series (2.57) obeys the uniqueness of the Taylor series, but now in a more general meaning. Let

$$(\alpha_1, \alpha_3, \alpha_5, \alpha_7, \cdots)$$

denote a point in a space $\mathcal{S}$, where α_k ($k = 1, 3, 5, \cdots$) is the coefficient in the perturbation solution (2.12). The perturbation solution (2.12) can be regarded as an approach to the point $(\alpha_1, \alpha_3, \alpha_5, \alpha_7, \cdots)$ along such a traditional path Γ_0 defined by:

$$(\alpha_1, 0, 0, 0, \cdots),$$
$$(\alpha_1, \alpha_3, 0, 0, \cdots),$$
$$(\alpha_1, \alpha_3, \alpha_5, 0, \cdots),$$
$$\vdots$$

However, the solution series (2.57) can be regarded as an approach to the same point

$$(\alpha_1, \alpha_3, \alpha_5, \alpha_7, \cdots)$$

but along such a more general path $\Gamma(\hbar)$ defined by:

$$(\alpha_1 \mu_0^{0,0}(\hbar), 0, 0, 0, \cdots),$$
$$(\alpha_1 \mu_0^{1,0}(\hbar), \alpha_3 \mu_0^{1,1}(\hbar), 0, 0, \cdots),$$
$$(\alpha_1 \mu_0^{2,0}(\hbar), \alpha_3 \mu_0^{2,1}(\hbar), \alpha_5 \mu_0^{2,2}(\hbar), 0, \cdots),$$
$$\vdots$$

Notice that the path $\Gamma(\hbar)$ depends on the auxiliary parameter $\hbar$. According to (2.59), the path $\Gamma(-1)$ (when $\hbar = -1$) is exactly the same as the traditional one Γ_0. When $|1 + \hbar| < 1$ but $\hbar \ne -1$, the path $\Gamma(\hbar)$ is different from the

Illustrative Description

traditional path Γ_0. Even in this case, according to the property (2.60), all of them approach to the same point $(\alpha_1, \alpha_3, \alpha_5, \alpha_7, \cdots)$. Therefore, the solution series (2.57) can be regarded as a kind of limit process along an infinite number of approaching paths $\Gamma(\hbar)$ to the same point $(\alpha_1, \alpha_3, \alpha_5, \alpha_7, \cdots)$. It is well known that the result of such kind of limit process often depends upon the approaching path. For example, consider the limit

$$\lim_{(x,y)\to(0,0)} \frac{\sqrt{x^2+y^2}}{|x|}.$$

There are a lot of approaching paths to $(0,0)$. For simplicity, let us consider the path $y = \beta x$, where β is a real number. It obviously holds that

$$\lim_{(x,y)\to(0,0)} \frac{\sqrt{x^2+y^2}}{|x|} = \sqrt{1+\beta^2}.$$

The limit is therefore dependent upon the approaching path to the point $(0,0)$. This clearly explains why the convergence region of the solution series (2.57) is dependent upon the auxiliary parameter $\hbar$, because the function $\mu_0^{m,n}(\hbar)$ defines different approaching paths by different values of $\hbar$.

According to above explanation, the function $\mu_\alpha^{m,n}(\hbar)$ can be used to define different approaching paths for a limit process by different values of α and $\hbar$. For the sake of this reason, the function $\mu_\alpha^{m,n}(\hbar)$ is called *the approach function of the first kind*. To generalize the definition (2.85) of $\sigma_0^{m,n}(\hbar)$, we define

$$\sigma_\alpha^{m,n,k}(\hbar) = \frac{1}{2}\left[\mu_\alpha^{m,n+k}(\hbar) + \mu_\alpha^{m,n+k-1}(\hbar)\right] \tag{2.96}$$

as *the approach function of the second kind*, where $|1+\hbar| < 1$ and $-\infty < \alpha < +\infty$. It is easy to prove that, for $\alpha \in (-\infty, +\infty)$ and $0 \le n \le m+1$, it holds

$$\sigma_\alpha^{m,n,k}(-1) = \begin{cases} 1, & \text{when } 0 \le k < m+1-n, \\ 1/2, & \text{when } k = m+1-n, \end{cases} \tag{2.97}$$

and

$$\lim_{m\to+\infty} \sigma_\alpha^{m,n,k}(\hbar) = \begin{cases} 1, & \text{when } |1+\hbar| < 1, \\ \infty, & \text{when } |1+\hbar| > 1, \end{cases} \tag{2.98}$$

where n and k are finite positive integers.

The approach functions $\mu_\alpha^{m,n}(\hbar)$ and $\sigma_\alpha^{m,n,k}(\hbar)$ have rather general meaning and therefore could be employed to greatly enlarge convergence regions of approximation series. For example, from the traditional Taylor series

$$\sum_{n=0}^{+\infty} \frac{f^{(n)}(z_0)}{n!}(z-z_0)^n$$

of a function $f(z)$, we can define *the generalized Taylor series of the first kind*

$$\lim_{m\to+\infty} \sum_{n=0}^{m} \mu_\alpha^{m,n}(\hbar)\left[\frac{f^{(n)}(z_0)}{n!}(z-z_0)^n\right]$$

and *the generalized Taylor series of the second kind*

$$\lim_{m\to+\infty} \sum_{n=0}^{m} \sigma_\alpha^{m,n,0}(\hbar) \left[\frac{f^{(n)}(z_0)}{n!}(z-z_0)^n \right],$$

where $\mu_\alpha^{m,n}(\hbar)$ and $\sigma_\alpha^{m,n,0}(\hbar)$ are defined by (2.91) and (2.96), respectively. The convergence regions of these generalized Taylor series might be greatly enlarged by choosing proper values of $\hbar$ and α. To illustrate this, we can generalize the mth-order approximations (2.57), (2.78), and (2.84) by

$$V(t) \approx \sum_{n=0}^{m} \mu_\alpha^{m,n}(\hbar) \left[\alpha_{2n+1}\, t^{2n+1} \right], \tag{2.99}$$

$$V(t) \approx 1 + 2 \sum_{n=1}^{m} [(-1)^n \exp(-nt)] \, \mu_\alpha^{m,n}\left(\frac{\hbar}{2}\right)$$
$$- \exp(-t) \left[\left(1 + \frac{\hbar}{2}\right) + \frac{\hbar}{2} \exp(-2t) \right]^m \tag{2.100}$$

and

$$V(t) \approx 1 + 2 \sum_{n=1}^{m+1} \sum_{k=0}^{m+1-n} \sigma_\alpha^{m,n,k}(\hbar) \left[(-1)^n \frac{(-nt)^k}{k!} \exp(-nt) \right], \tag{2.101}$$

respectively, where $|1 + \hbar| < 1$ and $\alpha \in (-\infty, +\infty)$. For example, when $\alpha = \pi/4$, the convergence region of the approximation (2.99) becomes larger and larger as $\hbar$ tends to zero from below, as shown in Figure 2.5. And when $\hbar = -1$, all of the 20th-order approximations of (2.100) given by $\alpha = \pm 1/2, \pm\pi/4$ agree with the exact solution, as shown in Table 2.8. When $\hbar = -1/2$, all of the 20th-order approximations of (2.101) given by $\alpha = \pm 1/2, \pm\pi/4$ agree with the exact solution, as shown in Table 2.9. So, the functions $\mu_\alpha^{m,n}(\hbar)$ and $\sigma_\alpha^{m,n,k}(\hbar)$ have indeed rather general meaning.

In this subsection we point out that the definition (2.58) of $\mu_0^{m,n}(\hbar)$, which is first obtained in the homotopy analysis method, can be independently deduced from the Newtonian binomial theorem. Furthermore, we prove that convergence region and rate of a series can be indeed adjusted and controlled by introducing an auxiliary parameter. We also point out that the functions $\mu_\alpha^{m,n}(\hbar)$ and $\sigma_\alpha^{m,n.k}(\hbar)$ define different approaching paths by different values of $\hbar$ and α. All of these provide us with a rational base for the validity of the homotopy analysis method.

2.3.7 Homotopy-Padé method

The homotopy analysis method is based on such an assumption that the series (2.36) of $\Phi(t; q)$ converges at $q = 1$ for the illustrative problem. Fortunately, as mentioned above, we have great freedom to choose the initial approximation

Illustrative Description

$V_0(t)$, the auxiliary linear operator $\mathcal{L}$, the auxiliary function $H(t)$, and the auxiliary parameter $\hbar$ in the frame of the homotopy analysis method. If all of them are properly chosen, the series (2.36) can be convergent at $q = 1$, as shown above. Besides, the convergence region and rate of the solution series given by the homotopy analysis method depend upon the auxiliary parameter $\hbar$. Therefore, the auxiliary parameter $\hbar$ provides us with a convenient way to adjust and control the convergence region and rate of solution series, as shown above.

There exist some techniques to accelerate the convergence of a given series. Among them, the so-called Padé technique is widely applied. For a given series

$$\sum_{n=0}^{+\infty} c_n x^n,$$

the corresponding $[m, n]$ Padé approximant is expressed by

$$\frac{\sum_{k=0}^{m} a_{m,k} x^k}{\sum_{k=0}^{n} b_{m,k} x^k},$$

where $a_{m,k}, b_{m,k}$ are determined by the coefficients c_j ($j = 0, 1, 2, 3, \cdots, m+n$). In many cases the traditional Padé technique can greatly increase the convergence region and rate of a given series. For example, employing the traditional Padé technique to the perturbation series (2.12), we have the $[1, 1], [2, 2]$ and $[3, 3]$ Padé approximants

$$t, \quad \frac{3t}{3+t^2}, \quad \frac{t(15+t^2)}{15+6t^2},$$

respectively. In general, the $[m, m]$ Padé approximant can be expressed by

$$\frac{\sum_{n=0}^{m} a_0^{m,n} t^n}{\sum_{n=0}^{m-1} b_0^{m,n} t^n}, \quad \text{when } m \text{ is an odd number,} \tag{2.102}$$

or

$$\frac{\sum_{n=0}^{m-1} a_0^{m,n} t^n}{\sum_{n=0}^{m} b_0^{m,n} t^n}, \quad \text{when } m \text{ is an even number,} \tag{2.103}$$

where $a_0^{m,n}$ and $b_0^{m,n}$ are coefficients. Note that all of these traditional Padé approximants tend to either infinity or zero as $t \to +\infty$. The $[4, 4]$ and $[10, 10]$ Padé approximants of the perturbation solution (2.12) are as shown in Figure 2.6.

The so-called homotopy-Padé technique [50] was proposed by means of combining the above-mentioned traditional Padé technique with the homotopy analysis method. To ensure that the series (2.36) is convergent at $q = 1$, we first employ the traditional $[m, n]$ Padé technique about the embedding parameter q to obtain the $[m, n]$ Padé approximant

$$\frac{\sum_{k=0}^{m} A_{m,k}(t)\, q^k}{\sum_{k=0}^{n} B_{m,k}(t)\, q^k}, \qquad (2.104)$$

where the coefficients $A_{m,k}(t)$ and $B_{m,k}(t)$ are determined by the first several approximations

$$V_0(t), V_1(t), V_2(t), \cdots, V_{m+n}(t).$$

Then, setting $q = 1$ in (2.104) and using (2.32), we have the so-called $[m, n]$ homotopy-Padé approximant

$$\frac{\sum_{k=0}^{m} A_{m,k}(t)}{\sum_{k=0}^{n} B_{m,k}(t)}. \qquad (2.105)$$

For the illustrative problem, the coefficients $A_{m,n}(t)$ and $B_{m,n}(t)$ are dependent of the base functions used to present the solution $V(t)$. Using the base functions denoted by (2.62), we have the corresponding $[1, 1]$ homotopy-Padé approximant

$$\frac{t(12 + 16t + 7t^2)}{(1+t)(12 + 4t + 7t^2)}$$

and the $[2, 2]$ homotopy-Padé approximant

$$\frac{t(168000 + 362880\,t + 238000\,t^2 + 14160\,t^3 - 47124\,t^4 - 36308\,t^5 - 13419\,t^6)}{3(1+t)(56000 + 64960\,t + 33040\,t^2 + 12000\,t^3 - 2508\,t^4 - 9076\,t^5 - 4473\,t^6)},$$

respectively. In general, the $[m, m]$ homotopy-Padé approximation can be expressed by

$$\frac{\sum_{n=1}^{m^2+m+1} a_2^{m,n}\, t^n}{\sum_{n=0}^{m^2+m+1} b_2^{m,n}\, t^n}, \qquad (2.106)$$

where $a_2^{m,n}$ and $b_2^{m,n}$ are coefficients. It is very interesting that $a_2^{m,n}$ and $b_2^{m,n}$ are found to be independent of the auxiliary parameter $\hbar$. Comparing (2.106) with (2.102) and (2.103), we find that in accuracy the $[m, m]$ homotopy-Padé approximant is equivalent to the traditional $[m^2 + m + 1, m^2 + m + 1]$ Padé approximant. Unlike the traditional Padé approximants (2.102) and (2.103)

Illustrative Description

which tend to either infinity or zero as $t \to +\infty$, all of the homotopy-Padé approximant (2.106) correctly tend to 1 as $t \to +\infty$. Thus, for a given m, the $[m, m]$ homotopy-Padé approximant (2.106) is much more accurate than the traditional $[m, m]$ Padé approximants (2.102) and (2.103). For example, the [4,4] homotopy-Padé approximant is more accurate and much better than the traditional [4,4] Padé approximant and is even better than the [10,10] traditional Padé approximant, as shown in Figure 2.6. In particular, using the base functions denoted by (2.70), we have the $[1, 1]$ homotopy-Padé approximant

$$\frac{1 - \exp(-2t)}{1 + \exp(-2t)}, \tag{2.107}$$

which is just the exact solution $V(t) = \tanh(t)$. Thus, the so-called homotopy-Padé method is indeed much more efficient than the traditional Padé technique.

Similarly, the so-called homotopy-Padé technique can be applied to accelerate the convergence of the related series. For example, to accelerate the series of $V''(0)$ and $V'''(0)$, we first apply the traditional Padé technique to the series

$$\left.\frac{\partial^2 \Phi(t; q)}{\partial t^2}\right|_{t=0} = \sum_{n=0}^{+\infty} V''_n(0)\, q^n$$

and

$$\left.\frac{\partial^3 \Phi(t; q)}{\partial t^3}\right|_{t=0} = \sum_{n=0}^{+\infty} V'''_n(0)\, q^n$$

to get their $[m, n]$ Padé approximants about the embedding parameter q, respectively, and then set $q = 1$ to obtain the corresponding $[m, n]$ homotopy-Padé approximants. The homotopy-Padé approximations of $V''(0)$ and $V'''(0)$, corresponding to the solution expression (2.69) expressed by the fractional functions, are listed in Table 2.10. The homotopy-Padé approximations of $V''(0)$ and $V'''(0)$, corresponding to the solution expression (2.78) expressed by the exponential functions are listed in Table 2.11. In both cases, the homotopy-Padé technique greatly accelerates the convergence of $V''(0)$ and $V'''(0)$.

For the illustrative problem, it is found that all of the $[m, m]$ homotopy-Padé approximants do not depend upon the auxiliary parameter $\hbar$. Thus, even if we choose a bad value of $\hbar$ such that the corresponding solution series diverges, we can still employ the homotopy-Padé technique to get a convergent result. As shown later in this book for other nonlinear problems, the $[m, m]$ homotopy-Padé approximants are often independent of the auxiliary parameter $\hbar$. However, up to now, we cannot give a mathematical proof about it in general cases.

All of these illustrate that the so-called homotopy-Padé technique can greatly enlarge the convergence region and rate of the solution series given by the homotopy analysis method.

In summary, we introduce in this chapter the basic ideas of the homotopy analysis method by means of a simple example. We show that, unlike all previous analytic techniques, the homotopy analysis method always gives a family of solution expressions in the auxiliary parameter $\hbar$, which may be expressed by different base functions. Using the freedom in choosing the initial guess, the auxiliary linear operator, and the auxiliary function, we can express the solution in many different base functions, and thus approximate a nonlinear problem more efficiently by choosing a better set of base functions. The *rule of solution expression*, the *rule of coefficient ergodicity*, and the *rule of solution existence* are proposed to direct the choice of the initial guess, the auxiliary linear operator, and the auxiliary function. These rules greatly simplify the application of the homotopy analysis method. We demonstrate that the convergence region and rate of the solution series may be adjusted and controlled by means of the auxiliary parameter $\hbar$. By plotting the so-called $\hbar$-curves, it is easy to find out a proper value of $\hbar$ to ensure that the solution series converge. Furthermore, the so-called homotopy-Padé technique is proposed to accelerate the convergence of solution series, which is often much more efficient than the traditional Padé technique.

Illustrative Description

TABLE 2.1
Approximations of $V''(0)$ given by (2.69) for different values of $\hbar$.

order	$\hbar = -1/2$	$\hbar = -3/4$	$\hbar = -1$	$\hbar = -5/4$	$\hbar = -3/2$
5	-0.062500	-0.001953	0	-0.001953	0.062500
10	-0.001953	-1.9×10^{-6}	0	-1.9×10^{-6}	-0.001953
15	-0.000061	-1.9×10^{-9}	0	1.9×10^{-9}	0.000061
20	-1.9×10^{-6}	-1.9×10^{-12}	0	-1.9×10^{-12}	-1.9×10^{-6}
25	-6.0×10^{-8}	-1.8×10^{-15}	0	1.8×10^{-15}	6.0×10^{-8}
30	-1.9×10^{-9}	-1.7×10^{-18}	0	-1.7×10^{-18}	-1.9×10^{-9}
35	-5.8×10^{-11}	-1.7×10^{-21}	0	1.7×10^{-21}	5.8×10^{-11}
40	-1.8×10^{-12}	-1.7×10^{-24}	0	-1.7×10^{-24}	-1.9×10^{-12}

TABLE 2.2
Approximations of $V'''(0)$ given by (2.69) for different values of $\hbar$.

order	$\hbar = -1/2$	$\hbar = -3/4$	$\hbar = -1$	$\hbar = -5/4$	$\hbar = -3/2$
5	-3.312500	-2.138672	-2	-2.251953	-6.937500
10	-2.089844	-2.000278	-2	-1.999516	-1.699219
15	-2.004333	-2.000000	-2	-2.000001	-2.013977
20	-2.000183	-2.000000	-2	-2.000000	-1.99942
25	-2.000007	-2.000000	-2	-2.000000	-2.000023
30	-2.000000	-2.000000	-2	-2.000000	-1.999999
35	-2.000000	-2.000000	-2	-2.000000	-2.000000
40	-2.000000	-2.000000	-2	-2.000000	-2.000000

TABLE 2.3
Comparison of the exact solution (2.8) with the mth-order approximations of $V(t)$ given by (2.69) when $\hbar = -1$.

t	10th-order approx.	20th-order approx.	40th-order approx.	60th-order approx.	exact result
1/4	0.2449	0.2449	0.2449	0.2449	0.2449
1/2	0.4621	0.4621	0.4621	0.4621	0.4621
3/4	0.6349	0.6351	0.6351	0.6351	0.6351
1	0.7516	0.7616	0.7616	0.7616	0.7616
3/2	0.9082	0.9053	0.9051	0.9051	0.9051
2	0.9720	0.9644	0.9640	0.9640	0.9640
5/2	0.9982	0.9870	0.9866	0.9866	0.9866
3	1.0082	0.9950	0.9950	0.9951	0.9951
4	1.0110	0.9979	0.9992	0.9993	0.9993
5	1.0082	0.9973	0.9997	0.9999	0.9999
10	0.9984	0.9968	1.0003	1.0001	1.0000
100	0.9987	0.9998	1.0001	1.0000	1.0000

TABLE 2.4
Approximations of $V''(0)$ given by (2.77) for different values of $\hbar$.

order	$\hbar = -1/2$	$\hbar = -3/4$	$\hbar = -1$	$\hbar = -5/4$	$\hbar = -3/2$
5	-0.031250	-0.000977	0	-0.000977	0.031250
10	-0.000977	-9.5 ×10^{-7}	0	-9.5 ×10^{-7}	-0.000977
15	-0.000031	-9.3×10^{-10}	0	9.3×10^{-10}	0.000031
20	-9.5×10^{-7}	-9.1×10^{-13}	0	-9.1×10^{-13}	-9.5×10^{-7}
25	-3.0×10^{-8}	-8.9×10^{-16}	0	8.8×10^{-16}	3.0×10^{-8}
30	-9.3×10^{-10}	-8.7×10^{-19}	0	-8.7×10^{-19}	-9.3×10^{-10}
35	-2.9×10^{-11}	-8.5×10^{-22}	0	8.5×10^{-22}	2.9×10^{-11}
40	-9.1×10^{-13}	-8.3×10^{-25}	0	-8.3×10^{-25}	-9.1×10^{-13}

TABLE 2.5
Approximations of $V'''(0)$ given by (2.77) for different values of $\hbar$.

order	$\hbar = -1/2$	$\hbar = -3/4$	$\hbar = -1$	$\hbar = -5/4$	$\hbar = -3/2$
5	-2.375000	-2.041016	-2	-2.076172	-3.500000
10	-2.026367	-2.000083	-2	-1.999854	-1.909180
15	-2.001282	-2.000000	-2	-2.000001	-2.004211
20	-2.000054	-2.000000	-2	-2.000000	-1.999825
25	-2.000002	-2.000000	-2	-2.000000	-2.000007
30	-2.000000	-2.000000	-2	-2.000000	-2.000000
35	-2.000000	-2.000000	-2	-2.000000	-2.000000
40	-2.000000	-2.000000	-2	-2.000000	-2.000000

TABLE 2.6
Comparison of the exact solution (2.8) with the approximations of $V(t)$ given by (2.77) when $\hbar = -1$.

t	5th-order approx.	10th-order approx.	15th-order approx.	20th-order approx.	exact result
1/4	0.2449	0.2449	0.2449	0.2449	0.2449
1/2	0.4619	0.4621	0.4621	0.4621	0.4621
3/4	0.6342	0.6351	0.6351	0.6351	0.6351
1	0.7596	0.7616	0.7616	0.7616	0.7616
3/2	0.9020	0.9051	0.9051	0.9051	0.9051
2	0.9612	0.9639	0.9640	0.9640	0.9640
5/2	0.9845	0.9866	0.9866	0.9866	0.9866
3	0.9937	0.9950	0.9951	0.9951	0.9951
4	0.9988	0.9993	0.9993	0.9993	0.9993
5	0.9997	0.9999	0.9999	0.9999	0.9999
10	1.0000	1.0000	1.0000	1.0000	1.0000
100	1.0000	1.0000	1.0000	1.0000	1.0000

TABLE 2.7
Comparison of the exact solution (2.8) with the approximations of $V(t)$ given by (2.84) when $\hbar = -1$.

t	10th-order approx.	20th-order approx.	40th-order approx.	50th-order approx.	exact result
1/4	0.2449	0.2449	0.2449	0.2449	0.2449
1/2	0.4621	0.4621	0.4621	0.4621	0.4621
3/4	0.6351	0.6351	0.6351	0.6351	0.6351
1	0.7616	0.7616	0.7616	0.7616	0.7616
3/2	0.9051	0.9051	0.9051	0.9051	0.9051
2	0.9640	0.9640	0.9640	0.9640	0.9640
5/2	0.9866	0.9866	0.9866	0.9866	0.9866
3	0.9953	0.9950	0.9951	0.9951	0.9951
4	0.9990	0.9993	0.9993	0.9993	0.9993
5	0.9975	0.9999	0.9999	0.9999	0.9999
10	1.0021	0.9982	0.9999	1.0000	1.0000
100	1.0000	1.0000	1.0000	1.0000	1.0000

TABLE 2.8
Comparison of the exact solution (2.8) with the 20th-order approximation of $V(t)$ given by (2.100) when $\hbar = -1$ and $\alpha = \pm 1/2, \pm \pi/4$.

t	when $\alpha = -\pi/4$	when $\alpha = -1/2$	when $\alpha = 1/2$	when $\alpha = \pi/4$	exact result
1/4	0.2449	0.2449	0.2449	0.2449	0.2449
1/2	0.4621	0.4621	0.4621	0.4621	0.4621
3/4	0.6351	0.6351	0.6351	0.6351	0.6351
1	0.7616	0.7616	0.7616	0.7616	0.7616
3/2	0.9051	0.9051	0.9051	0.9051	0.9051
2	0.9640	0.9640	0.9640	0.9640	0.9640
5/2	0.9866	0.9866	0.9866	0.9866	0.9866
3	0.9951	0.9951	0.9951	0.9951	0.9951
4	0.9993	0.9993	0.9993	0.9993	0.9993
5	0.9999	0.9999	0.9999	0.9999	0.9999
10	1.0000	1.0000	0.9999	1.0000	1.0000
100	1.0000	1.0000	1.0000	1.0000	1.0000

TABLE 2.9
Comparison of the exact solution (2.8) with the 20th-order approximation of $V(t)$ given by (2.101) when $\hbar = -1/2$ and $\alpha = \pm 1/2, \pm \pi/4$.

t	when $\alpha = -\pi/4$	when $\alpha = -1/2$	when $\alpha = 1/2$	when $\alpha = \pi/4$	exact result
1/4	0.2449	0.2449	0.2449	0.2449	0.2449
1/2	0.4621	0.4621	0.4621	0.4621	0.4621
3/4	0.6351	0.6351	0.6351	0.6351	0.6351
1	0.7616	0.7616	0.7616	0.7616	0.7616
3/2	0.9051	0.9051	0.9051	0.9051	0.9051
2	0.9640	0.9640	0.9640	0.9640	0.9640
5/2	0.9866	0.9866	0.9866	0.9866	0.9866
3	0.9951	0.9951	0.9951	0.9951	0.9951
4	0.9993	0.9993	0.9993	0.9993	0.9993
5	0.9999	0.9999	0.9999	0.9999	0.9999
10	1.0000	1.0000	0.9999	1.0000	1.0000
100	1.0000	1.0000	1.0000	1.0000	1.0000

TABLE 2.10
The $[m, m]$ homotopy-Padé approximation of $V''(0)$ and $V'''(0)$ corresponding to (2.69).

$[m, m]$	$V''(0)$	$V'''(0)$
[1, 1]	0	-3
[2, 2]	0	-2
[3, 3]	0	-2
[4, 4]	0	-2
[5, 5]	0	-2
[10, 10]	0	-2

TABLE 2.11
The $[m, m]$ homotopy-Padé approximation of $V''(0)$ and $V'''(0)$ corresponding to (2.78).

$[m, m]$	$V''(0)$	$V'''(0)$
[1, 1]	0	-5.57143
[2, 2]	0	-2
[3, 3]	0	-2
[4, 4]	0	-2
[5, 5]	0	-2
[10, 10]	0	-2

Illustrative Description

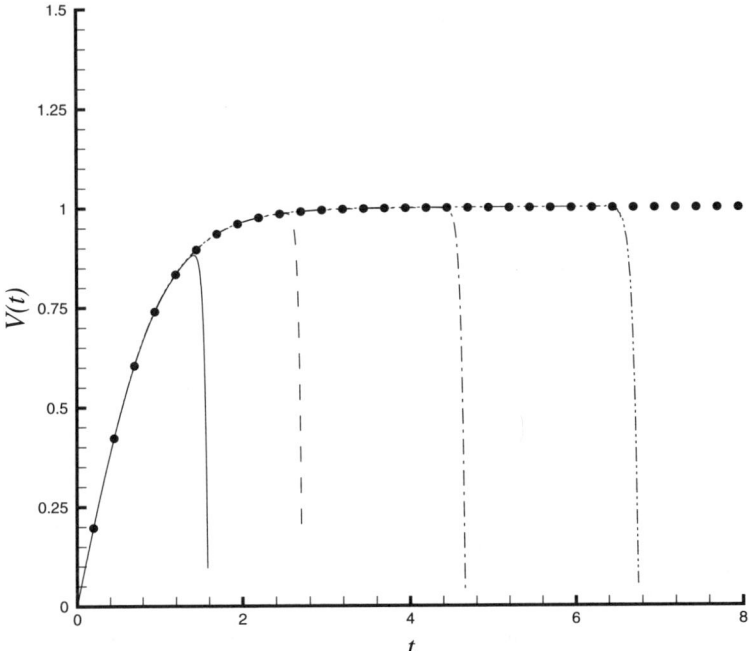

FIGURE 2.1
Comparison of the exact solution (2.8) with the solution expression (2.57). Symbols: exact solution; solid line: perturbation solution (2.12); dashed line: solution (2.57) when $\hbar = -1/2$; dash-dotted line: solution (2.57) when $\hbar = -1/5$; dash-dot-dotted line: solution (2.57) when $\hbar = -1/10$.

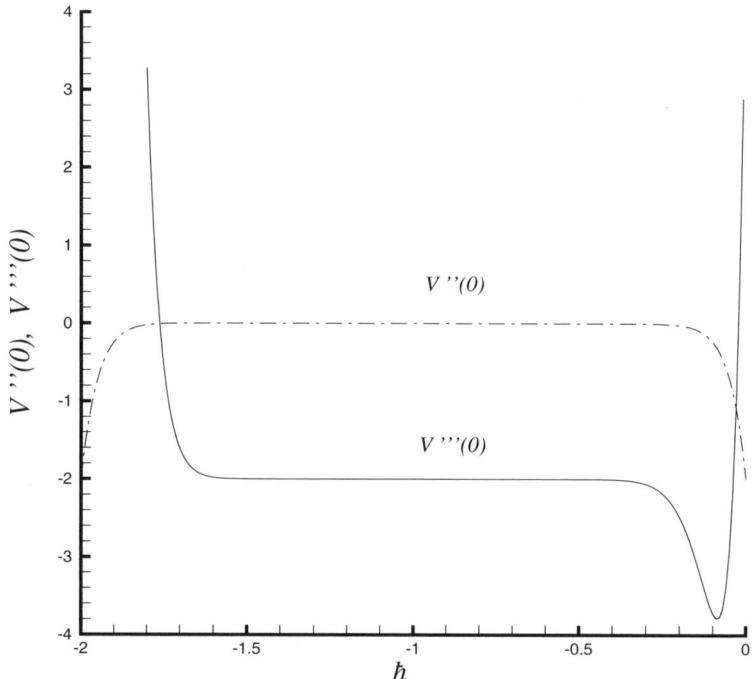

FIGURE 2.2
The $\hbar$-curve of $V''(0)$ and $V'''(0)$ given by (2.69) when $H(t) = 1/(1+t)$. Dash-dotted line: 20th-order approximation of $V''(0)$; solid line: 20th-order approximation of $V'''(0)$.

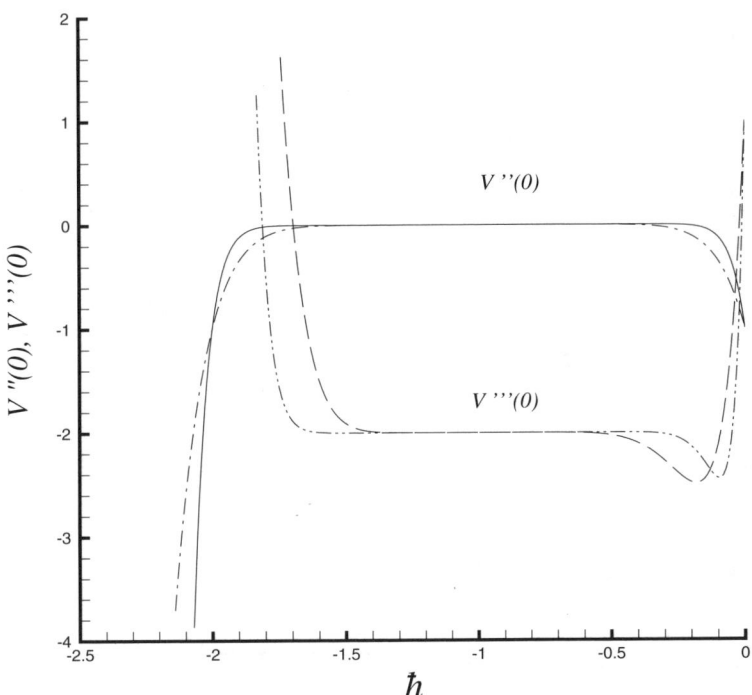

FIGURE 2.3
The $\hbar$-curves of $V''(0)$ and $V'''(0)$ given by (2.77) when $H(t) = \exp(-t)$. Dash-dotted line: 10th-order approximation of $V''(0)$; solid line: 20th-order approximation of $V''(0)$; dashed lined: 10th-order approximation of $V'''(0)$; dash-dot-dotted line: 20th-order approximation of $V'''(0)$.

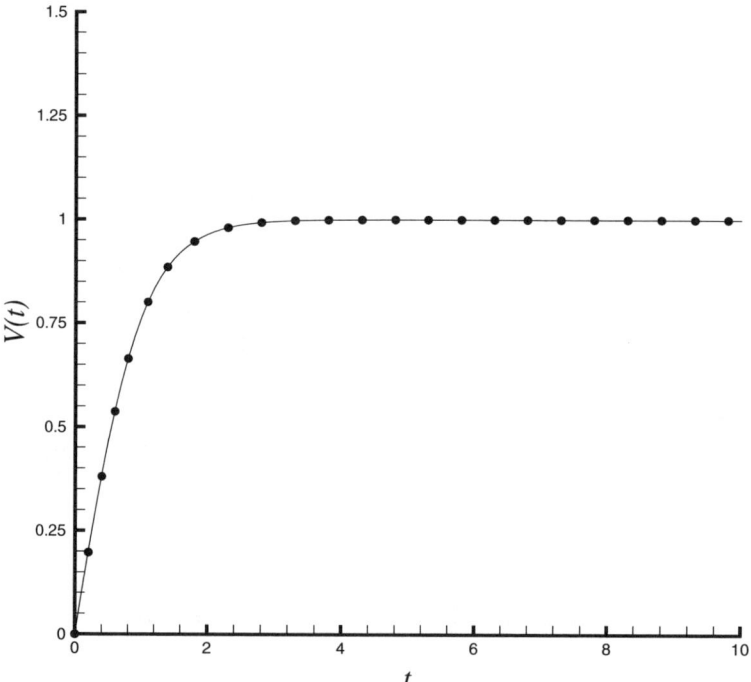

FIGURE 2.4
The comparison of the third-order approximation (2.81) of $V(t)$ with the exact solution (2.8). Solid line: third-order approximation (2.81); symbols: exact solution (2.8).

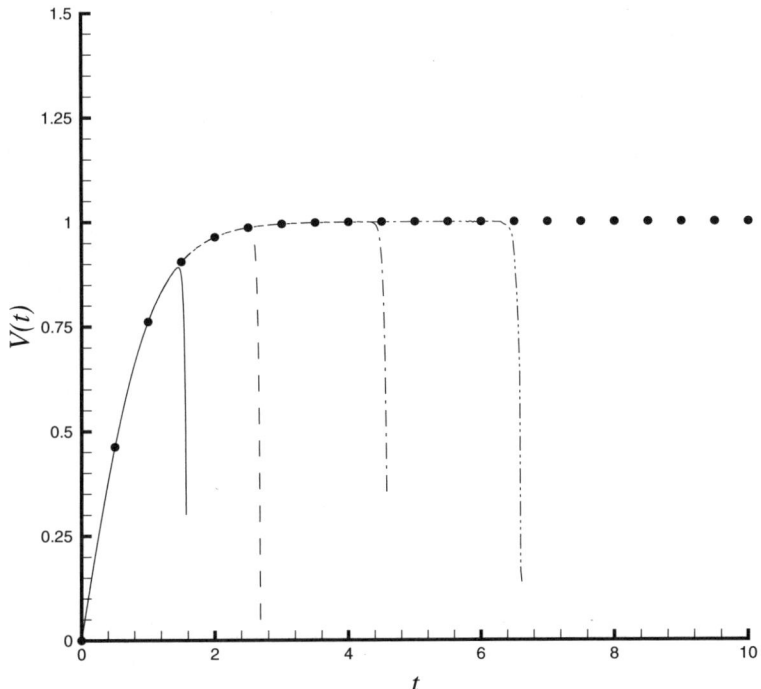

FIGURE 2.5
Comparison of the exact solution (2.8) with the solution expression (2.99) at the 31st-order of approximation when $\alpha = \pi/4$. Symbols: exact solution; solid line: (2.99) when $\hbar = -1$; dashed line: (2.99) when $\hbar = -1/2$; dash-dotted line: (2.99) when $\hbar = -1/5$; dash-dot-dotted line: (2.99) when $\hbar = -1/10$.

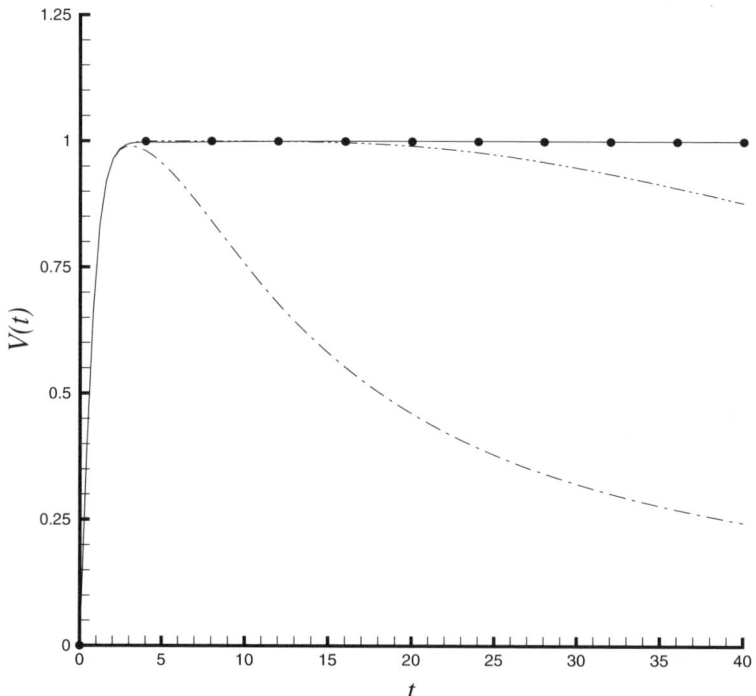

FIGURE 2.6
Comparison of the exact solution (2.8) with the homotopy-Padé approximation (2.106) and traditional Padé approximant (2.103) of $V(t)$. Symbols: exact solution; solid line: [4,4] homotopy-Padé approximant of $V(t)$; dash-dotted line: [4,4] traditional Padé approximant of $V(t)$; dash-dot-dotted line: [10,10] traditional Padé approximant.

3

Systematic description

In Chapter 2 the basic ideas of the homotopy analysis method are illustrated by a simple nonlinear problem. Here a systematic description is given for general nonlinear problems.

3.1 Zero-order deformation equation

In most cases a nonlinear problem can be described by a set of governing equations and initial and/or boundary conditions. For brevity, let us consider here only one nonlinear equation in a general form:

$$\mathcal{N}[u(\mathbf{r},t)] = 0, \qquad (3.1)$$

where $\mathcal{N}$ is a nonlinear operator, $u(\mathbf{r},t)$ is an unknown function, and $\mathbf{r}$ and t denote spatial and temporal independent variables, respectively.

Let $u_0(\mathbf{r},t)$ denote an initial guess of the exact solution $u(\mathbf{r},t)$, $\hbar \neq 0$ an auxiliary parameter, $H(\mathbf{r},t) \neq 0$ an auxiliary function, and $\mathcal{L}$ an auxiliary linear operator with the property

$$\mathcal{L}[f(\mathbf{r},t)] = 0 \qquad \text{when } f(\mathbf{r},t) = 0. \qquad (3.2)$$

Then, using $q \in [0,1]$ as an embedding parameter, we construct such a homotopy

$$\mathcal{H}[\Phi(\mathbf{r},t;q); u_0(\mathbf{r},t), H(\mathbf{r},t), \hbar, q]$$
$$= (1-q)\{\mathcal{L}[\Phi(\mathbf{r},t;q) - u_0(\mathbf{r},t)]\} - q\,\hbar\,H(\mathbf{r},t)\,\mathcal{N}[\Phi(\mathbf{r},t;q)]. \qquad (3.3)$$

It should be emphasized that the above homotopy contains the so-called auxiliary parameter $\hbar$ and the auxiliary function $H(\mathbf{r},t)$. To the best of the author's knowledge, the nonzero auxiliary parameter $\hbar$ and auxiliary function $H(\mathbf{r},t)$ are introduced for the first time in this way to construct a homotopy. So, such a kind of homotopy is more general than traditional ones. The auxiliary parameter $\hbar$ and the auxiliary function $H(\mathbf{r},t)$ play important roles within the frame of the homotopy analysis method. It should be emphasized that we have great freedom to choose the initial guess $u_0(\mathbf{r},t)$, the auxiliary linear operator $\mathcal{L}$, the nonzero auxiliary parameter $\hbar$, and the auxiliary function $H(\mathbf{r}, t)$.

Let $q \in [0, 1]$ denote an embedding parameter. Enforcing the homotopy (3.3) to be zero, i.e.,

$$\mathcal{H}[\Phi(\mathbf{r}, t; q); u_0(\mathbf{r}, t), H(\mathbf{r}, t), \hbar, q] = 0,$$

we have the so-called zero-order deformation equation

$$(1 - q) \{\mathcal{L}[\Phi(\mathbf{r}, t; q) - u_0(\mathbf{r}, t)]\} = q \hbar H(\mathbf{r}, t) \mathcal{N}[\Phi(\mathbf{r}, t; q)], \qquad (3.4)$$

where $\Phi(\mathbf{r}, t; q)$ is the solution which depends upon not only the initial guess $u_0(\mathbf{r}, t)$, the auxiliary linear operator $\mathcal{L}$, the auxiliary function $H(\mathbf{r}, t)$ and the auxiliary parameter $\hbar$ but also the embedding parameter $q \in [0, 1]$. When $q = 0$, the zero-order deformation equation (3.4) becomes

$$\mathcal{L}[\Phi(\mathbf{r}, t; 0) - u_0(\mathbf{r}, t)] = 0, \qquad (3.5)$$

which gives, using the property (3.2),

$$\Phi(\mathbf{r}, t; 0) = u_0(\mathbf{r}, t). \qquad (3.6)$$

When $q = 1$, since $\hbar \neq 0$ and $H(\mathbf{r}, t) \neq 0$, the zero-order deformation equation (3.4) is equivalent to

$$\mathcal{N}[\Phi(\mathbf{r}, t; 1)] = 0, \qquad (3.7)$$

which is exactly the same as the original equation (3.1), provided

$$\Phi(\mathbf{r}, t; 1) = u(\mathbf{r}, t). \qquad (3.8)$$

Thus, according to (3.6) and (3.8), as the embedding parameter q increases from 0 to 1, $\Phi(\mathbf{r}, t; q)$ varies (or deforms) continuously from the initial approximation $u_0(\mathbf{r}, t)$ to the exact solution $u(\mathbf{r}, t)$ of the original equation (3.1). Such a kind of continuous variation is called deformation in homotopy. This is the reason why we call (3.4) the zero-order deformation equation.

Define the so-called mth-order deformation derivatives

$$u_0^{[m]}(\mathbf{r}, t) = \left. \frac{\partial^m \Phi(\mathbf{r}, t; q)}{\partial q^m} \right|_{q=0}. \qquad (3.9)$$

By Taylor's theorem, $\Phi(\mathbf{r}, t; q)$ can be expanded in a power series of q as follows:

$$\Phi(\mathbf{r}, t; q) = \Phi(\mathbf{r}, t; 0) + \sum_{m=1}^{+\infty} \frac{u_0^{[m]}(\mathbf{r}, t)}{m!} q^m. \qquad (3.10)$$

Writing

$$u_m(\mathbf{r}, t) = \frac{u_0^{[m]}(\mathbf{r}, t)}{m!} = \frac{1}{m!} \left. \frac{\partial^m \Phi(\mathbf{r}, t; q)}{\partial q^m} \right|_{q=0} \qquad (3.11)$$

Systematic Description

and using (3.6), the power series (3.10) of $\Phi(\mathbf{r}, t; q)$ becomes

$$\Phi(\mathbf{r}, t; q) = u_0(\mathbf{r}, t) + \sum_{m=1}^{+\infty} u_m(\mathbf{r}, t) \, q^m. \tag{3.12}$$

Note that we have great freedom to choose the initial guess $u_0(\mathbf{r}, t)$, the auxiliary linear operator $\mathcal{L}$, the nonzero auxiliary parameter $\hbar$, and the auxiliary function $H(\mathbf{r}, t)$. Assume that all of them are properly chosen so that:

1. The solution $\Phi(\mathbf{r}, t; q)$ of the zero-order deformation equation (3.4) exists for all $q \in [0, 1]$.

2. The deformation derivative $u_0^{[m]}(\mathbf{r}, t)$ exists for $m = 1, 2, 3, \cdots, +\infty$.

3. The power series (3.12) of $\Phi(\mathbf{r}, t; q)$ converges at $q = 1$.

Then, from (3.8) and (3.12), we have under these assumptions the solution series

$$u(\mathbf{r}, t) = u_0(\mathbf{r}, t) + \sum_{m=1}^{+\infty} u_m(\mathbf{r}, t). \tag{3.13}$$

This expression provides us with a relationship between the exact solution $u(\mathbf{r}, t)$ and the initial approximation $u_0(\mathbf{r}, t)$ by means of the terms $u_m(\mathbf{r}, t)$ which are determined by the so-called high-order deformation equations described below.

3.2 High-order deformation equation

For brevity, define the vector

$$\vec{u}_n = \{u_0(\mathbf{r}, t), u_1(\mathbf{r}, t), u_2(\mathbf{r}, t), \cdots, u_n(\mathbf{r}, t)\}.$$

According to the definition (3.11), the governing equation of $u_m(\mathbf{r}, t)$ can be derived from the zero-order deformation equation (3.4). Differentiating the zero-order deformation equation (3.4) m times with respective to the embedding parameter q and then dividing it by $m!$ and finally setting $q = 0$, we have the so-called *mth-order deformation equation*

$$\mathcal{L}[u_m(\mathbf{r}, t) - \chi_m \, u_{m-1}(\mathbf{r}, t)] = \hbar \, H(\mathbf{r}, t) \, R_m(\vec{u}_{m-1}, \mathbf{r}, t), \tag{3.14}$$

where χ_m is defined by (2.42) and

$$R_m(\vec{u}_{m-1}, \mathbf{r}, t) = \frac{1}{(m-1)!} \left. \frac{\partial^{m-1} \mathcal{N}[\Phi(\mathbf{r}, t; q)]}{\partial q^{m-1}} \right|_{q=0}. \tag{3.15}$$

Substituting (3.12) into the above expression, we have

$$R_m(\vec{u}_{m-1}, \mathbf{r}, t) = \frac{1}{(m-1)!} \left\{ \frac{\partial^{m-1}}{\partial q^{m-1}} \mathcal{N} \left[\sum_{n=0}^{+\infty} u_n(\mathbf{r}, t) \, q^n \right] \right\} \bigg|_{q=0}. \quad (3.16)$$

Note that the high-order deformation equation (3.14) is governed by the same linear operator $\mathcal{L}$, and the term $R_m(\vec{u}_{m-1}, \mathbf{r}, t)$ can be expressed simply by (3.15) for any given nonlinear operator $\mathcal{N}$. According to the definition (3.15), the right-hand side of Equation (3.14) is only dependent upon $\vec{u}_{m-1}$. Thus, we gain $u_1(\mathbf{r}, t), u_2(\mathbf{r}, t), \cdots$ by means of solving the linear high-order deformation equation (3.14) one after the other in order. The mth-order approximation of $u(\mathbf{r}, t)$ is given by

$$u(\mathbf{r}, t) \approx \sum_{k=0}^{m} u_k(\mathbf{r}, t). \quad (3.17)$$

We can construct the zero-order deformation equation in a form even more general than (3.4). Let $A(q), B(q)$ be complex functions analytic in the region $|q| \leq 1$, called the embedding functions, which satisfy

$$A(0) = B(0) = 0, \quad A(1) = B(1) = 1, \quad (3.18)$$

respectively. Let

$$A(q) = \sum_{k=1}^{+\infty} \alpha_k \, q^k, \quad B(q) = \sum_{k=1}^{+\infty} \beta_k \, q^k \quad (3.19)$$

denote the Maclaurin series of $A(q)$ and $B(q)$, respectively. Because $A(q)$ and $B(q)$ are analytic in the region $|q| \leq 1$, we have from (3.18) that

$$\sum_{k=1}^{+\infty} \alpha_k = 1, \quad \sum_{k=1}^{+\infty} \beta_k = 1. \quad (3.20)$$

Then, we construct the zero-order deformation equation in a more general form

$$[1 - B(q)] \{\mathcal{L}[\Phi(\mathbf{r}, t; q) - u_0(\mathbf{r}, t)]\} = A(q) \, \hbar \, H(\mathbf{r}, t) \, \mathcal{N}[\Phi(\mathbf{r}, t; q)]. \quad (3.21)$$

All other related formulae are the same, except the high-order deformation equation which is now in a more general form

$$\mathcal{L}\left[u_m(\mathbf{r}, t) - \sum_{k=1}^{m-1} \beta_k \, u_{m-k}(\mathbf{r}, t) \right] = \hbar \, H(\mathbf{r}, t) \, R_m(\vec{u}_{m-1}, \mathbf{r}, t), \quad (3.22)$$

where

$$R_m(\vec{u}_{m-1}, \mathbf{r}, t) = \sum_{k=1}^{m} \alpha_k \, \delta_{m-k}(\mathbf{r}, t) \quad (3.23)$$

Systematic Description

under the definition

$$\delta_n(\mathbf{r}, t) = \frac{1}{n!} \left. \frac{\partial^n \mathcal{N}[\Phi(\mathbf{r}, t; q)]}{\partial q^n} \right|_{q=0}. \tag{3.24}$$

The zero-order deformation equation (3.4) and the high-order deformation equation (3.14) are clearly special cases of Equations (3.21) and (3.22) when $A(q) = B(q) = q$, respectively.

Note that, in general, a nonlinear problem might be described by a set of governing equations with related initial/boundary conditions. For the sake of brevity, only one equation (3.1) is employed here to systematically describe the basic ideas of the homotopy analysis method. However, the form of equation (3.1) is so general that it can denote either a governing equation or a boundary/initial condition. It may be a differential equation, an integral equation, an integro-differential equation, or an algebraic equation. All governing equations and boundary conditions can be treated in a similar way, although for different governing equations and intial/boundary conditions we should choose different initial approximations, different auxiliary linear operators, and different types of embedding functions $A(q), B(q)$. In addition, it is unnecessary for us to assume the existence of any small/large quantities in governing equations or initial/boundary conditions. Therefore, the analytic approach described above is very general.

3.3 Convergence theorem

The convergence of a series is important. A series is often of no use if it is convergent in a rather restricted region. In general cases, we can prove that, as long as the solution series (3.13) given by the homotopy analysis method is convergent, it must be the solution of the considered nonlinear problem.

THEOREM 3.1 Convergence theorem
As long as the series

$$u_0(\mathbf{r}, t) + \sum_{m=1}^{+\infty} u_m(\mathbf{r}, t)$$

is convergent, where $u_m(\mathbf{r}, t)$ is governed by the high-order deformation equation (3.22) under the definitions (3.23), (3.24), and (2.42), it must be a solution of Equation (3.1).

Proof: Let

$$s(\mathbf{r}, t) = u_0(\mathbf{r}, t) + \sum_{m=1}^{+\infty} u_m(\mathbf{r}, t)$$

denote the convergent series. Using (3.22) and (2.42), we have

$$\hbar\, H(\mathbf{r},t) \sum_{m=1}^{+\infty} R_m(\vec{u}_{m-1},\mathbf{r},t)$$

$$= \sum_{m=1}^{+\infty} \mathcal{L}\left[u_m(\mathbf{r},t) - \sum_{k=1}^{m-1} \beta_k\, u_{m-k}(\mathbf{r},t)\right]$$

$$= \mathcal{L}\left[\sum_{m=1}^{+\infty} u_m(\mathbf{r},t) - \sum_{m=1}^{+\infty}\sum_{k=1}^{m-1} \beta_k\, u_{m-k}(\mathbf{r},t)\right]$$

$$= \mathcal{L}\left[\sum_{m=1}^{+\infty} u_m(\mathbf{r},t) - \sum_{k=1}^{+\infty}\sum_{m=k+1}^{+\infty} \beta_k\, u_{m-k}(\mathbf{r},t)\right]$$

$$= \mathcal{L}\left[\sum_{m=1}^{+\infty} u_m(\mathbf{r},t) - \sum_{k=1}^{+\infty} \beta_k \sum_{n=1}^{+\infty} u_n(\mathbf{r},t)\right]$$

$$= \mathcal{L}\left[\left(1 - \sum_{k=1}^{+\infty} \beta_k\right) \sum_{m=1}^{+\infty} u_m(\mathbf{r},t)\right],$$

$$= \mathcal{L}\left[\left(1 - \sum_{k=1}^{+\infty} \beta_k\right) s(\mathbf{r},t)\right],$$

which gives, since $\hbar \neq 0$, $H(\mathbf{r},t) \neq 0$ and from (3.20) and (3.2),

$$\sum_{m=1}^{+\infty} R_m(\vec{u}_{m-1},\mathbf{r},t) = 0. \qquad (3.25)$$

On the other side, we have according to the definitions (3.23) and (3.24), that

$$\sum_{m=1}^{+\infty} R_m(\vec{u}_{m-1},\mathbf{r},t) = \sum_{m=1}^{+\infty}\sum_{k=1}^{m} \alpha_k\, \delta_{m-k}(\mathbf{r},t)$$

$$= \sum_{k=1}^{+\infty}\sum_{m=k}^{+\infty} \alpha_k\, \delta_{m-k}(\mathbf{r},t) = \left(\sum_{k=1}^{+\infty} \alpha_k\right) \sum_{n=0}^{+\infty} \delta_n(\mathbf{r},t),$$

which gives from (3.20), (3.24), and (3.25)

$$\sum_{m=1}^{+\infty} R_m(\vec{u}_{m-1},\mathbf{r},t) = \sum_{m=0}^{+\infty} \delta_m(\mathbf{r},t)$$

$$= \sum_{m=0}^{+\infty} \frac{1}{m!} \left.\frac{\partial^m \mathcal{N}[\Phi(\mathbf{r},t;q)]}{\partial q^m}\right|_{q=0} = 0. \qquad (3.26)$$

In general, $\Phi(\mathbf{r},t;q)$ does not satisfy the original nonlinear equation (3.1). Let

$$\mathcal{E}(\mathbf{r},t;q) = \mathcal{N}[\Phi(\mathbf{r},t;q)]$$

Systematic Description

denote the residual error of Equation (3.1). Clearly,

$$\mathcal{E}(\mathbf{r}, t; q) = 0$$

corresponds to the exact solution of the original equation (3.1). According to the above definition, the Maclaurin series of the residual error $\mathcal{E}(\mathbf{r}, t; q)$ about the embedding parameter q is

$$\sum_{m=0}^{+\infty} \frac{q^m}{m!} \frac{\partial^m \mathcal{E}(\mathbf{r}, t; q)}{\partial q^m}\bigg|_{q=0} = \sum_{m=0}^{+\infty} \frac{q^m}{m!} \frac{\partial^m \mathcal{N}[\Phi(\mathbf{r}, t; q)]}{\partial q^m}\bigg|_{q=0}.$$

When $q = 1$, the above expression gives, using (3.26),

$$\mathcal{E}(\mathbf{r}, t; 1) = \sum_{m=0}^{+\infty} \frac{1}{m!} \frac{\partial^m \mathcal{E}(\mathbf{r}, t; q)}{\partial q^m}\bigg|_{q=0} = 0. \tag{3.27}$$

This means, according to the definition of $\mathcal{E}(\mathbf{r}, t; q)$, that we gain the exact solution of the original equation (3.1) when $q = 1$. Thus, as long as the series

$$u_0(\mathbf{r}, t) + \sum_{m=1}^{+\infty} u_m(\mathbf{r}, t)$$

is convergent, it must be one solution of the original equation (3.1). This ends the proof.

THEOREM 3.2
As long as the series

$$u_0(\mathbf{r}, t) + \sum_{m=1}^{+\infty} u_m(\mathbf{r}, t)$$

is convergent, where $u_m(\mathbf{r}, t)$ is governed by the high-order deformation equation (3.22) under the definitions (3.23), (3.24), and (2.42), it holds that

$$\sum_{m=1}^{+\infty} R_m(\vec{u}_{m-1}, \mathbf{r}, t) = \sum_{m=0}^{+\infty} \delta_m(\mathbf{r}, t) = 0.$$

Proof: Using (3.26), the proof of this theorem is straightforward. This ends the proof.

Note that Equation (3.14) is only a special case of Equation (3.22) when $A(q) = B(q) = q$. We therefore have the following.

THEOREM 3.3
As long as the series

$$u_0(\mathbf{r}, t) + \sum_{m=1}^{+\infty} u_m(\mathbf{r}, t)$$

is convergent, where $u_m(\mathbf{r},t)$ is governed by the high-order deformation equation (3.14) under the definitions (2.42) and (3.15), it must be a solution of Equation (3.1). It holds therefore that

$$\sum_{m=1}^{+\infty} R_m(\vec{u}_{m-1}, \mathbf{r}, t) = 0.$$

According to Theorem 3.1 and Theorem 3.3, we need only focus on choosing the initial approximation $u_0(\mathbf{r},t)$, the auxiliary linear operator $\mathcal{L}$, the embedding functions $A(q), B(q)$, the auxiliary parameter $\hbar$, and the auxiliary function $H(\mathbf{r},t)$ to ensure that the solution series (3.13) converges. Theorem 3.2 provides us with an alternative method to estimate the convergence and accuracy of approximation series given by the homotopy analysis method.

3.4 Fundamental rules

Perturbation techniques and other nonperturbation methods for nonlinear problems are, more or less, based on some assumptions. Similarly, the homotopy analysis method is also based on the assumptions listed on page 55. Theoretically speaking, these assumptions impair the method. However, the homotopy analysis method provides us with great freedom to choose the initial approximation $u_0(\mathbf{r},t)$, the auxiliary linear operator $\mathcal{L}$, the auxiliary parameter $\hbar$, and the auxiliary function $H(\mathbf{r},t)$. Such freedom is so great that it is almost quite possible for us to satisfy all of these assumptions. This kind of freedom is therefore a cornerstone of the validity and flexibility of the homotopy analysis method.

However, from the view points of practical applications, the freedom seems too great. It is therefore better to have some fundamental rules to direct us to choose the initial approximation $u_0(\mathbf{r},t)$, the auxiliary linear operator $\mathcal{L}$, and the auxiliary function $H(\mathbf{r},t)$. We must first emphasize two facts. First, a solution of a nonlinear problem may be expressed by different sets of base functions, as illustrated in Chapter 2. Second, in many cases, from physical characteristics and boundary/initial conditions, it is often not very difficult to determine the type of base functions convenient to represent solutions of a given nonlinear problem, even without solving it. So, given a nonlinear problem, we can first choose a set of base function to present its solutions. This kind of presentation provides us with the *rule of solution expression*. For example, support that we choose a set of base functions

$$\{e_n(\mathbf{r},t) \mid n = 0, 1, 2, 3, \cdots\} \tag{3.28}$$

to represent the solution $u(\mathbf{r},t)$ of Equation (3.1) by

$$u(\mathbf{r},t) = \sum_{n=0}^{+\infty} c_n \, e_n(\mathbf{r},t), \qquad (3.29)$$

where c_n is a coefficient. The above expression provides the so-called *rule of solution expression* for Equation (3.1). To obey the *rule of solution expression*, the initial approximation $u_0(\mathbf{r},t)$ must be expressed by a sum of the base functions, i.e.,

$$u_0(\mathbf{r},t) = \sum_{n=0}^{M_0} a_n \, e_n(\mathbf{r},t), \qquad (3.30)$$

where a_n is a coefficient and M_0 is an integer. To obey the *rule of solution expression* denoted by (3.29), the auxiliary linear operator $\mathcal{L}$ must be chosen in such a way that the solution of the equation

$$\mathcal{L}[w(\mathbf{r},t)] = 0$$

must be expressed by a sum of the base functions, say,

$$w(\mathbf{r},t) = \sum_{n=0}^{M_1} b_n \, e_n(\mathbf{r},t), \qquad (3.31)$$

where b_n is a coefficient and the integer M_1 is determined by the highest order of the derivative of linear operator $\mathcal{L}$, which is generally the same as the highest order of the derivative of the original equation (3.1). This is because the solution of the high-order deformation equation (3.22) can be expressed by

$$u_m(\mathbf{r},t) = u_m^*(\mathbf{r},t) + w(\mathbf{r},t),$$

where $u_m^*(\mathbf{r},t)$ is a special solution of Equation (3.22). Furthermore, to obey the *rule of solution expression* denoted by (3.29), the auxiliary function $H(\mathbf{r},t)$ should be chosen so that the special solution $u_m^*(\mathbf{r},t)$ of the high-order deformation equation (3.22) must be expressed by a sum of the base functions, say,

$$u_m^*(\mathbf{r},t) = \mathcal{L}^{-1}[\hbar \, H(\mathbf{r},t) \, R_m(\vec{u}_{m-1},\mathbf{r},t)] = \sum_{n=0}^{M_2} d_n \, e_n(\mathbf{r},t), \qquad (3.32)$$

where d_n is a coefficient and $\mathcal{L}^{-1}$ is the inverse operator of the auxiliary linear operator $\mathcal{L}$. In this way, the *rule of solution expression* directs us to choose the initial approximation $u_0(\mathbf{r},t)$, the auxiliary linear operator $\mathcal{L}$, and the auxiliary function $H(\mathbf{r},t)$. Using the so-called *rule of solution expression*, we can easily avoid the appearance of the so-called secular terms in solution expressions, as illustrated in this book. Therefore, the *rule of solution expression* practically provides us with a starting point and therefore plays a very important role within the frame of the homotopy analysis method.

It is found that, in most cases, the auxiliary function $H(\mathbf{r},t)$ cannot be uniquely determined by above-mentioned *rule of solution expression*. Thus, more restrictions should be given to direct us to choose the auxiliary function $H(\mathbf{r},t)$. Note that, from the view point of the completeness, each base $e_n(\mathbf{r},t)$ of the set denoted by (3.28) should appear in the solution expression (3.29). In other words, each coefficient $c_{m,n}$ of the mth-order approximate solution

$$u(\mathbf{r},t) \approx \sum_{n=1}^{m} u_m(\mathbf{r},t) = \sum_{n=0}^{M_3} c_{m,n}\, e_n(\mathbf{r},t) \qquad (3.33)$$

can be modified as the order of approximation tends to infinity. This provides us with the so-called *rule of coefficient ergodicity*, i.e., *as the order of approximation tends to infinity, each base should appear in the solution expression and each coefficient can be modified*. This further restricts the choice of the auxiliary function. In many cases, using the *rule of solution expression* and the *rule of coefficient ergodicity*, we can uniquely determine the auxiliary function $H(\mathbf{r},t)$, as illustrated in this book. Thus, the *rule of coefficient ergodicity* also plays a very important role within the frame of the homotopy analysis method.

Using (3.13), the original nonlinear problem is transformed into an infinite number of linear subproblems governed by the high-order deformation equation (3.14) or (3.22). So, if the original nonlinear problem has a solution, all of these linear subproblems should have solutions too. Thus, we have the so-called *rule of solution existence*, i.e., the initial approximation $u_0(\mathbf{r},t)$, the auxiliary linear operator $\mathcal{L}$, and the auxiliary function $H(\mathbf{r},t)$ should be chosen so that all of the high-order deformation equation (3.14) or (3.22) are closed and have solutions, if the original nonlinear problem has a solution. This rule further restricts the choice of the initial approximation $u_0(\mathbf{r},t)$, the auxiliary linear operator $\mathcal{L}$, and the auxiliary function $H(\mathbf{r},t)$.

The above-mentioned *rule of solution expression* and *rule of coefficient ergodicity*, in addition to *rule of solution existence*, direct us to choose the initial approximation $u_0(\mathbf{r},t)$, the auxiliary linear operator $\mathcal{L}$, and the auxiliary function $H(\mathbf{r},t)$. These rules considerably simplify the application of the homotopy analysis method.

3.5 Control of convergence region and rate

It is important to ensure that a solution series is convergent in a large enough region. In general, the convergence region and rate of solution series are mainly determined by the base functions used to represent the solution series. Unlike previous analytic techniques, the homotopy analysis method provides us with great freedom to represent solutions of a given nonlinear problem

Systematic Description 63

by different base functions. Therefore, by means of the homotopy analysis method, we can gain solution series convergent in a whole region having physical meanings, as illustrated in this book. The so-called *rule of solution expression* is most important and the key, which determines the choice of the initial approximation $u_0(\mathbf{r},t)$, the auxiliary linear operator $\mathcal{L}$, and the auxiliary function $H(\mathbf{r},t)$.

Even if the initial approximation $u_0(\mathbf{r},t)$, the auxiliary linear operator $\mathcal{L}$, and the auxiliary function $H(\mathbf{r},t)$ are given, we still have great freedom to choose the value of the auxiliary parameter $\hbar$. Unlike all previous analytic techniques, the homotopy analysis method always provides us with a *family* of solution expressions in the auxiliary parameter $\hbar$. It is found that the auxiliary parameter $\hbar$ often affects convergence region and rate of solution series, as shown and proved in Chapter 2. The influence of $\hbar$ on the convergence region and rate becomes obvious, especially when a "bad" set of base function is chosen, as illustrated in Chapter 2. It is found that the convergence region and rate of solution series can be easily adjusted and controlled by means of setting $\hbar$ proper values. Thus, unlike all previous analytic techniques, the homotopy analysis method provides us with a convenient way to control and adjust convergence region and rate of solution series.

3.5.1 The $\hbar$-curve and the valid region of $\hbar$

Assume that we gain a family of solution series in the auxiliary parameter $\hbar$ by means of homotopy analysis method. How does one then to choose the value of $\hbar$ to ensure that the solution series converges fast enough in a large enough region?

Many nonlinear problems contain important physical quantities such as frequency of a nonlinear oscillator, wall skin friction of viscous flow, and so on. Because we have a family of solution expressions in the auxiliary parameter $\hbar$, those physical quantities also depend upon $\hbar$. So, regarding $\hbar$ as an independent variable, it is easy to plot curves of these kinds of quantities versus $\hbar$. For example, assume that

$$\gamma = \ddot{u}(\mathbf{r},t)|_{\mathbf{r}=0,t=0}$$

corresponds to a quantity having important physical meaning, where the dot denotes the derivative with respect to the time t. Then, γ is a function of $\hbar$ and thus can be plotted by a curve $\gamma \sim \hbar$. According to Theorem 3.1 or Theorem 3.3, all convergent series of γ given by different values of $\hbar$ converge to its exact value. So, if the solution is unique, all of them converge to the same value and therefore there exists a horizontal line segment in the figure of $\gamma \sim \hbar$ that corresponds to a region of $\hbar$ denoted by $\mathbf{R}_\hbar$. For the sake of brevity we call such a kind of curve *the $\hbar$-curve* and the corresponding region $\mathbf{R}_\hbar$ *the valid region of $\hbar$*, respectively. Thus, if we set $\hbar$ any value in the so-called valid region of $\hbar$, we are quite sure that the corresponding solution series converge. Certainly, if there exist many such kinds of quantities, we can plot corresponding $\hbar$-curves of them. And even if the term denoted by γ has no physical meanings, we can still plot the corresponding $\hbar$-curves. Obviously, the more the so-called $\hbar$-curves are plotted, the clearer it is to choose the value of $\hbar$. It is found that, for given initial approximation $u_0(\mathbf{r},t)$, the auxiliary linear operator $\mathcal{L}$, and the auxiliary function $H(\mathbf{r},t)$, the valid regions of $\hbar$ for different special quantities are often nearly the same for a given problem, although up to now we cannot give a mathematical proof in general. In most cases, using the same $\hbar$-curve gained by a special quantity such as γ mentioned above, we can find a proper value of $\hbar$ to ensure that the solution series of $u(\mathbf{r},t)$ converges in the whole spacial and temporal regions having physical meanings. So, the so-called $\hbar$-curve provides us with a convenient way to show the influence of $\hbar$ on the convergence region and rate of solution series.

3.5.2 Homotopy-Padé technique

The Padé technique is widely applied to enlarge the convergence region and rate of a given series. Traditionally, the $[m,n]$ Padé approximant of $u(\mathbf{r},t)$ is expressed by either

$$\frac{\sum_{k=0}^{m} F_k(\mathbf{r})\, t^k}{1 + \sum_{k=1}^{n} F_{m+1+k}(\mathbf{r})\, t^k}$$

or

$$\frac{\sum_{k=0}^{m} G_k(t)\, \mathbf{r}^k}{1 + \sum_{k=1}^{n} G_{m+1+k}(t)\, \mathbf{r}^k},$$

where $F_k(\mathbf{r})$ and $G_k(t)$ are functions. Note that the numerator and denominator are polynomial of either the spatial variable $\mathbf{r}$ or the temporal variable t.

The Padé technique can be employed within the frame of the homotopy analysis method. As mentioned before, the homotopy analysis method is based on such an assumption that the series (3.12) is convergent at $q = 1$ because the solution series (3.13) is obtained by setting $q = 1$ in (3.12). So, it is important to ensure that the series (3.12) is convergent at $q = 1$. We first employ the traditional Padé technique to the series (3.12) about the embedding parameter q to gain the $[m,n]$ Padé approximant

$$\frac{\sum_{k=0}^{m} W_k(\mathbf{r},t)\, q^k}{1 + \sum_{k=1}^{n} W_{m+1+k}(\mathbf{r},t)\, q^k},$$

Systematic Description

where $W_k(\mathbf{r}, t)$ is a function determined by the first several approximations

$$u_j(\mathbf{r}, t), \qquad j = 0, 1, 2, 3, \cdots, m+n.$$

Then, using (3.8), we set $q = 1$ to get the so-called $[m, n]$ homotopy-Padé approximant

$$\frac{\sum_{k=0}^{m} W_k(\mathbf{r}, t)}{1 + \sum_{k=1}^{n} W_{m+1+k}(\mathbf{r}, t)}.$$

It is found that the $[m, n]$ homotopy-Padé approximant often converges faster than the corresponding traditional $[m, n]$ Padé approximant. In many cases, the $[m, m]$ homotopy-Padé approximant does not depend upon the auxiliary parameter $\hbar$. In this case, we can gain convergent solution by means of the homotopy-Padé technique even if the solution series is divergent. However, up to now, we cannot prove it in general.

It is flexible to apply the homotopy-Padé technique to accelerate related solution series. For example, we can employ it to accelerate the convergence of the series

$$\dot{u}(\mathbf{r}, t) = \dot{u}_0(\mathbf{r}, t) + \sum_{n=1}^{+\infty} \dot{u}_n(\mathbf{r}, t).$$

First of all, from (3.12), we have the series

$$\frac{\partial \Phi(\mathbf{r}, t; q)}{\partial t} = \dot{u}_0(\mathbf{r}, t) + \sum_{n=1}^{+\infty} \dot{u}_n(\mathbf{r}, t) \, q^n.$$

Then, applying the Padé technique to the above series about the embedding parameter q, we have the traditional $[m, n]$ Padé approximation

$$\frac{\sum_{k=0}^{m} V_k(\mathbf{r}, t) \, q^k}{1 + \sum_{k=1}^{n} V_{m+1+k}(\mathbf{r}, t) \, q^k},$$

where $V_k(\mathbf{r}, t)$ is a function of $\mathbf{r}$ and t. Setting $q = 1$ in the above expression, we have using (3.8) the $[m, n]$ homotopy-Padé approximant

$$\dot{u}(\mathbf{r}, t) \approx \frac{\sum_{k=0}^{m} V_k(\mathbf{r}, t)}{1 + \sum_{k=1}^{n} V_{m+1+k}(\mathbf{r}, t)}.$$

In summary, the $\hbar$-curve provides us with a convenient way to determine the valid region of $\hbar$. In addition, the so-called homotopy-Padé technique can greatly enlarge the convergence region and rate of solution series. In many cases, the homotopy-Padé technique is more efficient than the traditional Padé method and is even independent of the auxiliary parameter $\hbar$. So, by means of choosing a proper set of base functions, selecting a proper value of $\hbar$, or employing the homotopy-Padé technique, we can gain accurate approximations convergent in a large enough region within the frame of the homotopy analysis method.

3.6 Further generalization

The homotopy analysis method can be further generalized by means of the zero-order deformation equation in the form

$$[1 - B(q)] \{\mathcal{L}[\Phi(\mathbf{r},t;q) - u_0(\mathbf{r},t)]\}$$
$$= A(q) \, \hbar \, H(\mathbf{r},t) \, \mathcal{N}[\Phi(\mathbf{r},t;q)] + \hbar_2 \, H_2(\mathbf{r},t) \, \Pi[\Phi(\mathbf{r},t;q);q], \quad (3.34)$$

where $u_0(\mathbf{r},t), \mathcal{L}, H(\mathbf{r},t), \hbar, A(q)$, and $B(q)$ are defined as before, $\hbar_2$ is the second auxiliary parameter, $H_2(\mathbf{r},t)$ is the second auxiliary function, and $\Pi[\Phi(\mathbf{r},t);q]$ is an auxiliary operator which equals to zero when $q = 0$ and $q = 1$, i.e.,

$$\Pi[\Phi(\mathbf{r},t;0);0] = \Pi[\Phi(\mathbf{r},t;1);1] = 0. \quad (3.35)$$

All other related formulae are the same, except the high-order deformation equation in a more general form

$$\mathcal{L}\left[u_m(\mathbf{r},t) - \sum_{k=1}^{m-1} \beta_k \, u_{m-k}(\mathbf{r},t)\right]$$
$$= \hbar \, H(\mathbf{r},t) \, R_m(\vec{u}_{m-1},\mathbf{r},t) + \hbar_2 \, H_2(\mathbf{r},t) \, \Delta_m(\mathbf{r},t), \quad (3.36)$$

where

$$\Delta_m(\mathbf{r},t) = \frac{1}{m!} \left. \frac{\partial^m \Pi\left[\Phi(\mathbf{r},t;q);q\right]}{\partial q^m} \right|_{q=0}. \quad (3.37)$$

In this way, we introduce the additional auxiliary parameter $\hbar_2$ and auxiliary function $H_2(\mathbf{r},t)$, and more importantly, an auxiliary operator $\Pi[\Phi(\mathbf{r},t;q);q]$. In this way the flexibility of the homotopy analysis method is further increased. Note that the solution series given by Equation (3.36) is now a family of two parameters, $\hbar$ and $\hbar_2$.

It is rather flexible to choose the auxiliary operator $\Pi[\Phi(\mathbf{r},t;q);q]$ which satisfies the property (3.35). For example, we can choose

$$\Pi[\Phi(\mathbf{r},t;q);q] = A(q)[1 - B(q)]F[\Phi(\mathbf{r},t;q)], \quad (3.38)$$

Systematic Description

where $F[\Phi(\mathbf{r},t;q)]$ is a function, or

$$\Pi[\Phi(\mathbf{r},t;q);q] = [1 - A(q)]\left\{[\Phi(\mathbf{r},t;q)]^{1+q} - \Phi(\mathbf{r},t;q)\right\}, \qquad (3.39)$$

and so on. However, it is under investigation how to choose the additional auxiliary parameter $\hbar_2$, the additional auxiliary function $H_2(\mathbf{r},t)$, and the auxiliary operator $\Pi[\Phi(\mathbf{r},t;q);q]$ for a given nonlinear problem in general. For the applications of the zero-order deformation equation in the form (3.34), the reader is referred to §4.3 and §12.1.

4

Relations to some previous analytic methods

In this chapter we reveal the relationships between the homtopy analysis method and other nonperturbation techniques such as Adomian's decomposition method, Lyapunov's artificial small parameter method, and the δ-expansion method. We show that these methods can be unified by the homotopy analysis method.

4.1 Relation to Adomian's decomposition method

Adomian's decomposition method [23, 24, 25] is a well-known, easy-to-use analytic tool for nonlinear problems and has been widely applied in science and engineering [63, 64, 65, 66, 67, 68, 69, 70, 71, 72, 73, 74, 75, 76, 77, 78, 79]. In Chapter 2 we show by an example that the solution expression (2.17) given by Adomian's decomposition method is just a special one of the solution expressions (2.57) given by the homotopy analysis method. In this section we prove that the homotopy analysis method logically contains Adomian's decomposition method in general.

To simply describe the basic ideas of Adomian's decomposition method, let us consider a nonlinear problem governed by

$$\mathcal{N}[u(\mathbf{r},t)] = f(\mathbf{r},t), \tag{4.1}$$

where $\mathcal{N}$ is a nonlinear operator, u is a dependent variable, $f(\mathbf{r},t)$ is a known function, and $\mathbf{r}$ and t denote the spatial and temporal variables, respectively. Assume that the nonlinear operator $\mathcal{N}$ can be divided into

$$\mathcal{N} = \mathcal{L}_0 + \mathcal{N}_0, \tag{4.2}$$

where $\mathcal{L}_0$ and $\mathcal{N}_0$ are linear and nonlinear operators, respectively. Under this assumption the original nonlinear equation becomes

$$\mathcal{L}_0[u(\mathbf{r},t)] + \mathcal{N}_0[u(\mathbf{r},t)] = f(\mathbf{r},t). \tag{4.3}$$

By means of Adomian's decomposition method we express $u(\mathbf{r},t)$ in such a series

$$u(\mathbf{r},t) = u_0(\mathbf{r},t) + \sum_{n=1}^{+\infty} u_n(\mathbf{r},t), \tag{4.4}$$

where
$$u_0(\mathbf{r}, t) = \mathcal{L}_0^{-1}[f(\mathbf{r}, t)] \qquad (4.5)$$
and
$$u_n(\mathbf{r}, t) = -\mathcal{L}_0^{-1}[A_{n-1}(\mathbf{r}, t)], \qquad n \geq 1, \qquad (4.6)$$
in which $\mathcal{L}_0^{-1}$ is the inverse operator of $\mathcal{L}_0$, and $A_n(\mathbf{r}, t)$ is the so-called Adomian polynomial defined by (see Cherruault [66] and Babolian et al. [75])
$$A_n(\mathbf{r}, t) = \frac{1}{n!} \left[\frac{d^n}{dq^n} \mathcal{N}_0 \left(u_0(\mathbf{r}, t) + \sum_{n=1}^{+\infty} u_n(\mathbf{r}, t) q^n \right) \right]\bigg|_{q=0}. \qquad (4.7)$$

Unlike Adomian's decomposition method, the homotopy analysis method is valid even without the assumption denoted by (4.2). Let $\mathcal{L}$ denote an auxiliary linear operator, $u_0(\mathbf{r}, t)$ an initial approximation that is unnecessary to be given by (4.5), $\hbar$ a nonzero auxiliary parameter, $H(\mathbf{r}, t)$ a nonzero auxiliary function, and $q \in [0, 1]$ an imbedding parameter, respectively. By means of the homotopy analysis method, we construct the so-called zero-order deformation equation

$$(1 - q) \mathcal{L} [\Phi(\mathbf{r}, t; q) - u_0(\mathbf{r}, t)] = \hbar\, q\, H(\mathbf{r}, t) \{\mathcal{N}[\Phi(\mathbf{r}, t; q)] - f(\mathbf{r}, t)\}, \qquad (4.8)$$

where $\Phi(\mathbf{r}, t; q)$ is a unknown dependent variable. It clearly holds

$$\Phi(\mathbf{r}, t; 0) = u_0(\mathbf{r}, t) \qquad (4.9)$$
and
$$\Phi(\mathbf{r}, t; 1) = u(\mathbf{r}, t) \qquad (4.10)$$

when $q = 0$ and $q = 1$, respectively. Thus, the unknown function $\Phi(\mathbf{r}, t; q)$ governed by Equation (4.8) deforms from the initial approximation $u_0(\mathbf{r}, t)$ to the exact solution $u(\mathbf{r}, t)$ of the original equation (4.1) as the embedding parameter q increases from 0 to 1. By Taylor's theorem and using (4.9) we expand $\Phi(\mathbf{r}, t; q)$ in a power series of q in the form

$$\Phi(\mathbf{r}, t; q) = u_0(\mathbf{r}, t) + \sum_{n=1}^{+\infty} u_n(\mathbf{r}, t)\, q^n, \qquad (4.11)$$

where
$$u_n(\mathbf{r}, t) = \frac{1}{n!} \frac{d^n \Phi(\mathbf{r}, t; q)}{dq^n}\bigg|_{q=0}. \qquad (4.12)$$

The zero-order deformation equation (4.8) contains the initial approximation $u_0(\mathbf{r}, t)$, the auxiliary linear operator $\mathcal{L}$, the auxiliary parameter $\hbar$, the auxiliary function $H(\mathbf{r}, t)$, and more importantly, we have great freedom to choose them. Assuming that all of them are properly chosen so that the series (4.11) converges at $q = 1$, we have, using (4.10), the solution series

$$u(\mathbf{r}, t) = u_0(\mathbf{r}, t) + \sum_{n=1}^{+\infty} u_n(\mathbf{r}, t). \qquad (4.13)$$

Note that in form this expression is the same as (4.4).

Differentiating the zero-order deformation equation (4.8) n times with respect to q and then dividing it by $n!$ and finally setting $q = 0$, we have the first-order deformation equation (when $n = 1$)

$$\mathcal{L}\left[u_1(\mathbf{r}, t)\right] = \hbar\, H(\mathbf{r}, t)\, \{\mathcal{N}\left[u_0(\mathbf{r}, t)\right] - f(\mathbf{r}, t)\} \tag{4.14}$$

and the nth-order deformation equation (when $n \geq 2$)

$$\mathcal{L}\left[u_n(\mathbf{r}, t) - u_{n-1}(\mathbf{r}, t)\right] = \hbar\, H(\mathbf{r}, t)\, R_n(\mathbf{r}, t), \tag{4.15}$$

where

$$R_n(\mathbf{r}, t) = \frac{1}{(n-1)!} \left.\frac{d^{n-1}\mathcal{N}\left[\Phi(\mathbf{r}, t; q)\right]}{dq^{n-1}}\right|_{q=0}. \tag{4.16}$$

We then can prove that Adomian's decomposition method is just a special case of the homotopy analysis method under the assumption (4.2). Because we have great freedom to choose the auxiliary linear operator $\mathcal{L}$ and the initial guess $u_0(\mathbf{r}, t)$, we certainly can choose

$$\mathcal{L} = \mathcal{L}_0, \quad u_0(\mathbf{r}, t) = \mathcal{L}_0^{-1}\left[f(\mathbf{r}, t)\right]. \tag{4.17}$$

Setting

$$\hbar = -1, \quad H(\mathbf{r}, t) = 1 \tag{4.18}$$

and substituting (4.2) and (4.17) into Equations (4.14) and (4.15), we have

$$\mathcal{L}_0\left[u_1(\mathbf{r}, t)\right] = f(\mathbf{r}, t) - \mathcal{L}_0\left[u_0(\mathbf{r}, t)\right] - \mathcal{N}_0\left[u_0(\mathbf{r}, t)\right] \tag{4.19}$$

and

$$\begin{aligned}
&\mathcal{L}_0\left[u_n(\mathbf{r}, t)\right] \\
&= \mathcal{L}_0\left[u_{n-1}(\mathbf{r}, t)\right] - \frac{1}{(n-1)!} \left.\frac{d^{n-1}\mathcal{L}_0\left[\Phi(\mathbf{r}, t; q)\right]}{dq^{n-1}}\right|_{q=0} \\
&\quad - \frac{1}{(n-1)!} \left.\frac{d^{n-1}\mathcal{N}_0\left[\Phi(\mathbf{r}, t; q)\right]}{dq^{n-1}}\right|_{q=0}, \quad n \geq 2,
\end{aligned} \tag{4.20}$$

respectively. From (4.17), it holds

$$f(\mathbf{r}, t) - \mathcal{L}_0\left[u_0(\mathbf{r}, t)\right] = 0$$

so that Equation (4.19) becomes, by the definition (4.7),

$$\mathcal{L}_0\left[u_1(\mathbf{r}, t)\right] = -A_0(\mathbf{r}, t), \tag{4.21}$$

where $A_0(\mathbf{r}, t)$ is an Adomian polynomial. According to definition (4.12), it holds

$$\mathcal{L}_0\left[u_{n-1}(\mathbf{r}, t)\right] - \frac{1}{(n-1)!} \left. \frac{d^{n-1}\mathcal{L}_0\left[\Phi(\mathbf{r}, t; q)\right]}{dq^{n-1}} \right|_{q=0}$$

$$= \mathcal{L}_0\left[u_{n-1}(\mathbf{r}, t)\right] - \mathcal{L}_0 \left[\frac{1}{(n-1)!} \left. \frac{d^{n-1}\Phi(\mathbf{r}, t; q)}{dq^{n-1}} \right|_{q=0} \right]$$

$$= \mathcal{L}_0\left[u_{n-1}(\mathbf{r}, t)\right] - \mathcal{L}_0\left[u_{n-1}(\mathbf{r}, t)\right]$$

$$= 0. \tag{4.22}$$

Thus, Equation (4.20) becomes

$$\mathcal{L}_0\left[u_n(\mathbf{r}, t)\right] = -\frac{1}{(n-1)!} \left. \frac{d^{n-1}\mathcal{N}_0\left[\Phi(\mathbf{r}, t; q)\right]}{dq^{n-1}} \right|_{q=0}. \tag{4.23}$$

Substituting (4.11) of $\Phi(\mathbf{r}, t; q)$ into the above expression, we have, according to the definition (4.7) of the Adomian polynomial,

$$\mathcal{L}_0\left[u_n(\mathbf{r}, t)\right]$$

$$= -\frac{1}{(n-1)!} \left[\frac{d^{n-1}}{dq^{n-1}} \mathcal{N}_0\left(u_0(\mathbf{r}, t) + \sum_{n=1}^{+\infty} u_n(\mathbf{r}, t)\, q^n \right) \right] \bigg|_{q=0}$$

$$= -A_{n-1}(\mathbf{r}, t). \tag{4.24}$$

So, the solution of Equation (4.21) and Equation (4.24) can be uniformly expressed by

$$u_n(\mathbf{r}, t) = -\mathcal{L}_0^{-1}\left[A_{n-1}(\mathbf{r}, t)\right], \qquad n \geq 1, \tag{4.25}$$

which is exactly the same as the solution (4.6) given by Adomian's decomposition method. Therefore, Adomian's decomposition method is just a special case of the homotopy analysis method under the assumption (4.2) when

$$u_0(\mathbf{r}, t) = \mathcal{L}_0^{-1}\left[f(\mathbf{r}, t)\right], \quad \mathcal{L} = \mathcal{L}_0, \quad H(\mathbf{r}, t) = 1, \quad \hbar = -1.$$

Some points should be emphasized here. First, we have great freedom to choose the initial guess $u_0(\mathbf{r}, t)$, the auxiliary linear operator $\mathcal{L}$, and the auxiliary function $H(\mathbf{r}, t)$ different from the above expressions so that the solution of high-order deformation equations (4.14) and (4.15) can be expressed by better base functions than those employed by Adomian's decomposition method that often uses polynomials. Second, it is unnecessary for us to assume that the nonlinear operator $\mathcal{N}$ should be divided into the form (4.2). Finally but most importantly, solutions given by the homotopy analysis method contain the auxiliary parameter $\hbar$, which provides us with a simply way to adjust and control convergence region and rate of solution series. Therefore, the homotopy analysis method is more general than Adomian's decomposition method.

4.2 Relation to artificial small parameter method

In 1892 Lyapunov [21] proposed the so-called artificial small parameter method. In Chapter 2 we illustrate that the solution expression (2.15) given by Lyapunov's artificial small parameter method is just a special one of the solution expressions (2.57) given by the homotopy analysis method. In this section we prove that Lyapunov's artificial small parameter method is in essence equivalent to Adomian's decomposition method and therefore is also a special case of the homotopy analysis method.

To simply describe the basic ideas of Lyapunov's artificial small parameter method, let us consider a nonlinear equation

$$\mathcal{N}[u(\mathbf{r},t)] = f(\mathbf{r},t), \tag{4.26}$$

where $\mathcal{N}$ is a nonlinear operator, u is a dependent variable, $f(\mathbf{r},t)$ is a known function, and $\mathbf{r}$ and t denote the spatial and temporal variables, respectively. Assume that the nonlinear operator $\mathcal{N}$ can be divided into

$$\mathcal{N} = \mathcal{L}_0 + \mathcal{N}_0, \tag{4.27}$$

where $\mathcal{L}_0$ and $\mathcal{N}_0$ are linear and nonlinear operators, respectively. Using the above expression and introducing the artificial small parameter ϵ, the original equation (4.26) becomes

$$\mathcal{L}_0[\phi(\mathbf{r},t;\epsilon)] + \epsilon \mathcal{N}_0[\phi(\mathbf{r},t;\epsilon)] = f(\mathbf{r},t), \tag{4.28}$$

where $\phi(\mathbf{r},t;\epsilon)$ is an unknown function. When $\epsilon = 1$, the above equation is clearly the same as Equation (4.26) so that

$$\phi(\mathbf{r},t;1) = u(\mathbf{r},t). \tag{4.29}$$

Expanding $\phi(\mathbf{r},t;\epsilon)$ in a power series of the artificial small parameter ϵ, we have

$$\phi(\mathbf{r},t;\epsilon) = u_0(\mathbf{r},t) + \sum_{n=1}^{+\infty} u_n(\mathbf{r},t)\,\epsilon^n. \tag{4.30}$$

Setting $\epsilon = 1$ in the above expression we have, using (4.29),

$$u(\mathbf{r},t) = u_0(\mathbf{r},t) + \sum_{n=1}^{+\infty} u_n(\mathbf{r},t), \tag{4.31}$$

which in form is exactly the same as the solution expression (4.4) given by Adomian's decomposition method.

Substituting (4.30) into Equation (4.28), we have

$$\mathcal{L}_0[u_0(\mathbf{r},t)] - f(\mathbf{r},t) + \sum_{n=1}^{+\infty} \epsilon^n \, \mathcal{L}_0\left[u_n(\mathbf{r},t)\right]$$

$$+ \epsilon \, \mathcal{N}_0\left[u_0(\mathbf{r},t) + \sum_{n=1}^{+\infty} u_n(\mathbf{r},t)\, \epsilon^n\right] = 0. \tag{4.32}$$

Write

$$\mathcal{N}_0\left[u_0(\mathbf{r},t) + \sum_{n=1}^{+\infty} u_n(\mathbf{r},t)\, \epsilon^n\right] = \sum_{n=0}^{+\infty} w_n(\mathbf{r},t)\, \epsilon^n.$$

Differentiating both sides of the above expression m times with respect to the artificial small parameter ϵ and then setting $\epsilon = 0$, we have

$$\left\{\frac{\partial^m}{\partial \epsilon^m}\mathcal{N}_0\left[u_0(\mathbf{r},t) + \sum_{n=1}^{+\infty} u_n(\mathbf{r},t)\, \epsilon^n\right]\right\}\bigg|_{\epsilon=0} = m!\, w_m(\mathbf{r},t),$$

which gives, using the definition (4.7), that

$$w_m(\mathbf{r},t) = \frac{1}{m!}\left\{\frac{\partial^m}{\partial \epsilon^m}\mathcal{N}_0\left[u_0(\mathbf{r},t) + \sum_{n=1}^{+\infty} u_n(\mathbf{r},t)\, \epsilon^n\right]\right\}\bigg|_{\epsilon=0} = A_m(\mathbf{r},t),$$

where $A_m(\mathbf{r},t)$ is the so-called Adomian polynomial. So, substituting

$$\mathcal{N}_0\left[u_0(\mathbf{r},t) + \sum_{n=1}^{+\infty} u_n(\mathbf{r},t)\, \epsilon^n\right] = \sum_{n=0}^{+\infty} A_n(\mathbf{r},t)\, \epsilon^n$$

into Equation (4.32), we have

$$\{\mathcal{L}_0[u_0(\mathbf{r},t)] - f(\mathbf{r},t)\} + \sum_{n=1}^{+\infty} \epsilon^n \left\{\mathcal{L}_0\left[u_n(\mathbf{r},t)\right] + A_{n-1}(\mathbf{r},t)\right\} = 0,$$

which gives

$$\mathcal{L}_0[u_0(\mathbf{r},t)] - f(\mathbf{r},t) = 0$$

and

$$\mathcal{L}_0\left[u_n(\mathbf{r},t)\right] + A_{n-1}(\mathbf{r},t) = 0, \quad n \geq 1.$$

Solving the above equations successively, we have

$$u_0(\mathbf{r},t) = \mathcal{L}_0^{-1}\left[f(\mathbf{r},t)\right]$$

and

$$u_n(\mathbf{r},t) = -\mathcal{L}_0^{-1}\left[A_{n-1}(\mathbf{r},t)\right], \quad n \geq 1,$$

which are exactly the same as the solutions (4.5) and (4.6) given by Adomian's decomposition method, respectively. So, Adomian's decomposition method is in essence equivalent to the artificial small parameter method.

In §4.1 we prove that Adomian's decomposition method is just a special case of the homotopy analysis method. Therefore, Lyapunov's artificial small parameter method is also a special case of the homotopy analysis method under the assumption

$$\mathcal{N} = \mathcal{L}_0 + \mathcal{N}_0$$

when

$$\hbar = -1, \quad H(\mathbf{r}, t) = 1, \quad \mathcal{L} = \mathcal{L}_0, \quad u_0(\mathbf{r}, t) = \mathcal{L}_0^{-1}[f(\mathbf{r}, t)].$$

This is easy to understand if we regard the so-called artificial small parameter ϵ as the embedding parameter and Equation (4.28) as a special zero-order deformation equation.

4.3 Relation to δ-expansion method

In Chapter 2 only the solution expression (2.21) given by the δ-expansion method is not among the four families of solution expressions given by means of the homotopy analysis method. However, using the generalized zero-order deformation equation (3.34) in §3.6, we can show that the δ-expansion method is also a special case of the homotopy analysis method. To illustrate this point, let us consider the same example in Chapter 2, i.e.,

$$\dot{V}(t) + V^2(t) = 1, \qquad V(0) = 0. \tag{4.33}$$

To solve this problem by means of the homotopy analysis method, we choose an auxiliary linear operator

$$\mathcal{L}\Phi = \frac{\partial \Phi}{\partial t} + \Phi - 1 \tag{4.34}$$

and an initial approximation $V_0(t)$ satisfying

$$\mathcal{L}[V_0(t)] = 0, \qquad V_0(0) = 0,$$

which gives

$$V_0(t) = 1 - \exp(-t). \tag{4.35}$$

From Equation (4.33), we define the nonlinear operator

$$\mathcal{N}[\Phi(t; q), q] = \frac{\partial \Phi(t; q)}{\partial t} + [\Phi(t; q)]^{q+1} - 1. \tag{4.36}$$

Define the auxiliary operator

$$\Pi\left[\Phi(t;q),q\right] = (1-q)\left\{[\Phi(t;q)]^{q+1} - \Phi(t;q)\right\} \quad (4.37)$$

which equals zero when $q = 0$ and $q = 1$. Let $\hbar, \hbar_2$ denote the auxiliary parameters, and $H(t), H_2(t)$ the auxiliary functions, respectively. According to (3.34), we construct the zero-order deformation equation

$$(1-q)\mathcal{L}[\Phi(t;q) - V_0(t)] = q\,\hbar\,H(t)\,\mathcal{N}\left[\Phi(t;q),q\right] + \hbar_2\,H_2(t)\,\Pi\left[\Phi(t;q),q\right], \quad (4.38)$$

subject to the initial condition

$$\Phi(0;q) = 0. \quad (4.39)$$

When $q = 0$, it is straightforward that

$$\Phi(t;0) = V_0(t) = 1 - \exp(-t). \quad (4.40)$$

When $q = 1$, Equation (4.38) is equivalent to the original equation (4.33), provided

$$\Phi(t;1) = V(t). \quad (4.41)$$

Expand $\Phi(t;q)$ in a power series

$$\Phi(t;q) = \Phi(t;0) + \sum_{n=1}^{+\infty} V_n(t)\,q^n, \quad (4.42)$$

where

$$V_n(t) = \frac{1}{n!}\frac{\partial^n \Phi(t;q)}{\partial q^n}\bigg|_{q=0}. \quad (4.43)$$

Assuming that the series (4.42) is convergent at $q = 1$, we have using Equations (4.40) and (4.41)

$$V(t) = V_0(t) + \sum_{m=1}^{+\infty} V_m(t). \quad (4.44)$$

The governing equation of $V_m(t)$ is deduced by means of the definition (4.43). Differentiating the zero-order deformation equation (4.38) m times with respect to the embedding parameter q and then dividing by $m!$ and finally setting $q = 0$, we have the high-order deformation equation

$$\mathcal{L}_0[V_m(t) - \chi_m\,V_{m-1}(t)] = \hbar\,H(t)\,R_m(t) + \hbar_2\,H_2(t)\,\Delta_m(t), \quad (4.45)$$

subject to the initial condition

$$V_m(0) = 0, \quad (4.46)$$

where χ_m is defined by (2.42) and

$$R_m(t) = \frac{1}{(m-1)!} \frac{\partial^{m-1} \mathcal{N}\left[\Phi(t;q),q\right]}{\partial q^{m-1}}\bigg|_{q=0}, \qquad (4.47)$$

$$\Delta_m(t) = \frac{1}{m!} \frac{\partial^m \Pi\left[\Phi(t;q),q\right]}{\partial q^m}\bigg|_{q=0} \qquad (4.48)$$

under the definition

$$\mathcal{L}_0 \Phi = \frac{\partial \Phi}{\partial t} + \Phi. \qquad (4.49)$$

Substituting Equations (4.36) and (4.37) into Equations (4.47) and (4.48), respectively, we have

$$R_1(t) = \dot{V}_0(t) + V_0(t) - 1,$$
$$R_2(t) = \dot{V}_1(t) + V_1(t) + V_0(t) \ln V_0(t),$$
$$\vdots$$

and

$$\Delta_1(t) = V_0(t) \ln V_0(t),$$
$$\Delta_2(t) = -V_0(t) \ln V_0(t) + V_1(t) \left[1 + \ln V_0(t)\right] + \frac{1}{2} V_0(t) \ln^2 V_0(t),$$
$$\vdots$$

In the special case

$$\hbar = \hbar_2 = -1, \qquad H(t) = H_2(t) = 1, \qquad (4.50)$$

we have the high-order deformation equations

$$\dot{V}_1 + V_1 = -V_0 \ln V_0 - R_1(t), \qquad V_1(0) = 0,$$
$$\dot{V}_2 + V_2 = -V_1(1 + \ln V_0) - \frac{1}{2} V_0 \ln^2 V_0 - R_2(t), \qquad V_2(0) = 0,$$
$$\vdots$$

Solving the above high-order deformation equations successively, we obtain

$$V_1(t) = \exp(-t)\left[t - \frac{\pi^2}{6} + P_2^L(e^{-t})\right] - (1 - e^{-t}) \ln(1 - e^{-t}),$$
$$\vdots$$

where

$$P_n^L(z) = \sum_{k=1}^{+\infty} \frac{z^k}{k^n}$$

is the nth polylogarithm function of z. So, the first-order approximation is

$$V(t) \approx 1 + \exp(-t)\left[t - \frac{\pi^2}{6} - 1 + P_2^L(e^{-t})\right] - (1 - e^{-t})\ln(1 - e^{-t}), \quad (4.51)$$

which is exactly the same as the approximation (2.21) given by the δ-expansion method in Chapter 2. It should be emphasized that the solution expression given by Equations (4.45) and (4.46) contains the two auxiliary parameters $\hbar$ and $\hbar_2$ and thus is more general than the solution expression (2.21) given by the δ-expansion method. In fact, from (4.35) it holds $R_1(t) = 0$. Furthermore, using the first-order deformation equation, we have $R_2(t) = 0$. Thus, the high-order deformation equations are exactly the same as those given by the δ-expansion method in Chapter 2. Substituting

$$\hbar = \hbar_2 = -1, \quad H(t) = H_2(t) = 1$$

into the zero-order deformation equation (4.38) we have

$$\frac{\partial \Phi(t;q)}{\partial t} + [\Phi(t;q)]^{1+q} = 1, \quad (4.52)$$

which is the same as the equation

$$\dot{V}(t) + V^{1+\delta}(t) = 1$$

in Chapter 2 used by the δ-expansion method, if δ and $V(t)$ are replaced by q and $\Phi(t;q)$, respectively. In general, we can regard δ as an embedding parameter and the corresponding equation as a special zero-order deformation equation. Therefore, the δ-expansion method is only a special case of the homotopy analysis method.

4.4 Unification of nonperturbation methods

As shown above, Adomian's decomposition method, Lyapunov's artificial small parameter method, and the δ-expansion method are only special cases of the homotopy analysis method. Therefore, these three nonperturbation methods can be unified in the frame of the homotopy analysis method. A unified theory is often believed to be closer to the truth. This, from another side, further indicates the validity of the homotopy analysis method.

5

Advantages, limitations, and open questions

As all things have their good and bad sides, so too does the homotopy analysis method. Here, we make some discussions about the advantages and limitations of this method and point out some open questions.

5.1 Advantages

Compared with perturbation techniques and nonperturbation methods such as Lyapunov's artificial small parameter method, the δ-expansion method, and Adomian's decomposition method, the homotopy analysis method has some or all of the following advantages.

Firstly, unlike all previous analytic techniques, the homotopy analysis method provides us with great freedom to express solutions of a given nonlinear problem by means of different base functions. Therefore, we can approximate a nonlinear problem more efficiently by choosing a proper set of base functions. This is because the convergence region and rate of a series are chiefly determined by the base functions used to express the solution.

Secondly, unlike all previous analytic techniques, the homotopy analysis method always provides us with a family of solution expressions in the auxiliary parameter $\hbar$, even if a nonlinear problem has a unique solution. The convergence region and rate of each solution expression among the family might be determined by the auxiliary parameter $\hbar$. So, the auxiliary parameter $\hbar$ provides us with an additional way to conveniently adjust and control the convergence region and rate of solution series. By means of the so-called $\hbar$-curves it is easy to find out the so-called valid regions of $\hbar$ to gain a convergent solution series. In addition, the so-called homotopy-Padé technique is often more efficient than the traditional Padé technique and is in some cases even independent of the auxiliary parameter $\hbar$.

Thirdly, unlike perturbation techniques, the homotopy analysis method is independent of any small or large quantities. So, the homotopy analysis method can be applied no matter if governing equations and boundary/initial conditions of a given nonlinear problem contain small or large quantities or not.

Finally, the homotopy analysis method logically contains Lyapunov's artificial small parameter method, the δ-expansion method, and Adomian's decomposition method, and therefore unifies these nonperturbation methods and is more general than them.

It should be pointed out that the homotopy analysis method is based on the following assumptions:

(A) There exists the solution of the zero-order deformation equation in the whole region of the embedding parameter $q \in [0, 1]$.

(B) All of the high-order deformation equations have solutions.

(C) All Taylor series expanded in the embedding parameter q converge at $q = 1$.

Fortunately, the homotopy analysis method provides us with great freedom to choose initial approximation, the auxiliary linear operator, the auxiliary function, and the auxiliary parameter $\hbar$. This kind of freedom provides us with the great possibility to ensure that all of these assumptions may be satisfied. So, the above assumptions do little damage to the homotopy analysis method. In fact, nearly all of the above-listed advantages of the homotopy analysis method come from such kinds of freedom. According to Theorem 3.1 and Theorem 3.3, as long as a solution series given by the homotopy analysis method converges, it must be one of the solutions of a given nonlinear problem. Thus, we need only focus on choosing proper initial approximations, auxiliary linear operators, auxiliary functions, and proper values of $\hbar$ to ensure that solution series converge. Therefore, it is this kind of freedom that establishes a cornerstone of the validity and the flexibility of the homotopy analysis method.

5.2 Limitations

However, such kinds of freedom seem too great for us. Up to now, there are no rigorous theories to direct us to choose the initial approximations, auxiliary linear operators, auxiliary functions, and auxiliary parameter $\hbar$. From the practical viewpoints, we propose some fundamental rules such as the rule of solution expression, the rule of coefficient ergodicity, and the rule of solution existence, which play important roles within the homotopy analysis method. The rule of solution expression provides us with a starting point. It is under the rule of solution expression that initial approximations, auxiliary linear operators, and the auxiliary functions are determined. The rule of coefficient ergodicity and the rule of solution existence play important roles in determining the auxiliary function and ensuring that the high-order deformation equations are closed and have solutions.

The rule of coefficient ergodicity is based on the completeness, and the rule of solution existence is straightforward. So, the rule of coefficient ergodicity and the rule of solution existence are reasonable. Unfortunately, the rule of solution expression implies such an assumption that we should have, more or less, some knowledge about a given nonlinear problem *a prior*. How can we get such kind of prior knowledge before we solve a problem that is completely new for us? How can we know that a set of base functions is better than others and is more efficient to approximate a nonlinear problem which we know nothing? So, theoretically, this assumption impairs the homotopy analysis method, although we can always attempt some base functions even if a given nonlinear problem is completely new for us. Fortunately, it seems that solutions of a nonlinear problem could be expressed by many different kinds of base functions, as illustrated in Chapter 2.

As mentioned in Chapter 2, the idea of avoiding the so-called secular term was proposed by a lot of researchers such as Lindstedt [52], Bohlin [53], Poincaré [54], Gyldén [55], and so on, and the rule of solution expression can be regarded as its generalization. However, for a completely new problem, how can we know that a term belongs to the so-called secular term or not? So, in fact, many previous analytic techniques also imply the assumption that some prior knowledge should be known. And this assumption also impairs these methods, although such kinds of damage seem tiny compared to other serious restrictions of these previous methods and thus are often neglected.

5.3 Open questions

To overcome the above-mentioned limitation of the homotopy analysis method, it is necessary to propose some pure mathematical theorems to direct us to choose the initial approximation, the auxiliary linear operator, and the auxiliary function. These mathematical theorems should be valid in rather general cases without any prior knowledge so that we can apply them without any physical backgrounds. Up to now, it is even an open question if such kinds of pure mathematical theorems exist or not.

Although the homotopy analysis method has been successfully applied to many nonlinear problems such as those illustrated in this book and published in some journals, it is unclear if this method is valid for nonlinear problems with discontinue or chaotic solutions. To the best of the author's knowledge, chaos is generally investigated by numerical techniques and hardly expressed analytically. Up to now, it seems not very clear what kind of base functions is efficient to analytically express a chaotic solution. Recently, Norden E. Huang et al. [80] developed the empirical mode decomposition method and showed that nonlinear and nonstationary time series can be expressed by the so-called "instrinic mode functions". However, up to now, we do not know how to use the "instrinic mode functions" to gain *analytic* expressions of chaotic solutions of a given nonlinear problem without numerically solving it.

PART II

APPLICATIONS

Great straightness seems bent;
Great skill seems awkward;
Great eloquence seems tongue-tied.

Lao Tzu, an ancient Chinese philosopher

6

Simple bifurcation of a nonlinear problem

Consider a nonlinear problem of the so-called Duffing oscillator in space (see Kahn and Zarmi [11], page 198) governed by

$$w''(x) + w(x) - w^3(x) = 0, \quad w(0) = w(L) = 0, \tag{6.1}$$

where x is a spatial variable, $w(x)$ is a real function of x defined in the region $0 \leq x \leq L$, and the prime denotes the derivation. Obviously,

$$w(x) = 0$$

satisfies all of the above equations and thus is one of its solutions. However, for some values of L, there exist nonzero solutions so that the so-called simple bifurcation occurs.

Under the transformation

$$x = \left(\frac{L}{\pi}\right)\xi, \quad \epsilon = \left(\frac{L}{\pi}\right)^2, \quad v(\xi) = w(x), \tag{6.2}$$

Equation (6.1) becomes

$$v'' + \epsilon(v - v^3) = 0, \quad v(0) = v(\pi) = 0, \tag{6.3}$$

where the prime denotes the derivation with respect to ξ.

For any $\epsilon \geq 0$, the above equation has the solution $v(\xi) = 0$. The so-called bifurcation occurs when a nonzero solution of Equation (6.3) exists for some values of ϵ. Thus, we focus on the nonzero solution of Equation (6.3) and the critical condition of its existence. Obviously, if $v(\xi)$ is a nonzero solution of Equation (6.3), then $-v(\xi)$ must be its solution as well. Without loss of any generality, define

$$A = v(\pi/2), \quad v(\xi) = A\,u(\xi). \tag{6.4}$$

Substituting the above expressions into Equation (6.3), we have

$$u'' + \epsilon(u - A^2 u^3) = 0, \quad u(0) = u(\pi) = 0. \tag{6.5}$$

Note that A is unknown in the above equation and it holds from (6.4) that

$$u(\pi/2) = 1. \tag{6.6}$$

According to Kahn and Zarmi [11], the exact relation between A and L is given by

$$L = 2 \int_0^A \frac{dz}{\sqrt{A^2 - z^2 - (A^4 - z^4)/2}},$$

which gives the exact solution

$$\frac{L}{\pi} = \frac{2}{\pi \sqrt{1 - A^2/2}} K\left(\frac{A^2}{2 - A^2}\right), \qquad (6.7)$$

where $K(\zeta)$ is the complete elliptic integral of the first kind. According to the above exact solution, $\epsilon = (L/\pi)^2$ tends to infinity as $|A|$ approaches to 1. By means of the method of normal forms, Kahn and Zarmi [11] gave the perturbation solution

$$A \approx \pm 2\sqrt{\frac{\epsilon - 1}{3}}, \qquad \epsilon \geq 1, \qquad (6.8)$$

which breaks down for large ϵ. In this chapter we employ the homotopy analysis method to solve the nonlinear boundary-value problem with simple bifurcations.

6.1 Homotopy analysis solution

6.1.1 Zero-order deformation equation

Using the boundary conditions $u(0) = u(\pi) = 0$ and considering the nonlinearity of Equation (6.5), it is straightforward to express the solution $u(\xi)$ by a set of base functions

$$\{\sin[(2m + 1)\xi] \mid m \geq 0\} \qquad (6.9)$$

such that

$$u(\xi) = \sum_{m=0}^{+\infty} c_m \sin[(2m + 1)\xi], \qquad (6.10)$$

where c_m is a coefficient. This provides us with the so-called *rule of solution expression*.

Under the *rule of solution expression* denoted by (6.10) and using (6.6), it is straightforward to choose

$$u_0(\xi) = \sin(\xi) \qquad (6.11)$$

Simple Bifurcation of a Nonlinear Problem

as the initial guess of $u(\xi)$. Under the *rule of solution expression* denoted by (6.10), we choose an auxiliary linear operator

$$\mathcal{L}[\Phi(\xi;q)] = \frac{\partial^2 \Phi(\xi;q)}{\partial \xi^2} + \Phi(\xi;q) \tag{6.12}$$

with the property

$$\mathcal{L}[C_1 \sin \xi + C_2 \cos \xi] = 0, \tag{6.13}$$

where C_1 and C_2 are coefficients. From Equation (6.5), we define a nonlinear operator

$$\mathcal{N}[\Phi(\xi;q), \alpha(q)] = \frac{\partial^2 \Phi(\xi;q)}{\partial \xi^2} + \epsilon \left[\Phi(\xi;q) - \alpha^2(q) \Phi^3(\xi;q) \right], \tag{6.14}$$

where $q \in [0,1]$ is the embedding parameter and $\alpha(q)$ is an unknown function dependent upon q. Let $\hbar \neq 0$ denote an auxiliary parameter and $H(\xi)$ an auxiliary function. We construct the so-called zero-order deformation equation

$$(1-q)\, \mathcal{L}[\Phi(\xi;q) - u_0(\xi)] = \hbar\, q\, H(\xi)\, \mathcal{N}[\Phi(\xi;q), \alpha(q)], \tag{6.15}$$

subject to the boundary conditions

$$\Phi(0;q) = \Phi(\pi;q) = 0. \tag{6.16}$$

Obviously, when $q = 0$, the solution of Equations (6.15) and (6.16) is

$$\Phi(\xi;0) = u_0(\xi). \tag{6.17}$$

When $q = 1$, Equations (6.15) and (6.16) are exactly the same as the original equations (6.5), provided

$$\Phi(\xi;1) = u(\xi), \quad \alpha(1) = A. \tag{6.18}$$

Thus, $\Phi(\xi;q)$ varies (or deforms) from the initial approximation $u_0(\xi) = \sin \xi$ to the exact solution $u(\xi)$ of Equations (6.5), as does $\alpha(q)$ from its initial approximation A_0 to the exact value $A = u(\pi/2)$.

Note that the zero-order deformation equation (6.15) contains the auxiliary parameter $\hbar$ and the auxiliary function $H(\xi)$. Assume that $\hbar$ and $H(\xi)$ are properly chosen so that the zero-order deformation equations (6.15) and (6.16) have solutions for all $q \in [0,1]$, and that there exist the derivatives

$$u_m(\xi) = \frac{1}{m!} \left.\frac{\partial^m \Phi(\xi;q)}{\partial q^m}\right|_{q=0}, \quad A_m = \frac{1}{m!} \left.\frac{d^m \alpha(q)}{dq^m}\right|_{q=0}. \tag{6.19}$$

Then, using Taylor's theorem and Equation (6.17), we can expand $\Phi(\xi;q)$ and $\alpha(q)$ in power series of q as follows

$$\Phi(\xi;q) = u_0(\xi) + \sum_{m=1}^{+\infty} u_m(\xi)\, q^m, \tag{6.20}$$

$$\alpha(q) = A_0 + \sum_{m=1}^{+\infty} A_m\, q^m, \tag{6.21}$$

respectively. Furthermore, assuming that $\hbar$ and $H(\xi)$ are so properly chosen that the power series (6.20) and (6.21) are convergent at $q = 1$, we have using (6.18) the solution series

$$u(\xi) = u_0(\xi) + \sum_{m=1}^{+\infty} u_m(\xi), \qquad (6.22)$$

$$A = A_0 + \sum_{m=1}^{+\infty} A_m. \qquad (6.23)$$

6.1.2 High-order deformation equation

For simplicity, define the vectors

$$\vec{u}_k = \{u_0(\xi), u_1(\xi), u_2(\xi), \cdots, u_k(\xi)\}, \qquad \vec{A}_k = \{A_0, A_1, A_2, \cdots, A_k\}.$$

Differentiating the zero-order deformation equations (6.15) and (6.16) m times with respect to the embedding parameter q and then dividing them by $m!$ and finally setting $q = 0$, we have the high-order deformation equation

$$\mathcal{L}\left[u_m(\xi) - \chi_m u_{m-1}(\xi)\right] = \hbar\, H(\xi)\, R_m(\vec{u}_{m-1}, \vec{A}_{m-1}), \qquad (6.24)$$

subject to the boundary conditions

$$u_m(0) = u_m(\pi) = 0, \qquad (6.25)$$

where χ_m is defined by (2.42) and

$$R_m(\vec{u}_{m-1}, \vec{A}_{m-1}) = \frac{1}{(m-1)!} \left.\frac{\partial^{m-1} \mathcal{N}[\Phi(\xi;q), \alpha(q)]}{\partial q^{m-1}}\right|_{q=0}$$
$$= u''_{m-1}(\xi) + \epsilon\, u_{m-1}(\xi)$$
$$- \epsilon \sum_{n=0}^{m-1}\left(\sum_{i=0}^{n} A_i\, A_{n-i}\right)\left[\sum_{j=0}^{m-1-n} u_j(\xi) \sum_{r=0}^{m-1-n-j} u_r(\xi) u_{m-1-n-j-r}(\xi)\right].$$
$$(6.26)$$

Note that there are two unknowns $u_m(\xi)$ and A_{m-1}, but we have only one differential equation for $u_m(\xi)$. The problem is therefore not closed and an additional algebraic equation is needed to determine A_{m-1}. Considering the *rule of solution expression* denoted by (6.10) and the property (6.13) of the auxiliary linear operator $\mathcal{L}$, $H(\xi)$ should be properly chosen so that the right-hand side term of the high-order deformation equation (6.24) can be expressed by

$$\hbar\, H(\xi)\, R_m(\vec{u}_{m-1}, \vec{A}_{m-1}) = \sum_{n=0}^{\mu_m} b_{m,n}(\vec{A}_{m-1})\, \sin[(2n+1)\xi], \qquad (6.27)$$

Simple Bifurcation of a Nonlinear Problem 89

where $b_{m,n}(\vec{A}_{m-1})$ is a coefficient and the positive integer μ_m depends upon m and the auxiliary function $H(\xi)$. Then, according to the property (6.13), if $b_{m,0}(\vec{A}_{m-1}) \neq 0$, the solution of the mth-order deformation equation (6.24) contains the term

$$\xi \sin \xi,$$

which disobeys the *rule of solution expression* denoted by (6.10). To avoid this, we had to enforce

$$b_{m,0}(\vec{A}_{m-1}) = 0, \tag{6.28}$$

which provides us with an additional algebraic equation for A_{m-1}. In this way, the problem is closed. Thereafter, it is easy to gain the solution of Equation (6.24)

$$u_m(\xi) = \chi_m \, u_{m-1}(\xi) - \sum_{n=1}^{\mu_m} \frac{b_{m,n}}{4n(n+1)} \sin[(2n+1)\xi]$$
$$+ C_1 \sin \xi + C_2 \cos \xi, \tag{6.29}$$

where C_1 and C_2 are coefficients. Under the *rule of solution expression* denoted by (6.10), C_2 must be zero. However, the coefficient C_1 cannot be determined by the boundary conditions (6.25), which is automatically satisfied when $C_2 = 0$. But, from Equation (6.6), it holds that

$$u_m(\pi/2) = 0, \tag{6.30}$$

which uniquely determines the value of C_1. In this way, we gain A_{m-1} and $u_m(\xi)$ successively. At the Nth-order of approximation, we have

$$u(\xi) \approx u_0(\xi) + \sum_{m=1}^{N} u_m(\xi), \tag{6.31}$$

$$A \approx A_0 + \sum_{m=1}^{N-1} A_m. \tag{6.32}$$

6.1.3 Convergence theorem

THEOREM 6.1
If the solution series (6.22) and (6.23) are convergent, where $u_k(\xi)$ is governed by Equations (6.24) and (6.25) under the definitions (6.26) and (2.42), they must be the exact solution of Equations (6.5).

Proof: If the solution series (6.22) is convergent, it is necessary that

$$\lim_{m \to +\infty} u_m(\xi) = 0, \qquad \xi \in [0, \pi].$$

Using (6.12) and (2.42) and from (6.24), we have

$$\hbar\, H(\xi) \sum_{k=1}^{+\infty} R_k(\vec{u}_{k-1}, \vec{A}_{k-1})$$

$$= \lim_{m\to+\infty} \sum_{k=1}^{m} \mathcal{L}[u_k(\xi) - \chi_k\, u_{k-1}(\xi)]$$

$$= \mathcal{L}\left\{ \lim_{m\to+\infty} \sum_{k=1}^{m} [u_k(\xi) - \chi_k\, u_{k-1}(\xi)] \right\}$$

$$= \mathcal{L}\left[\lim_{m\to+\infty} u_m(\xi) \right]$$

$$= 0,$$

which gives, since $\hbar \neq 0$ and $H(\xi) \neq 0$,

$$\sum_{k=1}^{+\infty} R_k(\vec{u}_{k-1}, \vec{A}_{k-1}) = 0.$$

Substituting (6.26) into the above expression and simplifying it, we have, due to the convergence of the series (6.22) and (6.23), that

$$\frac{d^2}{d\xi^2}\left[\sum_{k=0}^{+\infty} u_k(\xi)\right] + \epsilon \left\{ \left[\sum_{k=0}^{+\infty} u_k(\xi)\right] - \left(\sum_{m=0}^{+\infty} A_m\right)^2 \left[\sum_{k=0}^{+\infty} u_k(\xi)\right]^3 \right\} = 0.$$

From (6.11) and (6.25), it holds that

$$\sum_{k=0}^{+\infty} u_k(0) = \sum_{k=0}^{+\infty} u_k(\pi) = 0.$$

Thus, as long as the solution series (6.22) and (6.23) are convergent, they must be the exact solution of Equations (6.5). This ends the proof.

6.2 Result analysis

According to Theorem 6.1, we only need choose a proper auxiliary parameter $\hbar$ and a auxiliary function $H(\xi)$ to ensure that the solution series (6.22) and (6.23) are convergent. Note that, under the *rule of solution expression* denoted by (6.10) and without disobeying the *rule of coefficient ergodicity*, different auxiliary functions such as

$$H(\xi) = 1,\ H(\xi) = \sin^2(\xi),\ H(\xi) = \cos^2(\xi),\ H(\xi) = \cos(2\xi)$$

Simple Bifurcation of a Nonlinear Problem

and so on can be used. However, for the sake of simplicity, we choose here

$$H(\xi) = 1. \tag{6.33}$$

In this case, using (6.11) and (6.26), we have

$$\hbar\, H(\xi)\, R_1(\vec{u}_0, \vec{A}_0) = \hbar\left(\epsilon - 1 - \frac{3}{4}\epsilon\, A_0^2\right)\sin\xi + \frac{1}{4}\hbar\,\epsilon\, A_0^2 \sin(3\xi), \tag{6.34}$$

which gives from Equation (6.27) that

$$b_{1,0} = \hbar\left(\epsilon - 1 - \frac{3}{4}\epsilon\, A_0^2\right), \qquad b_{1,1} = \frac{1}{4}\hbar\,\epsilon\, A_0^2.$$

Thus, when $m = 1$, we have from Equation (6.28) an additional algebraic equation

$$\epsilon - 1 - \frac{3}{4}\epsilon\, A_0^2 = 0. \tag{6.35}$$

Since $\epsilon = (L/\pi)^2 \geq 0$, the above equation has no solution when $\epsilon < 1$. Thus, there does not exist a nonzero solution when $0 \leq \epsilon \leq 1$. However, when $\epsilon > 1$, Equation (6.35) has the solution

$$A_0 = \pm\frac{2}{\sqrt{3}}\sqrt{1 - \frac{1}{\epsilon}}, \qquad \epsilon > 1. \tag{6.36}$$

Therefore, the so-called simple bifurcation occurs at $\epsilon = 1$. The homotopy analysis method correctly provides us with the critical condition of the simple bifurcation of the considered nonlinear problem.

It should be emphasized that the homotopy analysis method provides us with two families of solution expressions in the auxiliary parameter $\hbar$, and $\hbar$ influences the convergence of the solution series (6.22) and (6.23). In particular, the series (6.23) of A is a power series of $\hbar$. To investigate the influence of $\hbar$ on the solution series (6.23), we plot the so-called $\hbar$-curve (see page 26 and §3.5.1) of A for any a given ϵ. For example, the $\hbar$-curves $A \sim \hbar$ when $\epsilon = 10$ and $\epsilon = 25$ are as shown in Figure 6.1, which clearly indicate the corresponding valid regions of $\hbar$. From Figure 6.1, the solution series (6.23) when $\epsilon = 10$ converges if $-3/4 \leq \hbar < 0$. When $\epsilon = 25$, it converges if $-1/4 \leq \hbar < 0$. So, by means of Theorem 6.1 and using the $\hbar$-curves, it is very clear that when $\epsilon = 10$ and $\epsilon = 25$ the solution series (6.23) is convergent to the exact value if we choose $\hbar$ in the corresponding valid region of $\hbar$, i.e., $-3/4 \leq \hbar < 0$ or $-1/4 \leq \hbar < 0$, respectively. For example, the approximations of A when $\epsilon = 10, \hbar = -1/2$ and $\epsilon = 25, \hbar = -1/5$ are listed in Table 6.1. It is found that, in general, as long as the solution series (6.23) is convergent, the corresponding solution series (6.22) of $u(\xi)$ given by the same auxiliary parameter $\hbar$ also converges in the whole region $0 \leq \xi \leq \pi$. For example, the 10th-order approximation of $u(\xi)$ when $\epsilon = 10, \hbar = -1/2$ and the 30th-order approximation of $u(\xi)$ when $\epsilon = 25, \hbar = -1/5$ agree well with the exact solution of $u(\xi)$, respectively, as shown in Figure 6.2. So, using the $\hbar$-curves, it is convenient to find out the valid region of $\hbar$ to ensure that the solution series (6.22) and (6.23) converge.

It is found that the mth-order approximation of A can be expressed by

$$A \approx \pm \sqrt{3(1-\epsilon^{-1})} \sum_{k=0}^{m} \beta_{m,k}(\hbar) \, \epsilon^k, \qquad (6.37)$$

where $\beta_{m,k}$ is a coefficient dependent upon $\hbar$. From Figure 6.1, it is clear that as ϵ enlarges the corresponding valid region of $\hbar$ becomes smaller. It is found that the convergence region of the solution series of A is governed by $\hbar$, as shown in Figure 6.3. Clearly, the closer the value of $\hbar$ is to zero from below ($\hbar < 0$), the larger the convergence region of A becomes. This implies that $\hbar$ should be a function of ϵ, whose absolution value should decrease as ϵ increases. It is found that, when

$$\hbar = -\frac{1}{1+\epsilon/3}, \qquad (6.38)$$

the 10th-order of approximation of A, i.e.,

$$A \approx \pm \frac{1}{(1+\epsilon/3)^{10}} \sqrt{1 - \frac{1}{\epsilon}} \left(1.1803 + 3.9075 \, \epsilon + 5.8128 \, \epsilon^2 + 5.1149 \, \epsilon^3 \right.$$
$$+ 2.9466 \, \epsilon^4 + 1.1603 \, \epsilon^5 + 0.31602 \, \epsilon^6 + 5.8726 \times 10^{-2} \, \epsilon^7$$
$$\left. + 7.1298 \times 10^{-3} \, \epsilon^8 + 5.1396 \times 10^{-4} \, \epsilon^9 + 1.7001 \times 10^{-5} \, \epsilon^{10} \right), \qquad (6.39)$$

agrees well with the exact result in the *whole* region $1 \leq \epsilon < +\infty$, as shown in Figure 6.3. The 10th-order approximation (6.39) of A gives

$$\lim_{\epsilon \to +\infty} |A| = 1.0039,$$

corresponding to a relative error of 0.39%. Using (6.38), even the third-order approximation of A, i.e.

$$A \approx \pm \frac{\left(7015 \, \epsilon^3 + 70251 \, \epsilon^2 + 220917 \, \epsilon + 226105\right)}{4096\sqrt{3}(\epsilon+3)^3} \sqrt{1 - \frac{1}{\epsilon}} \qquad (6.40)$$

agrees well with the exact result, as shown in Figure 6.4. So, it is the auxiliary parameter $\hbar$ which provides us with a convenient way to control and adjust the convergence region and rate of solutions series. Thus, the auxiliary parameter $\hbar$ indeed plays an important role within the frame of the homotopy analysis method.

It should be emphasized that the *rule of solution expression* denoted by (6.10) also plays an important role within the frame of the homotopy analysis method. It is under the *rule of solution expression* that the initial approximation (6.11) and the auxiliary linear operator (6.12) are chosen. Furthermore,

Simple Bifurcation of a Nonlinear Problem

it is under the *rule of solution expression* that Equation (6.28) is given to avoid the appearance of the term $\xi \sin \xi$ and to close the problem. Note that for the problem considered in this chapter, the term $\xi \sin \xi$ is not a traditional secular term because it is possible that $u(\xi)$ can be expressed by such a set of base functions

$$\{\xi^m \sin(n\xi), \xi^m \cos(n\xi) \mid m \geq 0, n \geq 1\}.$$

This is the reason why the auxiliary function $H(\xi)$ is not unique for the considered problem. However, it seems more efficient to use the set of base functions denoted by (6.9) to approximate $u(\xi)$.

Note that our approximations are much better than the perturbation result (6.8), as shown in Figures 6.3 and 6.4. Also note that even the third-order approximation (6.40) of A agrees well with the exact result in the whole region $1 \leq \epsilon < +\infty$ and correctly gives the bifurcation point $\epsilon = 1$ and two breaches of nonzero solutions.

Using the so-called homotopy-Padé technique (see page 38 and §3.5.2), the convergence of the solution series of A is greatly accelerated, as shown in Table 6.2 when $\epsilon = 10$ and $\epsilon = 25$. It is found that the $[m, m]$ homotopy-Padé approximant of A does not depend upon the auxiliary parameter $\hbar$. The $[4, 4]$ homotopy-Padé approximant

$$A \approx 2\sqrt{3}\sqrt{1 - \frac{1}{\epsilon} \frac{P(\epsilon)}{Q(\epsilon)}} \tag{6.41}$$

agrees well with the exact result in the whole region $1 \leq \epsilon < +\infty$, where

$$P(\epsilon) = 8665210296046039923 + 25009647825190573960\,\epsilon$$
$$+ 604034298653768562\,\epsilon^2 + 62408285303687028\,\epsilon^3$$
$$+ 3874319809940915\,\epsilon^4,$$
$$Q(\epsilon) = 25430938337575455089 + 7921677254280814588\,\epsilon$$
$$+ 1930521704826790758\,\epsilon^2 + 213027971364041596\,\epsilon^3$$
$$+ 13310678950379441\,\epsilon^4.$$

This example illustrates that the homotopy analysis method is valid for nonlinear problems with bifurcations.

TABLE 6.1
The 30th-order analytic approximations of A when
$\epsilon = 10, \hbar = -1/2$ and $\epsilon = 25, \hbar = -1/5$ by means of $H(\xi) = 1$.

Order of approximation	$\epsilon = 10, \hbar = -1/2$	$\epsilon = 25, \hbar = -1/5$
5	0.99833	1.01046
10	0.99588	1.00313
15	0.99624	1.00117
20	0.99644	1.00049
25	0.99644	1.00017
30	0.99644	1.00000

TABLE 6.2
The $[m, m]$ homotopy-Padé approximations of A when $\epsilon = 10$ and $\epsilon = 25$ by means of $H(\xi) = 1$.

$[m, m]$	$\epsilon = 10$	$\epsilon = 25$
$[2, 2]$	0.99914	1.01167
$[4, 4]$	0.99651	1.00113
$[6, 6]$	0.99644	1.00012
$[8, 8]$	0.99644	0.99996
$[10, 10]$	0.99644	0.99994
$[12, 12]$	0.99644	0.99994
$[15, 15]$	0.99644	0.99994

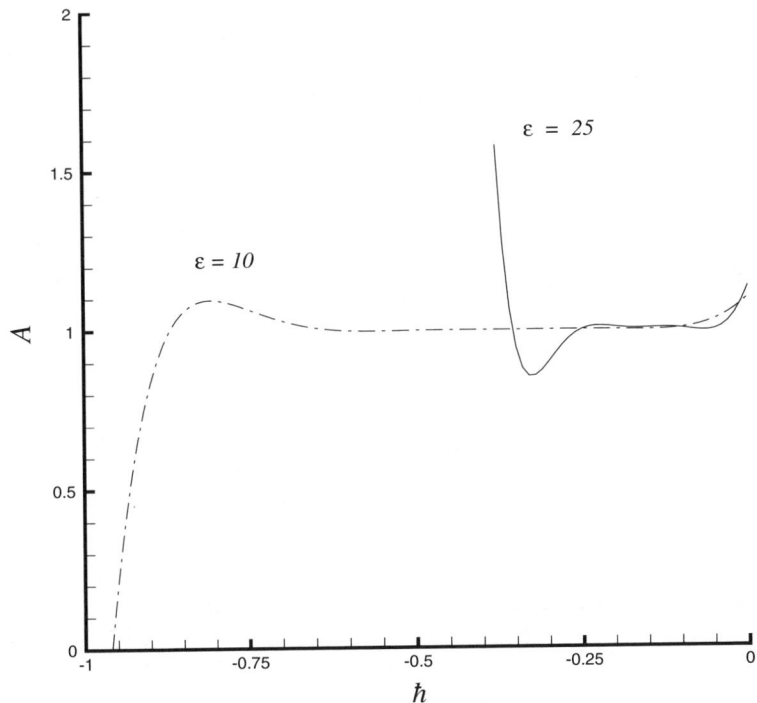

FIGURE 6.1
The $\hbar$-curves of A when $\epsilon = 10, 25$ by means of $H(\xi) = 1$. Dash-dotted line: 10th-order approximation of A when $\epsilon = 10$; solid line: 10th-order approximation of A when $\epsilon = 25$.

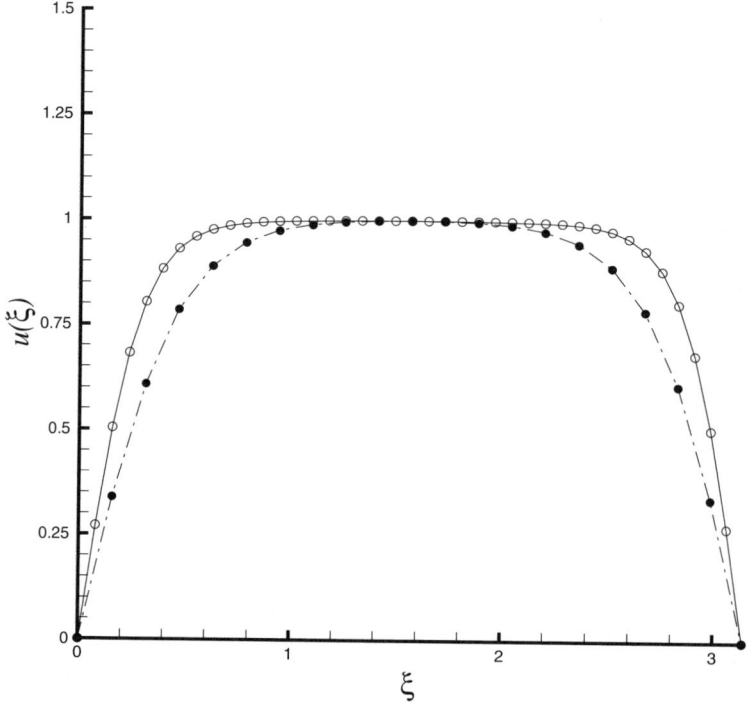

FIGURE 6.2
The analytic approximations of $u(\xi)$ by means of $H(\xi) = 1$. Dash-dotted line: 10th-order approximation when $\epsilon = 10$ and $\hbar = -1/2$; filled cycles: 20th-order approximation when $\epsilon = 10$ and $\hbar = -1/2$; solid line: 20th-order approximation of when $\epsilon = 25$ and $\hbar = -1/5$; open cycles: 30th-order approximation when $\epsilon = 25$ and $\hbar = -1/5$.

Simple Bifurcation of a Nonlinear Problem

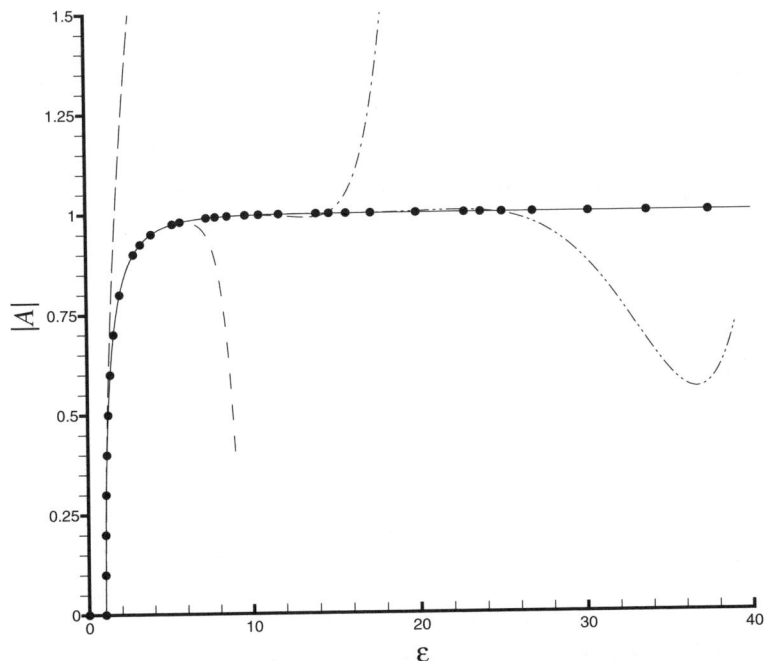

FIGURE 6.3
Comparison of the exact result (6.7) with the 10th-order approximation (6.37) of A when $H(\xi) = 1$. Symbols: exact result given by (6.7); long-dashed line: perturbation result (6.8); dashed line: approximation (6.37) when $\hbar = -1$; dash-dotted line: approximation (6.37) when $\hbar = -1/2$; dash-dot-dotted line: approximation (6.37) when $\hbar = -1/4$; solid line: approximation (6.39) when $\hbar = -1/(1 + \epsilon/3)$.

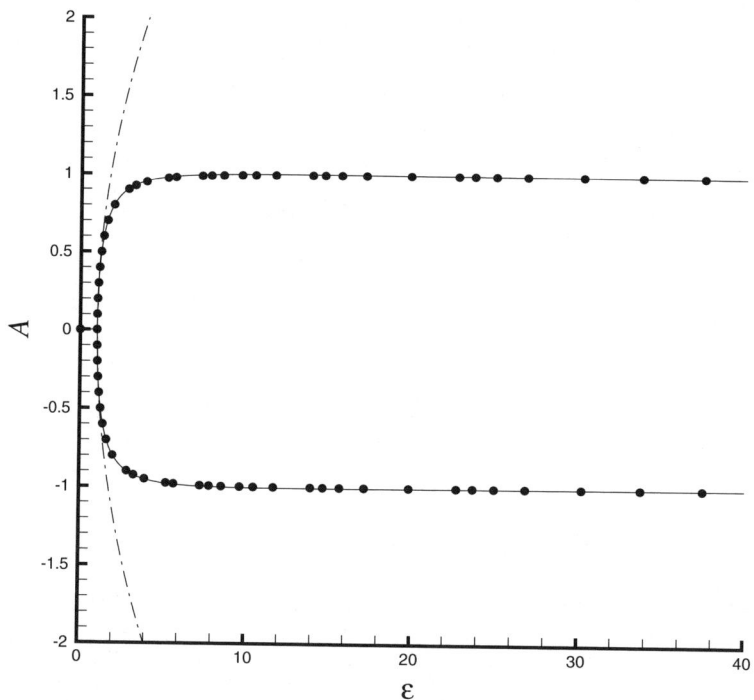

FIGURE 6.4
Comparison of the exact result (6.7) with the third-order analytic approximation (6.40) of A when $\hbar = -1/(1+\epsilon/3)$ and $H(\xi) = 1$. Symbols: exact result given by (6.7); dash-dotted line: perturbation result (6.8); solid line: analytic approximation (6.40).

7

Multiple solutions of a nonlinear problem

It is well known that many nonlinear problems have multiple solutions. For example, let us consider again the so-called Duffing oscillator in space, governed by

$$v'' + \epsilon(v - v^3) = 0, \tag{7.1}$$

subject to the boundary conditions

$$v(0) = v(\pi) = 0, \tag{7.2}$$

where the prime denotes the derivation with respect to ξ. In Chapter 6, we use the homotopy analysis method to solve the same problem and correctly discover its critical condition $\epsilon = 1$ for the simple bifurcation and express its solution by such a set of base functions

$$\{\sin[(2m+1)\xi] \mid m \geq 0\}. \tag{7.3}$$

Notice that there exist an infinite number of sets of base functions denoted by

$$\{\sin[(2m+1)\kappa\,\xi] \mid m \geq 0, \kappa \geq 1\}, \tag{7.4}$$

where $\kappa \geq 1$ is a positive integer, which can be used to express a real function satisfying the boundary conditions (7.2). This implies that Equations (7.1) and (7.2) might have multiple solutions. This is indeed true. We show in this chapter that, using base functions denoted by (7.4), we can gain all multiple solutions of Equations (7.1) and (7.2) by means of the homotopy analysis method.

Without loss of any generality, define

$$A = v(\pi/2\kappa), \quad v(\xi) = A\, u(\xi). \tag{7.5}$$

Then, Equation (7.1) becomes

$$u'' + \epsilon(u - A^2 u^3) = 0, \quad u(0) = u(\pi) = 0. \tag{7.6}$$

Note that A is unknown in the above equation. From (7.5), it holds

$$u(\pi/2\kappa) = 1. \tag{7.7}$$

7.1 Homotopy analysis solution

7.1.1 Zero-order deformation equation

Using the base functions (7.4) and the boundary conditions $u(0) = u(\pi) = 0$ and considering the nonlinearity of Equation (7.6), we express the solution $u(\xi)$ in the form

$$u(\xi) = \sum_{m=0}^{+\infty} c_m \sin[(2m+1)\kappa\xi], \quad (7.8)$$

where c_m is a coefficient. This provides us with the so-called *rule of solution expression*.

Under the *rule of solution expression* denoted by (7.8) and using (7.7), it is straightforward to choose

$$u_0(\xi) = \sin(\kappa\,\xi) \quad (7.9)$$

as an initial guess of $u(\xi)$, where $\kappa \geq 1$ is an integer. To obey the *rule of solution expression* denoted by (7.8), we choose an auxiliary linear operator

$$\mathcal{L}[\Phi(\xi;q)] = \frac{\partial^2 \Phi(\xi;q)}{\partial \xi^2} + \kappa^2\,\Phi(\xi;q) \quad (7.10)$$

such that

$$\mathcal{L}[C_1 \sin(\kappa\,\xi) + C_2 \cos(\kappa\,\xi)] = 0, \quad (7.11)$$

where C_1 and C_2 are coefficients. Furthermore, from Equation (7.6), we define the nonlinear operator

$$\mathcal{N}[\Phi(\xi;q),\alpha(q)] = \frac{\partial^2 \Phi(\xi;q)}{\partial \xi^2} + \epsilon\left[\Phi(\xi;q) - \alpha^2(q)\Phi^3(\xi;q)\right], \quad (7.12)$$

where $q \in [0,1]$ is the embedding parameter, $\Phi(\xi;q)$ is an unknown function of ξ and q, $\alpha(q)$ is an unknown function dependent upon q. Let $\hbar \neq 0$ denote an auxiliary parameter and $H(\xi) \neq 0$ an auxiliary function. We construct the so-called zero-order deformation equation

$$(1-q)\mathcal{L}[\Phi(\xi;q) - u_0(\xi)] = \hbar\,q\,H(\xi)\,\mathcal{N}[\Phi(\xi;q),\alpha(q)], \quad (7.13)$$

subject to the boundary conditions

$$\Phi(0;q) = \Phi(\pi;q) = 0. \quad (7.14)$$

When $q = 0$, the solution of Equations (7.13) and (7.14) is

$$\Phi(\xi;0) = u_0(\xi), \qquad \xi \in [0,\pi]. \quad (7.15)$$

Multiple Solutions of a Nonlinear Problem

When $q = 1$, Equations (7.13) and (7.14) are equivalent to Equations (7.6), provided
$$\Phi(\xi; 1) = u(\xi), \quad \alpha(1) = A. \tag{7.16}$$

Thus, $\Phi(\xi; q)$ varies (or deforms) from the initial approximation $u_0(\xi) = \sin(\kappa \xi)$ to the exact solution $u(\xi)$ of Equations (7.6), as does $\alpha(q)$ from its initial approximation A_0 to the exact value $A = u(\pi/2\kappa)$. Note that the zero-order deformation equation (7.13) contains the auxiliary parameter $\hbar$ and the auxiliary function $H(\xi)$. Assume that $\hbar$ and $H(\xi)$ are properly chosen so that the zero-order deformation equations (7.13) and (7.14) have solutions for all $q \in [0, 1]$ and that the terms

$$u_m(\xi) = \frac{1}{m!} \left. \frac{\partial^m \Phi(\xi; q)}{\partial q^m} \right|_{q=0}, \quad A_m = \frac{1}{m!} \left. \frac{d^m \alpha(q)}{dq^m} \right|_{q=0} \tag{7.17}$$

exist for $m \geq 1$. Then, by Taylor's theorem and using (7.15), we can expand $\Phi(\xi; q)$ and $\alpha(q)$ in power series of q as follows

$$\Phi(\xi; q) = u_0(\xi) + \sum_{m=1}^{+\infty} u_m(\xi) \, q^m, \tag{7.18}$$

$$\alpha(q) = A_0 + \sum_{m=1}^{+\infty} A_m \, q^m. \tag{7.19}$$

Furthermore, assuming that $\hbar$ and $H(\xi)$ are so properly chosen that the power series (7.18) and (7.19) are convergent at $q = 1$, we have from (7.16) the solution series

$$u(\xi) = u_0(\xi) + \sum_{m=1}^{+\infty} u_m(\xi), \tag{7.20}$$

$$A = A_0 + \sum_{m=1}^{+\infty} A_m. \tag{7.21}$$

7.1.2 High-order deformation equation

For brevity, write
$$\vec{u}_k = \{u_0(\xi), u_1(\xi), u_2(\xi), \cdots, u_k(\xi)\}, \quad \vec{A}_k = \{A_0, A_1, A_2, \cdots, A_k\}.$$

Differentiating the zero-order deformation equations (7.13) and (7.14) m times with respect to q and then dividing them by $m!$ and finally setting $q = 0$, we have the high-order deformation equation

$$\mathcal{L}[u_m(\xi) - \chi_m u_{m-1}(\xi)] = \hbar \, H(\xi) \, R_m(\vec{u}_{m-1}, \vec{A}_{m-1}), \tag{7.22}$$

subject to the boundary conditions
$$u_m(0) = u_m(\pi) = 0, \tag{7.23}$$
where χ_m is defined by (2.42) and

$$R_m(\vec{u}_{m-1}, \vec{A}_{m-1}) = \frac{1}{(m-1)!} \left. \frac{\partial^{m-1}\mathcal{N}[\Phi(\xi;q), \alpha(q)]}{\partial q^{m-1}} \right|_{q=0}$$
$$= u''_{m-1}(\xi) + \epsilon\, u_{m-1}(\xi)$$
$$- \epsilon \sum_{n=0}^{m-1} \left(\sum_{i=0}^{n} A_i\, A_{n-i} \right) \left[\sum_{j=0}^{m-1-n} u_j(\xi) \sum_{r=0}^{m-1-n-j} u_r(\xi) u_{m-1-n-j-r}(\xi) \right]. \tag{7.24}$$

Note that both $u_m(\xi)$ and A_{m-1} are unknown, but we have only one differential equation for $u_m(\xi)$. So, the problem is not closed and an additional algebraic equation is needed to determine A_{m-1}. Assume that $H(\xi)$ is properly chosen so that the right-hand side term of the high-order deformation equation (7.22) can be expressed by

$$\hbar\, H(\xi)\, R_m(\vec{u}_{m-1}, \vec{A}_{m-1}) = \sum_{n=0}^{\mu_m} b_{m,n}(\vec{A}_{m-1})\, \sin[(2n+1)\kappa\, \xi], \tag{7.25}$$

where $b_{m,n}(\vec{A}_{m-1})$ is a coefficient and the positive integer μ_m depends upon $H(\xi)$ and m. According to the property (7.11) of $\mathcal{L}$, when $b_{m,0}(\vec{A}_{m-1}) \neq 0$, the solution of the mth-order deformation equation (7.22) contains the term

$$\xi\, \sin(\kappa\, \xi),$$

which disobeys the *rule of solution expression* denoted by (7.8). To avoid this, we had to enforce
$$b_{m,0}(\vec{A}_{m-1}) = 0, \tag{7.26}$$
which provides us with an additional algebraic equation for A_{m-1}. In this way, the problem is closed. Thereafter, it is easy to gain the solution of Equation (7.22), say,

$$u_m(\xi) = \chi_m\, u_{m-1}(\xi) + \sum_{n=1}^{\mu_m} \frac{b_{m,n}}{[1-(2n+1)^2\kappa^2]}\, \sin[(2n+1)\kappa\, \xi]$$
$$+ C_1\, \sin(\kappa\xi) + C_2\, \cos(\kappa\xi), \tag{7.27}$$

where C_1 and C_2 are coefficients. Under the *rule of solution expression* denoted by (7.8), C_2 must be zero. Note that the coefficient C_1 cannot be determined by the boundary conditions (7.23), which is automatically satisfied when $C_2 = 0$. However, from (7.7), it should hold
$$u_m(\pi/2\kappa) = 0, \tag{7.28}$$

Multiple Solutions of a Nonlinear Problem

which uniquely determines C_1. In this way, we gain A_{m-1} and $u_m(\xi)$ successively.

At the Nth-order of approximation, we have

$$u(\xi) \approx u_0(\xi) + \sum_{m=1}^{N} u_m(\xi), \qquad (7.29)$$

$$A \approx A_0 + \sum_{m=1}^{N-1} A_m. \qquad (7.30)$$

7.1.3 Convergence theorem

THEOREM 7.1
If the solution series (7.20) and (7.21) are convergent, where $u_k(\xi)$ is governed by Equations (7.22) and (7.23) under the definitions (7.24) and (2.42), they must be the exact solution of Equations (7.6).

Proof: If the solution series (7.20) is convergent, it is necessary that

$$\lim_{m \to +\infty} u_m(\xi) = 0, \qquad \xi \in [0, \pi].$$

From (7.10), (2.42), and (7.22) and using the above expression, we have

$$\hbar\, H(\xi) \sum_{k=1}^{+\infty} R_k(\vec{u}_{k-1}, \vec{A}_{k-1})$$

$$= \lim_{m \to +\infty} \sum_{k=1}^{m} \mathcal{L}[u_k(\xi) - \chi_k\, u_{k-1}(\xi)]$$

$$= \mathcal{L}\left\{ \lim_{m \to +\infty} \sum_{k=1}^{m} [u_k(\xi) - \chi_k\, u_{k-1}(\xi)] \right\}$$

$$= \mathcal{L}\left[\lim_{m \to +\infty} u_m(\xi) \right]$$

$$= 0.$$

Since $\hbar \neq 0$ and $H(\xi) \neq 0$, the above expression gives

$$\sum_{k=1}^{+\infty} R_k(\vec{u}_{k-1}, \vec{A}_{k-1}) = 0.$$

Substituting (7.24) into the above expression and simplifying it, we have, due to the convergence of the series (7.20) and (7.21), that

$$\frac{d^2}{d\xi^2}\left[\sum_{k=0}^{+\infty} u_k(\xi)\right] + \epsilon\left\{ \left[\sum_{k=0}^{+\infty} u_k(\xi)\right] - \left(\sum_{m=0}^{+\infty} A_m\right)^2 \left[\sum_{k=0}^{+\infty} u_k(\xi)\right]^3 \right\} = 0.$$

Using (7.9) and (7.23), it holds

$$\sum_{k=0}^{+\infty} u_k(0) = \sum_{k=0}^{+\infty} u_k(\pi) = 0.$$

Thus, as long as the solution series (7.20) and (7.21) are convergent, they must be the exact solution of Equations (7.6). This ends the proof.

7.2 Result analysis

According to Theorem 7.1, we need only to properly choose an auxiliary function $H(\xi)$ and an auxiliary parameter $\hbar$ to ensure that the solution series (7.20) and (7.21) converge. As pointed out in Chapter 6, the auxiliary function $H(\xi)$ can be chosen in many different forms without disobeying the *rule of coefficient ergodicity*. For the sake of simplicity, we choose here

$$H(\xi) = 1. \tag{7.31}$$

Then, using (7.9) and (7.24), we have

$$\hbar\, H(\xi)\, R_1(\vec{u}_0, \vec{A}_0)$$
$$= \hbar\left(\epsilon - \kappa^2 - \frac{3}{4}\epsilon\, A_0^2\right)\sin(\kappa\,\xi) + \frac{1}{4}\hbar\,\epsilon\, A_0^2 \sin(3\kappa\,\xi), \tag{7.32}$$

which gives according to (7.25) that

$$b_{1,0} = \hbar\left(\epsilon - \kappa^2 - \frac{3}{4}\epsilon\, A_0^2\right), \quad b_{1,1} = \frac{1}{4}\hbar\,\epsilon\, A_0^2.$$

Thus, from Equation (7.26), we have an algebraic equation

$$\epsilon - \kappa^2 - \frac{3}{4}\epsilon\, A_0^2 = 0, \tag{7.33}$$

which has the nonzero solution

$$A_0 = \pm\frac{2}{\sqrt{3}}\sqrt{1 - \frac{\kappa^2}{\epsilon}} \tag{7.34}$$

when $\epsilon > \kappa^2$. Thus, for any a positive integer $\kappa \geq 1$, the so-called bifurcation occurs when

$$\epsilon = \kappa^2. \tag{7.35}$$

This critical condition of bifurcations indicates that there exist multiple bifurcation points for large ϵ. Note that κ determines the set of base functions denoted by (7.4). So, for large ϵ, there exist multiple solutions.

Multiple Solutions of a Nonlinear Problem

Without the loss of generality, we consider here the two cases of $\kappa = 2$ and $\kappa = 3$. Note that the convergence region and rate of the solution series (7.20) and (7.21) are determined by the auxiliary parameter $\hbar$. For a given ϵ and a positive integer κ, where $\epsilon > \kappa^2 \geq 1$, we can always find, by means of plotting the so-called $\hbar$-curves (see page 26 and §3.5.1) of A, a valid region of $\hbar$ to ensure that the solution series (7.21) converges. For example, the $\hbar$-curves of $A \sim \hbar$ when $\kappa = 2, \epsilon = 10, 25, 100$, and $\kappa = 3, \epsilon = 40, 90, 225$ are as shown in Figures 7.1 and 7.2, respectively. From these $\hbar$-curves, it is clear that the series (7.21) converges when $\epsilon = 10$ and $\kappa = 2, 3$ by means of $\hbar = -1$, or $\epsilon = 40, \kappa = 2$ and $\epsilon = 90, \kappa = 3$ by means of $\hbar = -1/2$, or $\epsilon = 100, \kappa = 2$ and $\epsilon = 225, \kappa = 3$ by means of $\hbar = -1/5$. This is indeed true, as shown in Tables 7.1 and 7.2.

From Figures 7.1 and 7.2, the so-called valid region of $\hbar$ for A decreases as ϵ increases for a given κ. So, the absolute value of $\hbar$ should decrease as ϵ increases. It is found that, for any a given ϵ and a given κ satisfying $\epsilon > \kappa^2 \geq 1$, the series (7.21) is always convergent in the region

$$\kappa^2 \leq \epsilon < +\infty,$$

when

$$\hbar = -\left(1 + \frac{\epsilon}{3\kappa^2}\right)^{-1}. \tag{7.36}$$

Besides, the corresponding 10th-order approximation

$$A \approx \pm \left(1 + \frac{\epsilon}{3\kappa^2}\right)^{-10} \sqrt{1 - \frac{\kappa^2}{\epsilon}} \left(1.1803 + 3.9075 \frac{\epsilon}{\kappa^2} + 5.8128 \frac{\epsilon^2}{\kappa^4}\right.$$

$$+ 5.1149 \frac{\epsilon^3}{\kappa^6} + 2.9466 \frac{\epsilon^4}{\kappa^8} + 1.1603 \frac{\epsilon^5}{\kappa^{10}} + 0.31602 \frac{\epsilon^6}{\kappa^{12}}$$

$$+ 5.8726 \times 10^{-2} \frac{\epsilon^7}{\kappa^{14}} + 7.1298 \times 10^{-3} \frac{\epsilon^8}{\kappa^{16}} + 5.1396 \times 10^{-4} \frac{\epsilon^9}{\kappa^{18}}$$

$$\left.+ 1.7001 \times 10^{-5} \frac{\epsilon^{10}}{\kappa^{20}}\right) \tag{7.37}$$

agrees well in the whole region $\kappa^2 \leq \epsilon < +\infty$ with the exact analytic result given by the implicit formula

$$\epsilon = \frac{8\kappa^2}{\pi^2(2-A^2)} K\left(\frac{A^2}{2-A^2}\right), \tag{7.38}$$

where K denotes the complete elliptic integral of the first kind, as shown in Figure 7.3. In fact, Figure 7.3 provides us with a complete bifurcation diagram of the so-called Duffing oscillator in space problem.

Using the so-called homotopy-Padé technique (see page 38 and §3.5.2), we can greatly accelerate the convergence of the series (7.21), as shown in Tables 7.3 and 7.4. It is found that the $[m, m]$ homotopy-Padé approximant does not depend upon the auxiliary parameter $\hbar$. The [4,4] homotopy-Padé approximant

$$A \approx 2\sqrt{3}\sqrt{1 - \frac{\kappa^2}{\epsilon}}\,\frac{P(\epsilon)}{Q(\epsilon)} \tag{7.39}$$

gives an accurate approximation of A in the whole region $1 \leq \epsilon/\kappa^2 < +\infty$, where

$$P(\epsilon) = 8665210296046039923 + 2500964782519057396 \left(\frac{\epsilon}{\kappa^2}\right)$$
$$+ 6040342986 53768562 \left(\frac{\epsilon}{\kappa^2}\right)^2 + 62408285303687028 \left(\frac{\epsilon}{\kappa^2}\right)^3$$
$$+ 3874319809940915 \left(\frac{\epsilon}{\kappa^2}\right)^4,$$

$$Q(\epsilon) = 25430938337575455089 + 7921677254280814588 \left(\frac{\epsilon}{\kappa^2}\right)$$
$$+ 1930521704826790758 \left(\frac{\epsilon}{\kappa^2}\right)^2 + 213027971364041596 \left(\frac{\epsilon}{\kappa^2}\right)^3$$
$$+ 13310678950379441 \left(\frac{\epsilon}{\kappa^2}\right)^4.$$

It is found that, as long as the series (7.21) of A is convergent, the corresponding series (7.20) of $u(\xi)$ given by the same value of $\hbar$ converges in the whole region $\xi \in [0, \pi]$, as shown in Figures 7.4 and 7.5. Due to the odd nonlinearity of Equation (7.6), if $u(\xi)$ is a solution, $-u(\xi)$ must be also a solution. However, for brevity, we do not give this kind of solution in Figures 7.4 and 7.5. The nonlinear problem of the so-called Duffing oscillator in space has multiple solutions for large ϵ. For example, when $\epsilon = 10$, there exist two nonzero solutions corresponding to $\kappa = 1$, two nonzero solutions to $\kappa = 2$, and two nonzero solutions to $\kappa = 3$, respectively, so that there are six nonzero solutions. In general, for any given $\epsilon \geq 1$, the problem of the so-called Duffing oscillator in space has $2[\sqrt{\epsilon}]$ nonzero solutions, where $[x]$ denotes the integer part of x. Therefore, the larger the value of ϵ, the more the multiple solutions, as shown in Figure 7.3. As ϵ tends to infinity, there exists an infinite number of solutions. Therefore, the nonlinear equation (7.1) with boundary conditions (7.2) contains rather rich mathematical structure and a complicated bifurcation diagram.

The *rule of solution expression* plays an important role in finding these multiple solutions. This example clearly indicates that, by means of different base functions, we can employ the homotopy analysis method to gain all multiple solutions of some nonlinear problems. Indeed, the so-called *rule of solution expression* of the homotopy analysis method provides us with a new viewpoint and a different starting point to investigate nonlinear problems.

TABLE 7.1
The analytic approximations of A when $\epsilon = 10$ and $\kappa = 2, 3$ by means of $\hbar = -1$ and $H(\xi) = 1$.

Order of approximation	$\kappa = 2$	$\kappa = 3$
2	0.8694142054	0.3643175731
4	0.8696932532	0.3643100899
6	0.8696857656	0.3643100194
8	0.8696860265	0.3643100187
10	0.8696860164	0.3643100187
12	0.8696860168	0.3643100187
14	0.8696860168	0.3643100187
16	0.8696860168	0.3643100187
18	0.8696860168	0.3643100187
20	0.8696860168	0.3643100187

TABLE 7.2
The analytic approximations of A when $\epsilon = 40, \kappa = 2$ and $\epsilon = 90, \kappa = 3$ by means of $\hbar = -1/2$ and $H(\xi) = 1$.

Order of approximation	$\epsilon = 40, \kappa = 2$	$\epsilon = 90, \kappa = 3$
2	0.98070	0.98171
4	0.99912	0.99854
6	0.99613	0.99639
8	0.99634	0.99625
10	0.99656	0.99658
12	0.99635	0.99635
14	0.99649	0.99648
16	0.99641	0.99642
18	0.99645	0.99644
20	0.99643	0.99644
22	0.99644	0.99644
24	0.99644	0.99644
26	0.99644	0.99644
28	0.99644	0.99644
30	0.99644	0.99644

TABLE 7.3
The $[m,m]$ homotopy-Padé approximant of A when $\epsilon = 10$ and $\kappa = 2, 3$ by means of $H(\xi) = 1$.

$[m,m]$	$\kappa = 2$	$\kappa = 3$
[1, 1]	0.8694029457	0.3643104636
[2, 2]	0.8696902377	0.3643100178
[3, 3]	0.8696859569	0.3643100187
[4, 4]	0.8696860176	0.3643100187
[5, 5]	0.8696860168	0.3643100187
[6, 6]	0.8696860168	0.3643100187
[7, 7]	0.8696860168	0.3643100187
[8, 8]	0.8696860168	0.3643100187
[9, 9]	0.8696860168	0.3643100187
[10, 10]	0.8696860168	0.3643100187

TABLE 7.4
The $[m,m]$ homotopy-Padé approximant of A when $\epsilon = 40, \kappa = 2$ and $\epsilon = 90, \kappa = 3$ by means of $H(\xi) = 1$.

$[m,m]$	$\epsilon = 40, \kappa = 2$	$\epsilon = 90, \kappa = 3$
[1, 1]	0.9747449855	0.9753179745
[2, 2]	0.9988803766	0.9988250895
[3, 3]	0.9960551840	0.9960761350
[4, 4]	0.9964957766	0.9964921829
[5, 5]	0.9964304420	0.9964305571
[6, 6]	0.9964370766	0.9964368336
[7, 7]	0.9964352860	0.9964352709
[8, 8]	0.9964353707	0.9964353614
[9, 9]	0.9964353314	0.9964353314
[10, 10]	0.9964353352	0.9964353355
[11, 11]	0.9964353351	0.9964353363
[12, 12]	0.9964353362	0.9964353363
[13, 13]	0.9964353363	0.9964353363
[14, 14]	0.9964353363	0.9964353363
[15, 15]	0.9964353363	0.9964353363

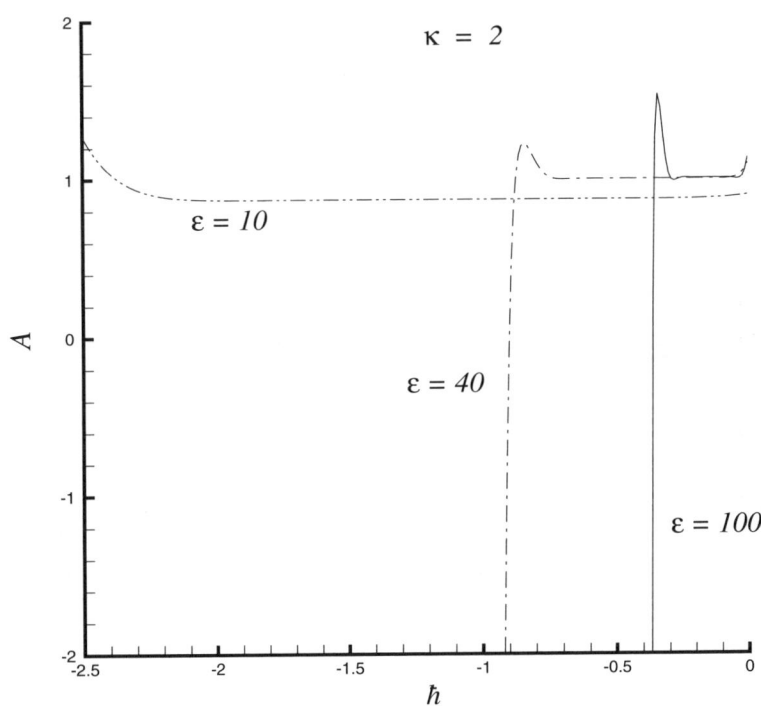

FIGURE 7.1
The $\hbar$-curves of A when $\kappa = 2$ and $\epsilon = 10, 40, 100$ by means of $H(\xi) = 1$. Dash-dot-dotted line: 20th-order approximation of A when $\epsilon = 10$; dash-dotted line: 20th-order approximation of A when $\epsilon = 40$; solid line: 20th-order approximation of A when $\epsilon = 100$.

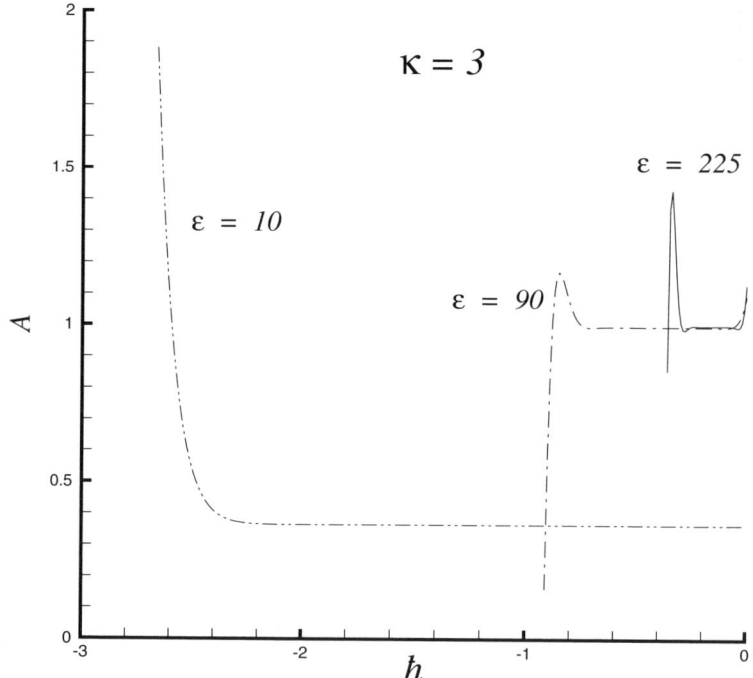

FIGURE 7.2
The $\hbar$-curves of A when $\kappa = 3$ and $\epsilon = 10, 90, 225$ by means of $H(\xi) = 1$. Dash-dot-dotted line: 20th-order approximation of A when $\epsilon = 10$; dash-dotted line: 20th-order approximation of A when $\epsilon = 90$; solid line: 20th-order approximation of A when $\epsilon = 225$.

Multiple Solutions of a Nonlinear Problem

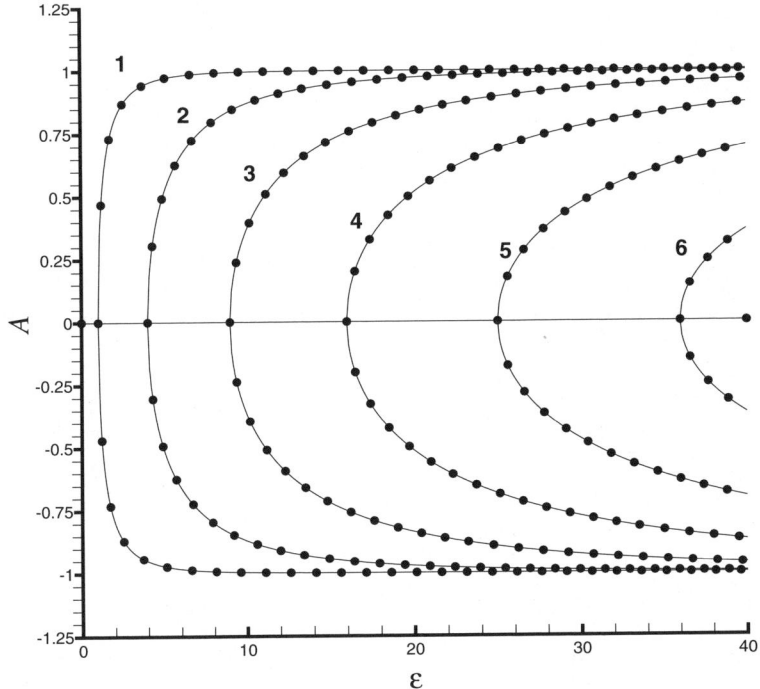

FIGURE 7.3
The comparison of the 10th-order approximation (7.37) of A with the exact implicit solution (7.38). Symbols: exact result; curve 1: $\kappa = 1$; curve 2: $\kappa = 2$; curve 3: $\kappa = 3$; curve 4: $\kappa = 4$; curve 5: $\kappa = 5$; curve 6: $\kappa = 6$.

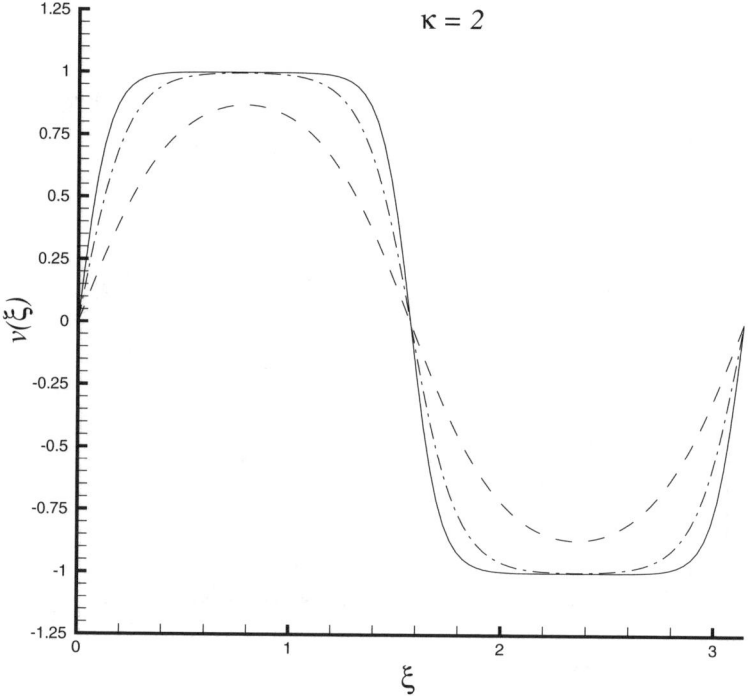

FIGURE 7.4
The convergent analytic result of $v(\xi) = Au(\xi)$ when $\kappa = 2$ and $\epsilon = 10, 40, 100$ by means of $H(\xi) = 1$. Dashed line: fifth-order approximation of $v(\xi)$ when $\epsilon = 10$ by means of $\hbar = -1$; dash-dotted line: 10th-order approximation of $v(\xi)$ when $\epsilon = 40$ by means of $\hbar = -1/2$; solid line: 20th-order approximation of $v(\xi)$ when $\epsilon = 100$ by means of $\hbar = -1/5$.

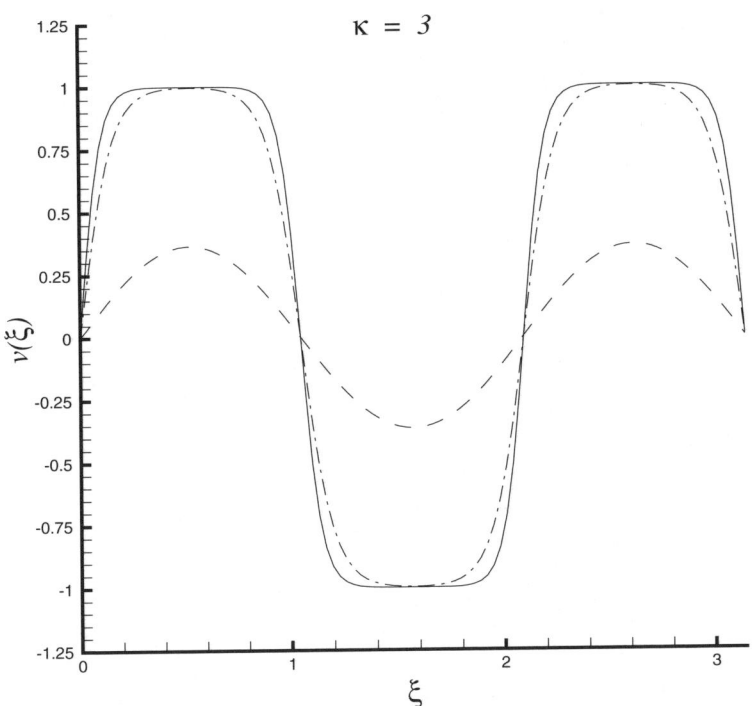

FIGURE 7.5
The convergent analytic result of $v(\xi) = Au(\xi)$ when $\kappa = 3$ and $\epsilon = 10, 90, 225$ by means of $H(\xi) = 1$. Dashed line: fifth-order approximation of $v(\xi)$ when $\epsilon = 10$ by means of $\hbar = -1$; dash-dotted line: 10th-order approximation of $v(\xi)$ when $\epsilon = 90$ by means of $\hbar = -1/2$; solid line: 20th-order approximation of $v(\xi)$ when $\epsilon = 225$ by means of $\hbar = -1/5$.

8
Nonlinear eigenvalue problem

It is often necessary to know eigenvalues and eigenfunctions of a nonlinear problem. In this chapter we illustrate how to obtain eigenvalues and eigenfunctions of a given nonlinear problem by means of the homotopy analysis method.

For example, let us consider an eigenvalue problem governed by

$$u''(x) + \lambda\, u(x) + \epsilon\, u^3(x) = 0, \tag{8.1}$$

subject to the boundary conditions

$$u(0) = u(1) = 0, \tag{8.2}$$

where the prime denotes differentiation with respect to x, ϵ is a parameter. Our object is to find such an eigenvalue λ_n and a normalized eigenfunction $u_n(x)$ such that

$$u_n''(x) + \lambda_n\, u_n(x) + \epsilon\, u_n^3(x) = 0, \tag{8.3}$$

subject to the boundary conditions

$$u_n(0) = u_n(1) = 0 \tag{8.4}$$

and the normalization condition

$$\int_0^1 u_n^2(x)\,dx = 1, \tag{8.5}$$

where the subscript $n \geq 1$ is an integer. Nayfeh [12] described a perturbation approach to the same problem and gave the first-order perturbation approximation

$$u_n(x) = \sqrt{2}\sin(n\pi x) - \frac{\epsilon\sqrt{2}}{16n^2\pi^2}\sin(3n\pi x) + O(\epsilon^2), \tag{8.6}$$

$$\lambda_n = n^2\pi^2 - \frac{3}{2}\epsilon + O(\epsilon^2), \tag{8.7}$$

valid for small ϵ.

115

8.1 Homotopy analysis solution

8.1.1 Zero-order deformation equation

From the boundary conditions (8.4) and considering the nonlinearity of Equation (8.3), it is straightforward that the eigenfunction $u_n(x)$ can be expressed by the set of base functions

$$\{\sin[(2k+1)n\pi x] \mid n \geq 1, k = 0, 1, 2, 3, \cdots\} \tag{8.8}$$

in the form

$$u_n(x) = \sum_{k=0}^{+\infty} a_{n,k} \sin[(2k+1)n\pi x], \tag{8.9}$$

where $a_{n,k}$ is a coefficient. This provides us with the so-called *rule of solution expression*.

Under the *rule of solution expression* denoted by (8.9) and from (8.4) and (8.5), it is straightforward to choose

$$u_{n,0}(x) = \sqrt{2} \sin(n\pi x) \tag{8.10}$$

as an initial guess of $u_n(x)$. Furthermore, under the *rule of solution expression* denoted by (8.9) and from Equation (8.3), we choose an auxiliary linear operator

$$\mathcal{L}\Phi = \frac{\partial^2 \Phi}{\partial x^2} + (n\pi)^2 \Phi \tag{8.11}$$

with the property

$$\mathcal{L}[C_1 \sin(n\pi x) + C_2 \cos(n\pi x)] = 0, \tag{8.12}$$

where C_1 and C_2 are constant coefficients. From Equation (8.3), we define the nonlinear operator

$$\mathcal{N}[\Phi(x;q), \Lambda(q)] = \frac{\partial^2 \Phi(x;q)}{\partial x^2} + \Lambda(q)\, \Phi(x;q) + \epsilon\, \Phi^3(x;q), \tag{8.13}$$

where $q \in [0,1]$ is an embedding parameter, $\Phi(x;q)$ is a function of x and q, and $\Lambda(q)$ is a function of q, corresponding to $u_n(x)$ and λ_n, respectively. Let $\hbar \neq 0$ denote a nonzero auxiliary parameter and $H(x) \neq 0$ a nonzero auxiliary function. Then, we construct the so-called zero-order deformation equation

$$(1-q)\,\mathcal{L}[\Phi(x;q) - u_{n,0}(x)] = q\,\hbar\, H(x)\, \mathcal{N}[\Phi(x;q), \Lambda(q)], \tag{8.14}$$

subject to the boundary conditions

$$\Phi(0;q) = \Phi(1;q) = 0. \tag{8.15}$$

Nonlinear Eigenvalue Problem

When $q = 0$, it is straightforward to show that the zero-order deformation equations (8.14) and (8.15) have the solution

$$\Phi(x; 0) = u_{n,0}(x). \tag{8.16}$$

When $q = 1$, they are equivalent to the original equations (8.3) and (8.4), respectively, provided

$$\Phi(x; 1) = u_n(x), \quad \Lambda(1) = \lambda_n. \tag{8.17}$$

Thus, as the embedding parameter q increases from 0 to 1, $\Phi(x; q)$ varies from the initial guess $u_{n,0}(x)$ to the exact eigenfunction $u_n(x)$, so does $\Lambda(q)$ from the initial guess $\lambda_{n,0}$ to the exact eigenvalue λ_n, respectively. Note that the zero-order deformation equations contain the auxiliary parameter $\hbar$ and the auxiliary function $H(x)$, therefore $\Phi(x; q)$ and $\Lambda(q)$ are dependent of $\hbar$ and $H(x)$. Assume that $\hbar$ and $H(x)$ are properly chosen so that the zero-order deformation equations (8.14) and (8.15) have solutions for all $q \in [0, 1]$, that the terms

$$u_{n,k}(x) = \frac{1}{k!} \left.\frac{\partial^k \Phi(x; q)}{\partial q^k}\right|_{q=0}, \quad \lambda_{n,k} = \frac{1}{k!} \left.\frac{\partial^k \Lambda(q)}{\partial q^k}\right|_{q=0} \tag{8.18}$$

exist for $k \geq 1$, and that the Taylor series

$$\Phi(x; q) = \sum_{k=0}^{+\infty} u_{n,k}(x) \, q^k, \tag{8.19}$$

$$\Lambda(q) = \sum_{k=0}^{+\infty} \lambda_{n,k} \, q^k \tag{8.20}$$

are convergent at $q = 1$. Then, using (8.16) and (8.17), we have

$$u_n(x) = u_{n,0}(x) + \sum_{k=1}^{+\infty} u_{n,k}(x), \tag{8.21}$$

$$\lambda_n = \lambda_{n,0} + \sum_{k=1}^{+\infty} \lambda_{n,k}, \tag{8.22}$$

which provide us with the relationships between the exact eigenfunction, eigenvalue, and their initial guesses by $u_{n,k}(x)$ and $\lambda_{n,k}$, respectively.

8.1.2 High-order deformation equation

For brevity, define the vector

$$\vec{u}_{n,m} = \{u_{n,0}(x), u_{n,1}(x), u_{n,2}(x), \cdots, u_{n,m}(x)\}$$

and
$$\vec{\lambda}_{n,m} = \{\lambda_{n,0}, \lambda_{n,1}, \lambda_{n,2}, \cdots, \lambda_{n,m}\}.$$

Differentiating the zero-order deformation equations (8.14) and (8.15) k times with respect to q and then dividing by $k!$ and finally setting $q = 0$, we have the high-order deformation equation

$$\mathcal{L}[u_{n,k}(x) - \chi_k u_{n,k-1}(x)] = \hbar H(x) R_{n,k}(\vec{u}_{n,k-1}, \vec{\lambda}_{n,k-1}), \qquad (8.23)$$

subject to the boundary conditions

$$u_{n,k}(0) = u_{n,k}(1) = 0, \qquad (8.24)$$

where χ_k is defined by (2.42) and

$$R_{n,k}(\vec{u}_{n,k-1}, \vec{\lambda}_{n,k-1})$$
$$= u''_{n,k-1}(x) + \sum_{m=0}^{k-1} \lambda_m u_{n,k-1-m}(x)$$
$$+ \epsilon \sum_{m=0}^{k-1} u_{n,k-1-m}(x) \sum_{j=0}^{m} u_{n,j}(x) u_{n,m-j}(x). \qquad (8.25)$$

Note that, for any a given integer $n \geq 1$, both $u_{n,k}(x)$ and $\lambda_{n,k-1}$ are unknown for $k \geq 1$, but we have only one differential equation (8.23) for $u_{n,k}(x)$. Thus, the problem is not closed and an additional algebraic equation is needed to determine $\lambda_{n,k-1}$. Note that we have great freedom to choose the auxiliary function $H(x)$. Under the *rule of solution expression* denoted by (8.9), $R_{n,k}(\vec{u}_{n,k-1}, \vec{\lambda}_{n,k-1})$ can be expressed in the form

$$R_{n,k}(\vec{u}_{n,k-1}, \vec{\lambda}_{n,k-1}) = \sum_{m=0}^{M_{n,k}} d_{n,m} \sin[(2m+1)n\pi x],$$

where $d_{n,m}$ is a coefficient, and $M_{n,k}$ is an integer dependent on both n and k. Under the *rule of solution expression* denoted by (8.9) and from (8.11) and (8.23), $H(x)$ can be in the form

$$H(x) = \sin^2[(2m-1)n\pi x], \qquad (m \geq 1) \qquad (8.26)$$

or

$$H(x) = \sin[(2m)n\pi x], \qquad (m \geq 1) \qquad (8.27)$$

or

$$H(x) = \cos^2[(2m-1)n\pi x], \qquad (m \geq 1) \qquad (8.28)$$

or

$$H(x) = \cos[(2m)n\pi x], \qquad (m \geq 1) \qquad (8.29)$$

or even simply
$$H(x) = 1, \tag{8.30}$$
where $m \geq 1$ is an integer. So, it holds
$$H(x)\, R_{n,k}(\vec{u}_{n,k-1}, \vec{\lambda}_{n,k-1}) = \sum_{m=0}^{\mu_{n,k}} b_m^{n,k}(\vec{\lambda}_{k-1})\, \sin[(2m+1)n\pi x], \tag{8.31}$$
where $b_m^{n,k}(\vec{\lambda}_{k-1})$ is a coefficient, and $\mu_{n,k}$ is an integer determined by $H(x)$ and the values of n and k. Thus, there exists the term
$$\hbar\, b_0^{n,k}(\vec{\lambda}_{k-1})\, \sin(n\pi x)$$
on the right-hand side of the high-order deformation equation (8.23). If
$$b_0^{n,k}(\vec{\lambda}_{k-1}) \neq 0,$$
due to the property (8.12), the solution $u_{n,k}(x)$ contains the term
$$x\, \sin(n\pi x),$$
which disobeys the *rule of solution expression* denoted by (8.9). To avoid this, we had to enforce
$$b_0^{n,k}(\vec{\lambda}_{k-1}) = 0, \tag{8.32}$$
which provides us with an additional equation to determine $\lambda_{n,k-1}$. In this way, the problem is closed.

After solving the above algebraic equation to gain $\lambda_{n,k-1}$, it is easy to get the solution
$$u_{n,k}(x) = \chi_k\, u_{n,k-1}(x) + \sum_{m=1}^{\mu_{n,k}} \frac{\hbar\, b_m^{n,k}(\vec{\lambda}_{k-1})}{n^2\pi^2\,[1-(2m+1)^2]}\, \sin[(2m+1)n\pi x]$$
$$+ C_1 \sin(n\pi x) + C_2 \cos(n\pi x), \tag{8.33}$$
where C_1 and C_2 are coefficients. Using the *rule of solution expression* denoted by (8.9), we have
$$C_2 = 0.$$
Note that the above solution automatically satisfies the boundary conditions (8.24) so that the coefficient C_1 cannot be determined. However, from the normalization condition (8.5), we have
$$\int_0^1 \left(\sum_{m=0}^k u_{n,m}(x) \right)^2 dx = 1, \tag{8.34}$$
which gives an algebraic equation
$$C_1^2 + 4\alpha C_1 + 2\beta = 1, \tag{8.35}$$

where

$$\alpha = \int_0^1 w_{n,k}(x)\sin(n\pi x)dx, \qquad \beta = \int_0^1 w_{n,k}^2(x)dx \qquad (8.36)$$

and

$$w_{n,k}(x) = \sum_{j=0}^{k-1} u_{n,j}(x) + \chi_k\, u_{n,k-1}(x)$$
$$+ \sum_{m=1}^{\mu_{n,k}} \frac{\hbar\, b_m^{n,k}(\vec{\lambda}_{k-1})}{n^2\pi^2[1-(2m+1)^2]}\sin[(2m+1)n\pi x]. \qquad (8.37)$$

Solving Equation (8.35), we have two solutions

$$C_1 = -2\alpha + \sqrt{4\alpha^2 - 2\beta} \qquad (8.38)$$

and

$$C_1 = -2\alpha - \sqrt{4\alpha^2 - 2\beta}, \qquad (8.39)$$

which correspond to two different eigenfunctions, respectively. In this way, we can gain $\lambda_{n,0}, u_{n,1}(x), \lambda_{n,1}, u_{n,2}(x)$, and so on, successively.

For any given n, the mth-order approximation of the eigenfunction and eigenvalue are given by

$$u_n(x) \approx u_{n,0} + \sum_{k=1}^m u_{n,k}(x), \qquad (8.40)$$

$$\lambda_n \approx \lambda_{n,0} + \sum_{k=1}^m \lambda_{n,k}, \qquad (8.41)$$

respectively.

8.1.3 Convergence theorem

THEOREM 8.1
If the series

$$u_{n,0}(x) + \sum_{k=1}^{+\infty} u_{n,k}(x)$$

and

$$\lambda_{n,0} + \sum_{k=1}^{+\infty} \lambda_{n,k}$$

are convergent, where $u_{n,k}(x)$ is governed by Equations (8.23), (8.24), and (8.34) under the definitions (8.11), (8.25), and (2.42), they must be the eigenfunction and eigenvalue of Equations (8.1) and (8.2), respectively.

Nonlinear Eigenvalue Problem

Proof: If the series of the eigenfunction is convergent, it is necessary that

$$\lim_{m \to +\infty} u_{n,m}(x) = 0.$$

Then, using (8.11), (8.23), and (2.42), we have

$$\hbar H(x) \sum_{k=1}^{+\infty} R_k(\vec{u}_{n,k-1}, \vec{\lambda}_{n,k-1})$$

$$= \lim_{m \to +\infty} \sum_{k=1}^{m} \mathcal{L}[u_{n,k}(x) - \chi_k u_{n,k-1}(x)]$$

$$= \mathcal{L} \left\{ \lim_{m \to +\infty} \sum_{k=1}^{m} [u_{n,k}(x) - \chi_k u_{n,k-1}(x)] \right\}$$

$$= \mathcal{L} \left[\lim_{m \to +\infty} u_{n,m}(x) \right]$$

$$= 0,$$

which gives, since $\hbar \neq 0$ and $H(x) \neq 0$,

$$\sum_{k=1}^{+\infty} R_k(\vec{u}_{n,k-1}, \vec{\lambda}_{n,k-1}) = 0.$$

Substituting (8.25) into the above expression and simplifying it, we obtain, due to the convergence of the series of the eigenfunction and eigenvalue, that

$$\frac{d^2}{dx^2} \left[\sum_{k=0}^{+\infty} u_{n,k}(x) \right] + \left(\sum_{m=0}^{+\infty} \lambda_{n,m} \right) \left[\sum_{k=0}^{+\infty} u_{n,k}(x) \right] + \epsilon \left[\sum_{k=0}^{+\infty} u_{n,k}(x) \right]^3 = 0.$$

From (8.10) and (8.24) it holds that

$$\sum_{k=0}^{+\infty} u_{n,k}(0) = \sum_{k=0}^{+\infty} u_{n,k}(1) = 0.$$

Furthermore, from (8.34), the normalization condition (8.5) is satisfied. Thus, as long as the two series are convergent, they must be the eigenfunction $u_n(x)$ and the eigenvalue λ_n of the nonlinear problem governed by Equations (8.3), (8.4), and (8.5). This ends the proof.

8.2 Result analysis

Note that the series (8.21) and (8.22) contain the auxiliary parameter $\hbar$ and the auxiliary function $H(x)$. In particular, for any given values of n and ϵ, the series (8.22) for the eigenvalue is a power series of $\hbar$ so that its convergence region and rate are dependent on $\hbar$. According to Theorem 8.1, we need only focus on the choice of the auxiliary parameter $\hbar$ and the auxiliary function $H(x)$ to ensure that the two series converge. Note that, under the *rule of solution expression* denoted by (8.9) and the *rule of coefficient ergodicity*, the auxiliary functions $H(x)$ can be chosen in many different forms such as those expressed by (8.26) to (8.30). For the sake of simplicity, we first consider the case of $H(x) = 1$. For any given values of n and ϵ, we can investigate the influence of $\hbar$ on the convergence region of the series (8.22) for the eigenvalue by plotting the so-called $\hbar$-curves (see page 26 and §3.5.1) of λ_n versus $\hbar$. For example, the $\hbar$-curves of the eigenvalue λ_1 when $\epsilon = 5, 25, -50$ are as shown in Figure 8.1. Using the $\hbar$-curves, we can easily find out the valid regions of $\hbar$ which ensure that the corresponding series (8.22) converge. From Figure 8.1, it is clear that the series (8.22) of λ_1 when $\epsilon = -50$ converges by means of $\hbar = -1/2$ or $\hbar = -2/5$. This is indeed true, as shown in Table 8.1, and the convergence rate can be accelerated by means of the homotopy-Padé technique (see page 38 and §3.5.2), as shown in Table 8.2. It is found that, as long as the series (8.22) for the eigenvalue is convergent, the series (8.21) for the corresponding eigenfunction also converges in the whole region $0 \le x \le 1$. For example, the approximations of the eigenfunction $u_1(x)$ when $\epsilon = -50$ are as shown in Figure 8.2. In this way, we can gain the convergent eigenvalue and eigenfunction for any given values of n and ϵ. For instance, some convergent analytic results of the eigenvalues are listed in Table 8.3 and some eigenfunctions are as shown in Figures 8.3 and 8.4.

It is found that the valid region of $\hbar$ becomes shorter as the nonlinearity of the problem is stronger, as shown in Figure 8.1. So, as the absolute value of ϵ increases, the value of $\hbar$ had to be chosen closer to zero from the below. It is found that, for $\epsilon < 0$, we can always gain convergent results by means of

$$\hbar = -\frac{1}{\sqrt{1+|\epsilon|}}. \tag{8.42}$$

Using this expression, we can investigate the eigenvalues and eigenfunctions when the nonlinearity becomes very strong. Some eigenvalues for negative ϵ far away from zero are listed in Table 8.4 and some eigenfunctions are as shown in Figures 8.5 to 8.7, respectively. According to these analytic results, it seems that

$$\lim_{\epsilon \to -\infty} \frac{\lambda_n}{\epsilon} = -1 \tag{8.43}$$

and

$$\lim_{\epsilon \to -\infty} u_n(x) = \begin{cases} 1, & 2k/n < x < (2k+1)/n, \\ -1, & (2k+1)/n < x < (2k+2)/n, \end{cases} \tag{8.44}$$

where $n \ge 2$, $k = 0, 1, 2, \cdots, [(n-1)/2]$ and $[x]$ denotes the integer part of x. Note that, when $\epsilon = -10000$, we had to choose a negative value of $\hbar$ with a small enough absolute value so as to ensure that the series (8.21) and (8.22) converge. This indicates once again that the auxiliary parameter $\hbar$ plays an important role in the homotopy analysis method.

It is found that, if (8.38) is used to calculate the coefficient C_1, the corresponding eigenfunction is positive in the region

$$0 < x < 1/n.$$

Let $u_n^+(x)$ denote such kind of eigenfunction. If (8.39) is used, the corresponding eigenfunction is negative in the same region, denoted by $u_n^-(x)$. It also is found that these two kinds of eigenfunctions are symmetrical about the x axis. Thus, for given ϵ and n, there exists a unique eigenvalue λ_n but two eigenfunctions $u_n^+(x)$ and $u_n^-(x)$ satisfying

$$u_n^-(x) = -u_n^+(x).$$

However, the series of the eigenfunction $u_n^-(x)$ converges more slowly than that of the eigenfunction $u_n^+(x)$. This is mainly because $u_n^+(x)$ is closer to the initial guess (8.10). It is found that, if we use the initial guess

$$u_{n,0}(x) = -\sqrt{2}\sin(n\pi x)$$

and employ the formula (8.39), it is easier to get convergent eigenfunctions $u_n^-(x)$. So, by means of the homotopy analysis method we can gain the multiple eigenfunctions of the considered nonlinear problem.

All of the above results are given by means of the auxiliary function $H(x) = 1$. It is found that the other four types of the auxiliary functions denoted by (8.26) to (8.29) can also give convergent results. Using different auxiliary functions, we gain the same eigenvalue and eigenfunction, however, the solution series given by the other four types of the auxiliary functions converge more slowly than those by $H(x) = 1$. It seems that $H(x) = 1$ might be the best auxiliary function for the considered problem, although we cannot prove it.

This example illustrates that the homotopy analysis method can be employed to gain all eigenvalues and eigenfunctions of nonlinear boundary-value problems with very strong nonlinearity.

TABLE 8.1
The analytic approximations of λ_1/π^2 when $\epsilon = -50$ by means of $H(x) = 1$.

order of approximation	$\hbar = -1/2$	$\hbar = -2/5$
5	7.5375343842	7.5399457051
10	7.5384488341	7.5384600578
15	7.5384471198	7.5384473078
20	7.5384471141	7.5384471261
25	7.5384471141	7.5384471146
30	7.5384471141	7.5384471141
35	7.5384471141	7.5384471141
40	7.5384471141	7.5384471141

TABLE 8.2
The $[m, m]$ homotopy-Padé approximant of λ_1/π^2 when $\epsilon = -50$ by means of $H(x) = 1$.

$[m, m]$	$\hbar = -1/2$	$\hbar = -2/5$
[2, 2]	7.5407539111	7.5410810211
[4, 4]	7.5384474321	7.5384485282
[6, 6]	7.5384473644	7.5384480394
[8, 8]	7.5384471141	7.5384471141
[10, 10]	7.5384471141	7.5384471141
[12, 12]	7.5384471141	7.5384471141
[14, 14]	7.5384471141	7.5384471141
[16, 16]	7.5384471141	7.5384471141
[18, 18]	7.5384471141	7.5384471141
[20, 20]	7.5384471141	7.5384471141

TABLE 8.3
The analytic value of $\lambda_n/(n\pi)^2$ by means of $\hbar = -1$ and $H(x) = 1$.

ϵ	$n=1$	$n=2$	$n=3$
-25	4.43277	1.91746	1.41524
-20	3.78508	1.73857	1.33324
-15	3.12328	1.55758	1.25074
-10	2.44317	1.37430	1.16771
-5	1.73857	1.18852	1.08414
0	1	1	1
5	0.212582	0.808470	0.915264
10	-0.647567	0.613626	0.829909
15	-1.61838	0.415125	0.743906
20	-2.75608	0.212582	0.657228
25	-4.13061	0.005561	0.569843

TABLE 8.4
The analytic value of λ_n/ϵ by means of $H(x) = 1$.

ϵ	λ_1/ϵ	λ_2/ϵ	λ_3/ϵ
-200	-1.221	-1.488	-1.810
-400	-1.152	-1.325	-1.524
-600	-1.122	-1.272	-1.412
-1000	-1.093	-1.196	-1.310
-2000	-1.065	-1.135	-1.215
-5000	-1.041	-1.083	-1.133
-10000	-1.029	-1.059	-1.090

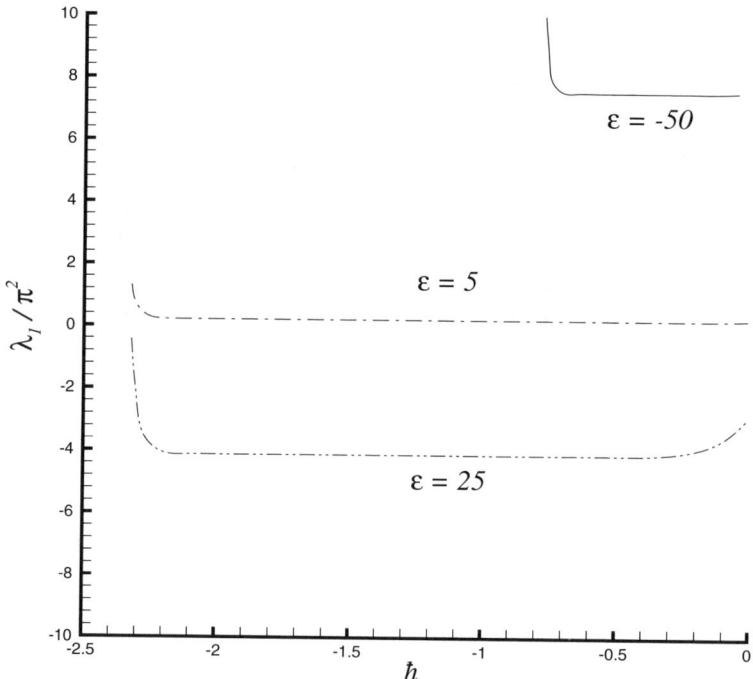

FIGURE 8.1
The $\hbar$-curves of λ_1/π^2 by means of $H(x) = 1$. Dash-dotted line: 20th-order approximation when $\epsilon = 5$; dash-dot-dotted line: 20th-order approximation when $\epsilon = 25$; solid line: 30th-order approximation when $\epsilon = -50$.

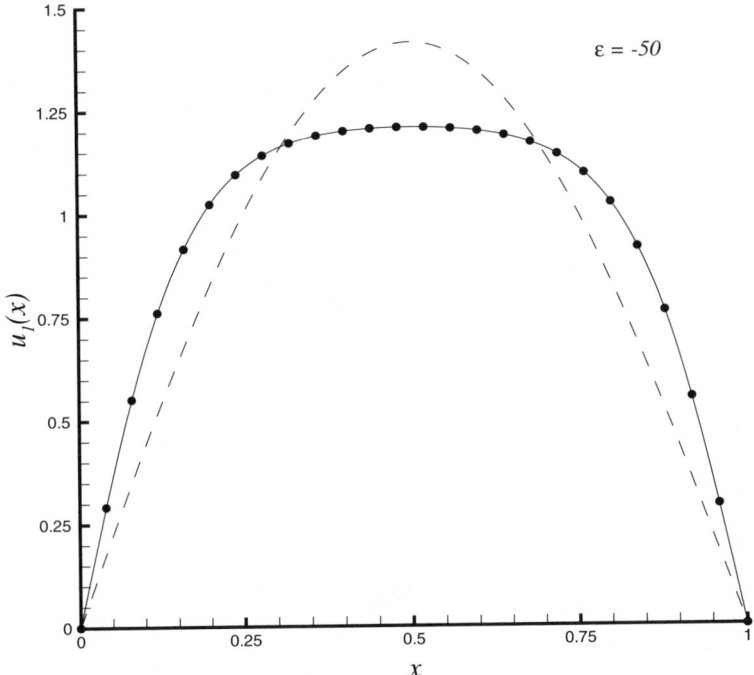

FIGURE 8.2
The analytic approximations of the eigenfunction $u_1(x)$ when $\epsilon = -50$ by means of $\hbar = -1/2$ and $H(x) = 1$. Dashed line: zero-order approximation; solid line: fifth-order approximation; symbols: 10th-order approximation.

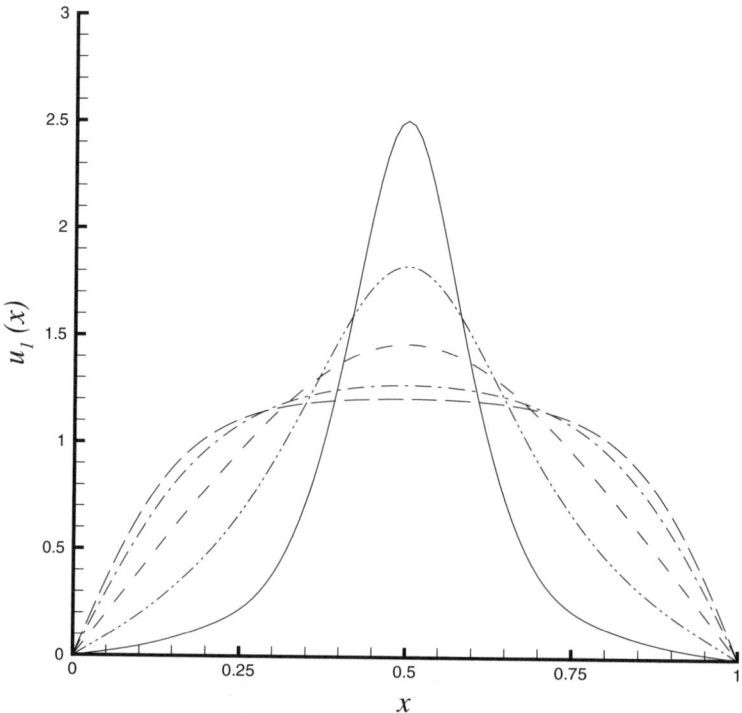

FIGURE 8.3
The convergent analytic results of the eigenfunction $u_1(x)$ by means of $H(x) = 1$. Solid line: 30th-order approximation when $\epsilon = 50$ and $\hbar = -1/2$; dash-dot-dotted line: 10th-order approximation when $\epsilon = 25$ and $\hbar = -1/2$; dashed line: fifth-order approximation when $\epsilon = 5$ and $\hbar = -1$; dash-dotted line: 20th-order approximation when $\epsilon = -25$ and $\hbar = -1/2$; long-dashed line: 20th-order approximation when $\epsilon = -50$ and $\hbar = -1/2$.

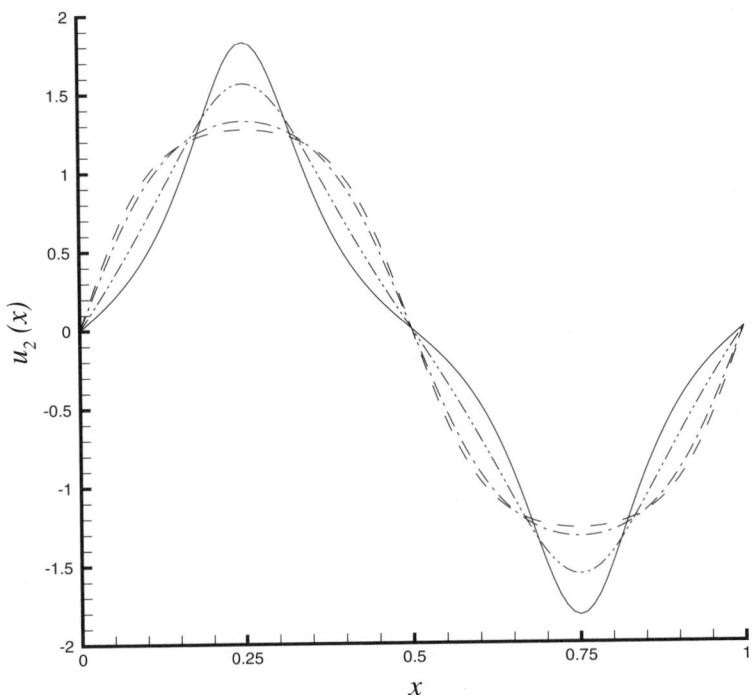

FIGURE 8.4
The convergent analytic results of the eigenfunction $u_2(x)$ by means of $\hbar = -1$ and $H(x) = 1$. Solid line: 10th-order approximation when $\epsilon = 100$; dash-dot-dotted line: fifth-order approximation when $\epsilon = 50$; dash-dotted line: 10th-order approximation when $\epsilon = -50$; dashed line: 20th-order approximation when $\epsilon = -100$.

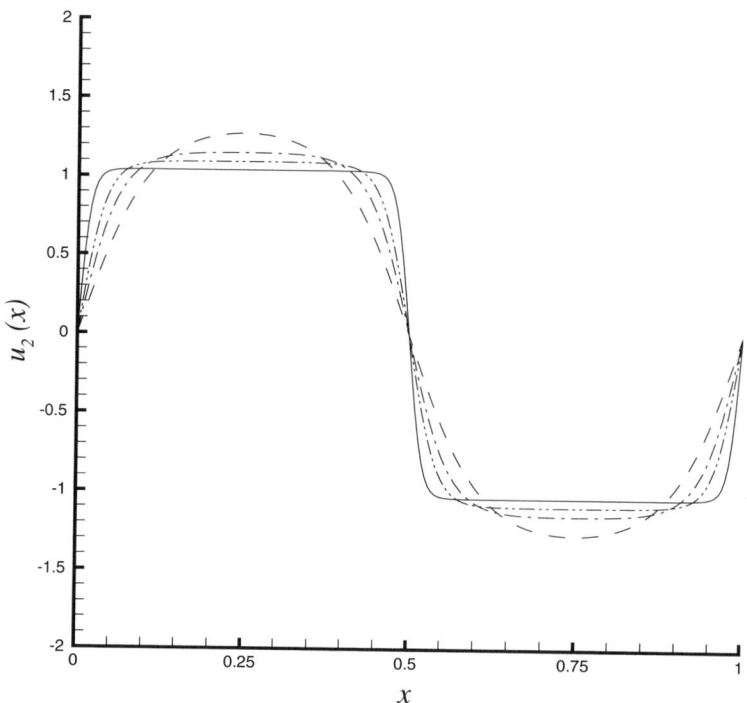

FIGURE 8.5
The convergent analytic result of the eigenfunction $u_2(x)$ by means of $H(x) = 1$. Solid line: 100th-order approximation when $\epsilon = -5000$ and $\hbar = -1/50$; dash-dot-dotted line: 20th-order approximation when $\epsilon = -1000$ and $\hbar = -1/10$; dash-dotted line: 20th-order approximation when $\epsilon = -400$ and $\hbar = -1/4$; dashed line: 20th-order approximation when $\epsilon = -100$ and $\hbar = -1$.

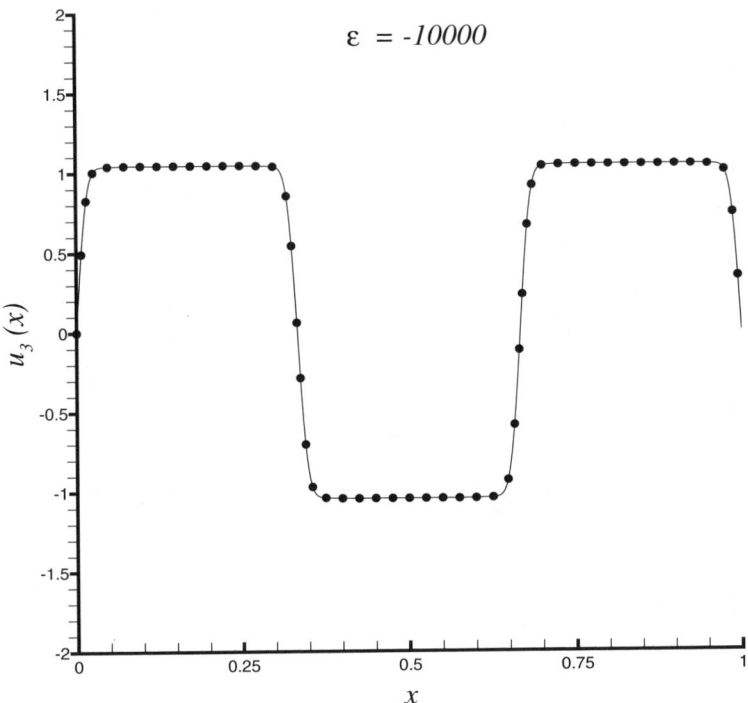

FIGURE 8.6
The analytic result of the eigenfunction $u_3(x)$ when $\epsilon = -10000$ by means of $H(x) = 1$ and $\hbar = -1/50$. Solid line: 70th-order approximation; symbols: 90th-order approximation

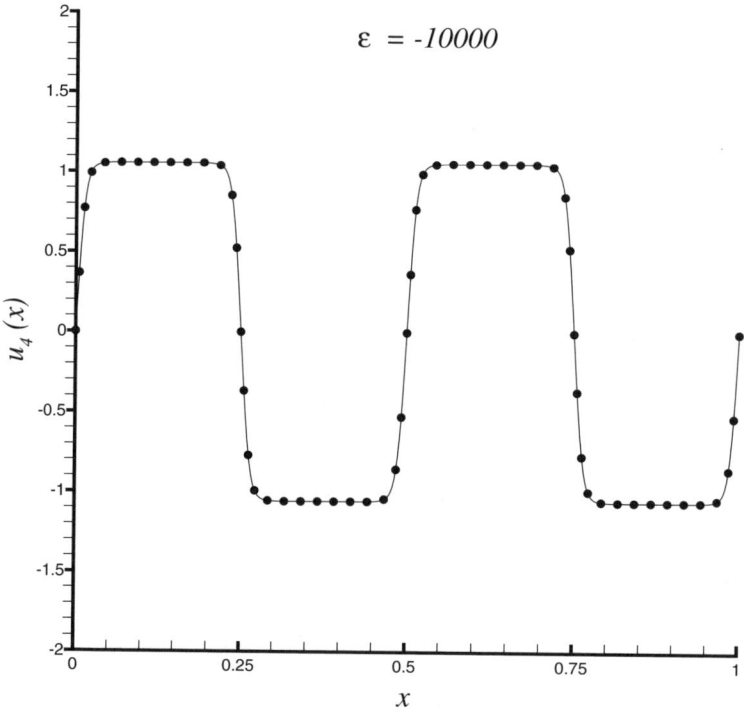

FIGURE 8.7
The analytic result of the eigenfunction $u_4(x)$ when $\epsilon = -10000$ by means of $H(x) = 1$ and $\hbar = -1/20$. Solid line: 40th-order approximation; symbols: 60th-order approximation

9

Thomas-Fermi atom model

In the Thomas-Fermi atom model [81, 82] there exists the so-called Thomas-Fermi equation

$$u''(x) = \sqrt{\frac{u^3(x)}{x}}, \qquad (9.1)$$

subject to the boundary conditions

$$u(0) = 1, u(+\infty) = 0 \qquad (9.2)$$

in the common case. The Thomas-Fermi atom model views the electrons in an atom as a gas and derives atomic structure in terms of the electrostatic potential and the electron density in the ground state. Equation (9.1) describes the spherically symmetric charge distribution concerning a multi-electron atom.

From (9.1) and (9.2) it holds that $u''(0) \to +\infty$. So, there exists a singularity at $x = 0$. The analytic approximations of the Thomas-Fermi equation were given by the variational approach [83, 84], the δ-expansion method [85, 86, 87], Adomian's decomposition method [88, 89, 90, 91], and so on [92, 93, 94, 95, 96, 97]. However, all of these results are analytic-numerical because numerical techniques had to be employed to gain the value of $u'(0)$. Recently, Liao [51] applied the homotopy analysis method to give, for the first time, an explicit, purely analytic solution of the Thomas-Fermi equation by means of the base functions

$$\{(1+x)^{-n} \mid n \geq 1\}. \qquad (9.3)$$

Although Liao's [51] solution is valid in the whole region, its convergence rate is slow for large x. In this chapter, using a set of base functions better than the above ones, we apply the homotopy analysis method to give a more efficient analytic expansion of Thomas-Fermi equations.

9.1 Homotopy analysis solution

9.1.1 Asymptotic property

Equation (9.1) can be rewritten by

$$x [u''(x)]^2 - u^3(x) = 0. \qquad (9.4)$$

Under the transformation
$$\tau = 1 + \lambda\, x, \tag{9.5}$$
where λ is a constant parameter to be determined later, Equation (9.4) becomes
$$\lambda^3 (\tau - 1) \left(\frac{d^2 u}{d\tau^2}\right)^2 - u^3(\tau) = 0, \tag{9.6}$$
subject to the boundary conditions
$$u(1) = 1, \quad u(+\infty) = 0. \tag{9.7}$$

Note that Equation (9.6) contains neither linear terms nor small/large parameters. So, its nonlinearity is very strong. According to (9.7), as $\tau \to +\infty$, $u(\tau) \to 0$ either algebraically or exponentially. However, from Equations (9.6) and (9.7), it is hard to determine the asymptotic property of $u(x)$ at infinity. So, let us first assume that $u(\tau) \to 0$ algebraically and that $u(\tau)$ has the asymptotic expression
$$u(\tau) \sim \tau^\kappa \quad \text{as} \quad \tau \to +\infty,$$
where κ is an unknown constant. Substituting it into Equation (9.6) and then balancing the main terms, we have
$$\kappa = -3. \tag{9.8}$$
Therefore, $u(\tau)$ can be expressed by the set of base functions
$$\{\tau^{-m} \mid m \geq 3\} \tag{9.9}$$
in the form
$$u(\tau) = \sum_{m=3}^{+\infty} c_m\, \tau^{-m}, \tag{9.10}$$
where c_m is a coefficient. This provides us with the so-called *rule of solution expression*.

9.1.2 Zero-order deformation equation

Under the *rule of solution expression* denoted by (9.10) and using the boundary conditions (9.7), it is straightforward to choose
$$u_0(\tau) = \tau^{-3} \tag{9.11}$$
as the initial guess of $u(\tau)$. From (9.6) and using the *rule of solution expression* denoted by (9.10), we choose the auxiliary linear operator
$$\mathcal{L}[\Phi(\tau; q)] = \left(\frac{\tau}{4}\right) \frac{\partial^2 \Phi(\tau; q)}{\partial \tau^2} + \frac{\partial \Phi(\tau; q)}{\partial \tau} \tag{9.12}$$

with the property

$$\mathcal{L}\left(\frac{C_1}{\tau^3} + C_2\right) = 0, \tag{9.13}$$

where C_1 and C_2 are coefficients. From Equation (9.6), we define the nonlinear operator

$$\mathcal{N}[\Phi(\tau;q)] = \lambda^3 \, (\tau - 1) \left[\frac{\partial^2 \Phi(\tau;q)}{\partial \tau^2}\right]^2 - \Phi^3(\tau;q), \tag{9.14}$$

where $\Phi(\tau;q)$ is an unknown function of τ and q. Let $\hbar$ denote a nonzero auxiliary parameter and $H(\tau)$ a nonzero auxiliary function, respectively. Then, we construct the zero-order deformation equation

$$(1-q)\,\mathcal{L}\left[\Phi(\tau;q) - u_0(\tau)\right] = \hbar\, H(\tau)\, q\, \mathcal{N}[\Phi(\tau;q)], \tag{9.15}$$

subject to the boundary conditions

$$\Phi(1;q) = 1,\, \Phi(+\infty;q) = 0, \tag{9.16}$$

where $q \in [0,1]$ is an embedding parameter.

From (9.11), it is straightforward to show that when $q = 0$ the solution of Equations (9.15) and (9.16) is

$$\Phi(\tau;0) = u_0(\tau). \tag{9.17}$$

Since $\hbar \neq 0$ and $H(\tau) \neq 0$, when $q = 1$, Equations (9.15) and (9.16) are equivalent to Equations (9.6) and (9.7), respectively, provided

$$\Phi(\tau;1) = u(\tau). \tag{9.18}$$

Thus, as q increases from 0 to 1, $\Phi(\tau;q)$ varies from the initial guess $u_0(\tau)$ to the exact solution $u(\tau)$ of Equations (9.6) and (9.7).

By Taylor's theorem and using (9.17), we can expand $\Phi(\tau;q)$ in the series of q in the form

$$\Phi(\tau;q) = u_0(\tau) + \sum_{k=1}^{+\infty} u_k(\tau)\, q^k, \tag{9.19}$$

where

$$u_k(\tau) = \frac{1}{k!} \left.\frac{\partial^k \Phi(\tau;q)}{\partial q^k}\right|_{q=0}. \tag{9.20}$$

$\Phi(\tau;q)$ is also dependent upon the auxiliary parameter $\hbar$ and the auxiliary function $H(x)$. Assuming that $\hbar$ and $H(x)$ are properly chosen so that the series (9.19) converges at $q = 1$, we have, using (9.18),

$$u(\tau) = u_0(\tau) + \sum_{k=1}^{+\infty} u_k(\tau). \tag{9.21}$$

It provides us with a relationship between the initial guess $u_0(x)$ and the exact solution $u(x)$ by the terms $u_k(x)$ $(k \geq 1)$.

9.1.3 High-order deformation equations

For brevity, define
$$\vec{u}_n = \{u_0(\tau), u_1(\tau), u_2(\tau), \cdots, u_n(\tau)\}.$$

Differentiating the zero-order deformation equations (9.15) and (9.16) k times with respect to q and then setting $q = 0$ and finally dividing them by $k!$, we have the so-called high-order deformation equation

$$\mathcal{L}[u_k(\tau) - \chi_k u_{k-1}(\tau)] = \hbar\, H(\tau)\, R_k(\vec{u}_{k-1}, \tau), \qquad (9.22)$$

subject to the boundary conditions

$$u_k(1) = 0,\, u_k(+\infty) = 0, \qquad (9.23)$$

where χ_k is defined by (2.42) and

$$R_k(\vec{u}_{k-1}, \tau)$$
$$= \sum_{j=0}^{k-1} \left[\lambda^3 (\tau - 1)\, u_j''(\tau)\, u_{k-1-j}''(\tau) - u_{k-1-j}(\tau) \sum_{i=0}^{j} u_i(\tau)\, u_{j-i}(\tau) \right]. \qquad (9.24)$$

Note that $u_k(\tau)$ ($k \geq 1$) is governed by the linear equation (9.22) and the linear boundary conditions (9.23). Thus, according to (9.21), the homotopy analysis method in essence transfers the original nonlinear problem, governed by Equations (9.6) and (9.7), to an infinite number of linear sub-problems, governed by Equations (9.22) and (9.23). Note that such a kind of transformation does not need any small or large parameters at all.

Let $u_k^*(\tau)$ denote a special solution of the equation

$$\mathcal{L}[u_k^*(\tau)] = \hbar\, H(\tau)\, R_k(\vec{u}_{k-1}, \tau).$$

Then, using (9.13), the general solution of Equation (9.22) is

$$u_k(\tau) = \chi_k\, u_{k-1}(\tau) + u_k^*(\tau) + C_1\, \tau^{-3} + C_2, \qquad (9.25)$$

where the coefficients C_1 and C_2 are determined by the boundary conditions (9.23). In this way we can successively solve the high-order deformation equations (9.22) and (9.23), provided $H(\tau)$ is known. Under the *rule of solution expression* denoted by (9.10), $H(\tau)$ should be in the form

$$H(\tau) = \tau^\sigma, \qquad (9.26)$$

where σ is an integer to be determined. It is found that when

$$\sigma > 4,$$

the solution contains the term

$$\tau \ln \tau$$

Thomas-Fermi Atom Model

that disobeys the *rule of solution expression* denoted by (9.10). When

$$\sigma < 4,$$

the coefficient of the term τ^{-4} is always zero and cannot be improved even if the order of approximation tends to infinity. This disobeys the *rule of coefficient ergodicity*. Thus, to obey the *rule of solution expression* denoted by (9.10) and the *rule of coefficient ergodicity*, we had to choose $\sigma = 4$ which uniquely determines the auxiliary function

$$H(\tau) = \tau^4. \tag{9.27}$$

Thereafter, it is straightforward to solve the high-order deformation equations (9.22) and (9.23), successively.

9.1.4 Recursive expressions

Considering the importance of Thomas-Fermi atom model, it is worthwhile to give an explicit analytic expression of its solution. It is found that $u_k(\tau)$ can be expressed by

$$u_k(\tau) = \sum_{n=0}^{2k} \frac{\alpha_{k,n}}{\tau^{n+3}}, \tag{9.28}$$

where $\alpha_{k,n}$ is a coefficient. Substituting this expression into the high-order deformation equations (9.22) and (9.23), we gain the recurrence formulae

$$\alpha_{k,j} = \chi_k \chi_{2k-j}\, \alpha_{k-1,j}$$
$$+ \frac{4\hbar \left[\chi_{2k+1-j}\left(\lambda^3 \beta_{k,j+1} - \gamma_{k,j+1}\right) - \chi_j\, \lambda^3 \beta_{k,j}\right]}{j(j+3)}, \tag{9.29}$$

$$\beta_{k,i} = \sum_{j=0}^{k-1} \sum_{n=\max\{0,i+2j-2k\}}^{\min\{2j,i-2\}} (n+3)(n+4)(i+1-n)$$
$$\times (i+2-n)\, \alpha_{j,n}\, \alpha_{k-1-j,i-n-2}, \tag{9.30}$$

$$\gamma_{k,i} = \sum_{j=0}^{k-1} \sum_{n=\max\{0,i+2j-2k\}}^{\min\{2j,i-2\}} \delta_{j,n}\, \alpha_{k-1-j,i-n-2}, \tag{9.31}$$

and

$$\delta_{j,n} = \sum_{i=0}^{j} \sum_{r=\max\{0,n+2i-2j\}}^{\min\{2i,n\}} \alpha_{i,r}\, \alpha_{j-i,n-r}, \tag{9.32}$$

respectively. From (9.23), we have

$$\alpha_{k,0} = -\sum_{n=1}^{2k} \alpha_{k,n}. \tag{9.33}$$

From (9.11) we gain the first coefficient

$$\alpha_{0,0} = 1. \tag{9.34}$$

Thus, using the above recurrence formulae and from the first coefficient $\alpha_{0,0} = 1$, we can calculate successively all other coefficients $\alpha_{k,n}$. Therefore, we obtain an explicit analytic solution of the Thomas-Fermi atom model in the form:

$$u(x) = \sum_{k=0}^{+\infty} \sum_{n=0}^{2k} \frac{\alpha_{k,n}}{(1+\lambda\,x)^{n+3}}. \tag{9.35}$$

The corresponding mth-order approximation is expressed by

$$u(x) \approx \sum_{k=0}^{m} \sum_{n=0}^{2k} \frac{\alpha_{k,n}}{(1+\lambda\,x)^{n+3}}, \tag{9.36}$$

which gives

$$u'(0) \approx -\lambda \sum_{k=0}^{m} \sum_{n=0}^{2k} (n+3)\alpha_{k,n} \tag{9.37}$$

and

$$u''(0) \approx \lambda^2 \sum_{k=0}^{m} \sum_{n=0}^{2k} (n+3)(n+4)\alpha_{k,n}. \tag{9.38}$$

9.1.5 Convergence theorem

THEOREM 9.1
If the series

$$u_0(\tau) + \sum_{k=1}^{+\infty} u_k(\tau)$$

is convergent, where $u_k(\tau)$ is governed by Equations (9.22) and (9.23) under the definitions (9.12), (9.24), and (2.42), it must be an exact solution of the Thomas-Fermi equation.

Proof: If the series is convergent, it holds

$$\lim_{m \to +\infty} u_m(\tau) = 0$$

and we can express it by

$$s(\tau) = u_0(\tau) + \sum_{k=1}^{+\infty} u_k(\tau).$$

Then, using (9.12), (9.22), and (2.42), we have

$$\hbar\, H(\tau) \sum_{k=1}^{+\infty} R_k(\vec{u}_{k-1}, \tau) = \lim_{m \to +\infty} \sum_{k=1}^{m} \mathcal{L}[u_k(\tau) - \chi_k u_{k-1}(\tau)]$$

$$= \mathcal{L}\left\{\lim_{m \to +\infty} \sum_{k=1}^{m} [u_k(\tau) - \chi_k u_{k-1}(\tau)]\right\}$$

$$= \mathcal{L}\left[\lim_{m \to +\infty} u_m(\tau)\right]$$

$$= 0,$$

which gives, since $\hbar \neq 0$ and $H(\tau) = \tau^4$,

$$\sum_{k=1}^{+\infty} R_k(\vec{u}_{k-1}, \tau) = 0$$

for any $\tau \geq 1$. Substituting (9.24) into the above expression and simplifying it, we obtain

$$\sum_{k=1}^{+\infty} R_k(\vec{u}_{k-1}, \tau)$$

$$= \sum_{k=1}^{+\infty} \sum_{j=0}^{k-1} \left[\lambda^3(\tau-1)\, u_j''(\tau)\, u_{k-1-j}''(\tau) - u_{k-1-j}(\tau) \sum_{i=0}^{j} u_i(\tau)\, u_{j-i}(\tau)\right]$$

$$= \lambda^3\,(\tau-1) \left[\sum_{k=0}^{+\infty} u_k''(\tau)\right]^2 - \left[\sum_{k=0}^{+\infty} u_k(\tau)\right]^3$$

$$= \lambda^3\,(\tau-1) \left[\frac{d^2 s(\tau)}{d\tau^2}\right]^2 - s^3(\tau)$$

$$= 0.$$

From (9.23) and (9.11), it holds

$$s(1) = 1,\, s(+\infty) = 0.$$

So, $s(\tau)$ satisfies Equations (9.6) and (9.7), and therefore is an exact solution of the original Thomas-Fermi equations (9.1) and (9.2). This ends the proof.

9.2 Result analysis

According to Theorem 9.1, we should ensure that the solution series (9.35) converges. Note that this series contains the auxiliary parameter $\hbar$ and the parameter λ, which influence its convergence region and rate. We should therefore focus on the choice of $\hbar$ and λ.

The energy of a neutral atom in the Thomas-Fermi model is determined by

$$E = \frac{6}{7}\left(\frac{4\pi}{3}\right)^{2/3} Z^{7/3}\, u'(0),$$

where Z is the unclear charge. So, the initial slope $u'(0)$ has an important physical meaning. Instead of investigating the influence of $\hbar$ and λ on the convergence of $u(x)$ in the whole region $0 \le x < +\infty$, first we consider the series of $u'(0)$. Clearly, $u'(0)$ is dependent of both $\hbar$ and λ. For any a given $\hbar$, we can investigate the influence of λ on the convergence of the series $u'(0)$ by regarding it as a function of λ and plotting the corresponding curves of $u'(0)$ versus λ, as shown in Figure 9.1 for $\hbar = -1, \hbar = -3/4$, and $\hbar = -1/2$. According to Theorem 9.1, $u'(0)$ should converge to the same value, corresponding to a nearly horizontal line segment in Figure 9.1. From this figure, it is clear that $u'(0)$ is convergent when $0.2 < \lambda < 0.3$ and $-1 \le \hbar \le -1/2$. Then, it is natural to choose

$$\lambda = 1/4.$$

To investigate the influence of $\hbar$ on the convergence region and rate of series $u'(0)$ in the case of $\lambda = 1/4$, we plot the corresponding $\hbar$-curves (see page 26 and §3.5.1) of $u'(0)$, as shown in Figure 9.2. $u'(0)$ is convergent in the region $-2 < \hbar < 0$ when $\lambda = 1/4$. Furthermore, it is found that, as long as the series of $u'(0)$ converges, the corresponding series of $u(x)$ is convergent in the whole region $0 \le x < +\infty$. So, the series (9.35) is convergent in the whole region $0 \le x < +\infty$ when $\lambda = 1/4$ and $-2 < \hbar < 0$. For example, when $\hbar = -1$ and $\lambda = 1/4$, the 10th-order approximation of (9.36) agrees well with the 100th-order approximation, as shown in Figure 9.3, clearly indicating the convergence of the corresponding solution series. The convergent analytic results of $u(x)$ by means of $\hbar = -1$ and $\lambda = 1/4$ are listed in Table 9.1. According to Theorem 9.1, it must be the exact solution of the Thomas-Fermi equation. The explicit analytic expression (9.35) when $\hbar = -1$ and $\lambda = 1/4$ may be regarded as a definition of the solution of the Thomas-Fermi equation.

Kobayashi [98] gave the numerical result

$$u'(0) = -1.588071. \tag{9.39}$$

Thomas-Fermi Atom Model

The approximations of the initial slope $u'(0)$ given by (9.36) when $\hbar = -1$ and $\lambda = 1/4$ are listed in Table 9.2. Clearly, the error decreases as the order of approximation increases. However, the convergence rate of $u'(0)$ is much slower than that of $u(x)$, possibly due to the singularity at $x = 0$. We employ the homotopy-Padé method (see page 38 and §3.5.2) to gain more accurate approximations of the initial slope $u'(0)$, as shown in Table 9.3. The approximations of $u''(0)$ when $\hbar = -1$ and $\lambda = 1/4$ are listed in Table 9.4. The homotopy-Padé approximations of $u''(0)$ are listed in Table 9.5. Obviously, $u''(0)$ given by (9.35) tends to infinity; this indicates that the homotopy analysis method may handle nonlinear problems with singularity and strong nonlinearity.

At the beginning of this chapter we assume that $u(x)$ tends to zero algebraically as $x \to +\infty$. Under this assumption we obtain the convergent results of the original Thomas-Fermi equation in the whole region $0 \leq x < +\infty$. Thus, this assumption seems to be reasonable. So, we have many reasons to believe that the solution of the Thomas-Fermi equation behaves algebraically at infinity. Note that it is hard to get this kind of conclusion by numerical techniques. This example also illustrates that we may employ the homotopy analysis method to get an accurate approximation of a nonlinear problem by means of assuming a set of base functions even if we know a little about its properties, and such a kind of assumption damages the method a little in practice.

Having not realized the importance of the asymptotic property at infinity and expressing $u(x)$ in the form

$$u(x) = \sum_{n=1}^{+\infty} \frac{a_n}{(1+x)^n},$$

Liao [51] employed the homotopy analysis method to give an explicit analytic solution of the Thomas-Fermi equation, which converges slowly for large x. So, for nonlinear problems in an infinite domain, it seems better to investigate asymptotic properties of solutions at infinity. This can considerably increase the convergence rate of approximate series. Note that the solution of the Thomas-Fermi equation can be expressed, respectively, by the base functions (9.3) and (9.10), and the approximation series given by the latter converges faster than that by the former. This indicates that, although the solution of the Thomas-Fermi equation seems unique, it can be expressed by different base functions, and there might even exist the best one among them.

TABLE 9.1
The convergent analytic results of $u(x)$ given by (9.36) when $\hbar = -1$ and $\lambda = 1/4$.

x	$u(x)$	x	$u(x)$
0.25	0.755202	4.25	0.0996979
0.50	0.606987	4.50	0.0919482
0.75	0.502347	4.75	0.0850218
1.00	0.424008	5.00	0.0788078
1.25	0.363202	6.00	0.0594230
1.50	0.314778	7.00	0.0460978
1.75	0.275451	8.00	0.0365873
2.00	0.243009	9.00	0.0295909
2.25	0.215895	10.0	0.0243143
2.50	0.192984	15.0	0.0108054
2.75	0.173441	20.0	0.00578494
3.00	0.156633	25.0	0.00347375
3.25	0.142070	50.0	0.000632255
3.50	0.129370	75.0	0.000218210
3.75	0.118229	100	0.000100243
4.00	0.108404	1000	1.3513×10^{-7}

TABLE 9.2
The initial slope $u'(0)$ given by (9.37) when $\hbar = -1$ and $\lambda = 1/4$ compared with Kobayashi's numerical result.

Order of approximation	$u'(0)$	Error (%)
10	-1.28590	19.03
20	-1.40932	11.26
30	-1.46306	7.87
40	-1.49236	6.03
50	-1.51063	4.88
60	-1.52309	4.09
70	-1.53211	3.52
80	-1.53895	3.09
90	-1.54430	2.76
100	-1.54860	2.49
110	-1.55214	2.26
120	-1.55509	2.07

TABLE 9.3
The $[m,m]$ Homotopy-Padé approximation of the initial slope $u'(0)$ given by (9.37) when $\hbar = -1$ and $\lambda = 1/4$ compared with Kobayashi's numerical result.

$[m,m]$	$u'(0)$	Error (%)
[5, 5]	-1.50419	5.28
[10, 10]	-1.54600	2.65
[15, 15]	-1.56437	1.49
[20, 20]	-1.56474	1.47
[25, 25]	-1.57666	0.72
[30, 30]	-1.558032	0.49
[35, 35]	-1.58187	0.39
[40, 40]	-1.58301	0.32
[45, 45]	-1.58388	0.26
[50, 50]	-1.58469	0.21
[55, 55]	-1.58538	0.17
[60, 60]	-1.58605	0.13

TABLE 9.4
The analytic approximations of $u''(0)$ given by (9.38) when $\hbar = -1$ and $\lambda = 1/4$.

Order of approximation	$u''(0)$
10	3.79
20	6.41
30	8.96
40	11. 49
50	14.01
60	16.52
70	19.03
80	21.54
90	24.04
100	26.55
110	29.05
120	31.56

TABLE 9.5
The $[m, m]$ Homotopy-Padé approximations of $u''(0)$ given by (9.38) when $\hbar = -1$ and $\lambda = 1/4$.

$[m, m]$	$u''(0)$
$[5, 5]$	122.7
$[15, 15]$	6087.7
$[30, 30]$	168917
$[40, 40]$	643063
$[50, 50]$	2.15707×10^6
$[60, 60]$	8.78329×10^6

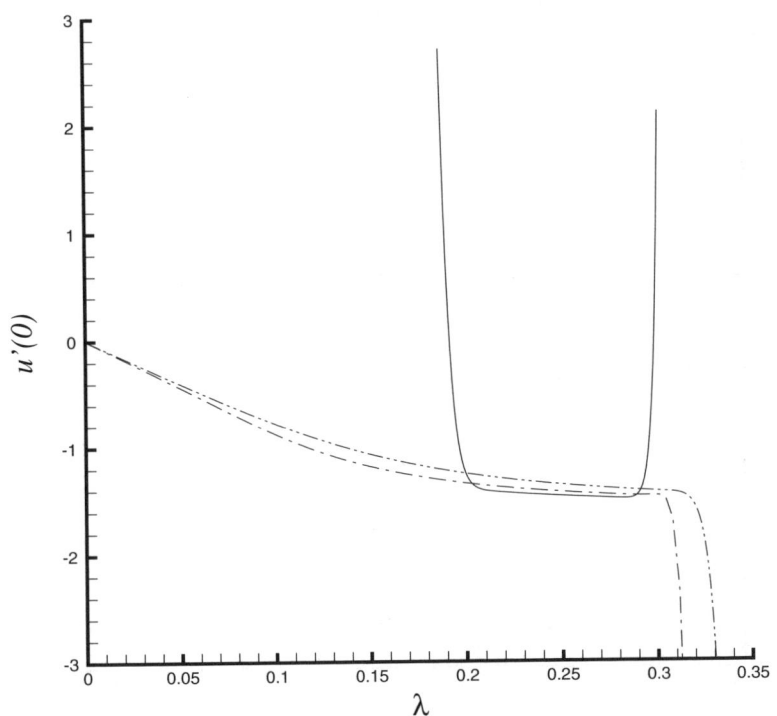

FIGURE 9.1
The 30th-order approximation of $u'(0)$ versus λ. Solid line: $\hbar = -1$; dash-dotted line: $\hbar = -3/4$; dash-dot-dotted line: $\hbar = -1/2$.

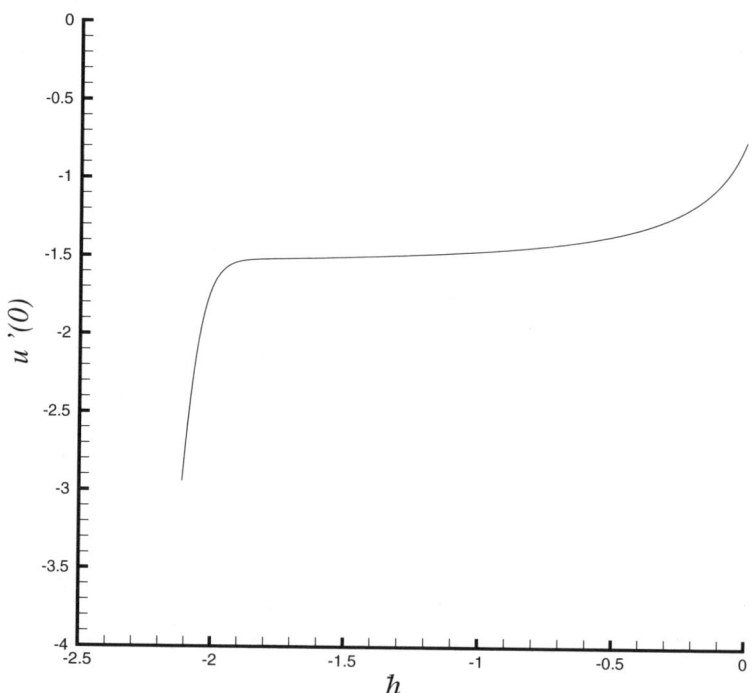

FIGURE 9.2
The $\hbar$-curve of $u'(0)$ at the 30th order of approximation when $\lambda = 1/4$.

Thomas-Fermi Atom Model

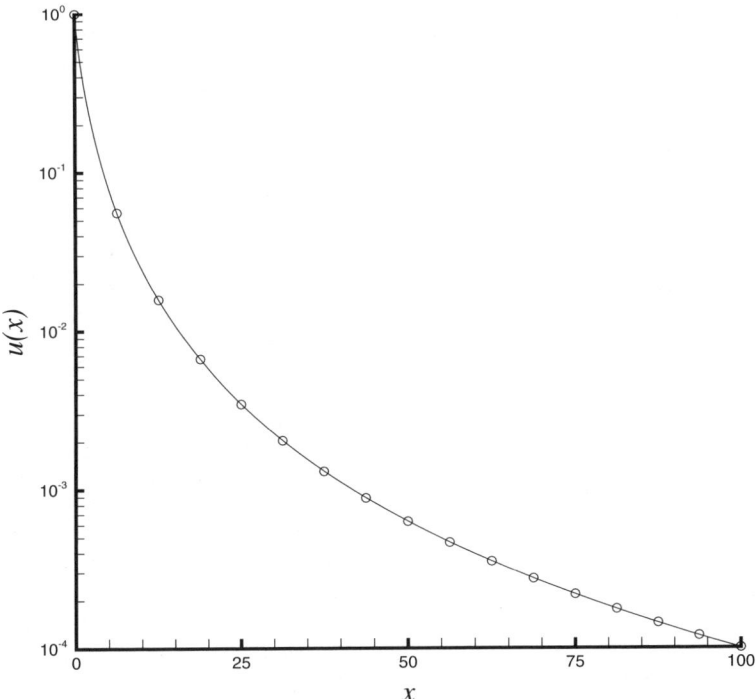

FIGURE 9.3
The analytic approximations of the Thomas-Fermi equation given by (9.36) when $\hbar = -1$ and $\lambda = 1/4$. Symbols: 10th-order approximation; solid line: 100th-order approximation.

10

Volterra's population model

Consider the Volterra model for the population growth [99] of a species within a closed system governed by a nonlinear integro-differential equation

$$\beta \frac{du(t)}{dt} = u(t) - u^2(t) - u(t) \int_0^t u(x)dx, \qquad (10.1)$$

subject to the initial condition

$$u(0) = \alpha, \qquad (10.2)$$

where $u(t)$ is the scaled population of identical individuals, t denotes the time, and $\beta = c/(ab)$ is a nondimensional parameter in which $a > 0$ is the birth rate coefficient, $b > 0$ is the crowding coefficient, and $c > 0$ is the toxicity coefficient, respectively. For details the reader is referred to Scudo [99], Small [100], TeBeest [101], and Wazwaz [102].

10.1 Homotopy analysis solution

10.1.1 Zero-order deformation equation

Let $\lambda > 0$ denote a so-called time-scale parameter. Under the transformation

$$\tau = \lambda t, \quad w(\tau) = u(t) \qquad (10.3)$$

Equation (10.1) becomes

$$(\beta \lambda^2) \frac{dw(\tau)}{d\tau} = \lambda \left[w(\tau) - w^2(\tau) \right] - w(\tau) \int_0^\tau w(x)dx, \qquad (10.4)$$

subject to the initial condition

$$w(0) = \alpha. \qquad (10.5)$$

It was shown by Small [100] that a rise occurs along the solution curve that will reach a peak and then followed by an exponential decay. So, it is reasonable to express $w(\tau)$ by a set of base functions

$$\{\exp(-n\tau) \mid n \geq 1\} \qquad (10.6)$$

in the form
$$w(\tau) = \sum_{n=1}^{+\infty} a_n \exp(-n\tau), \tag{10.7}$$

where a_n is a coefficient. This provides us with the so-called *rule of solution expression* of $w(\tau)$. Under *the rule of solution expression* and using (10.5), it is straightforward to choose the initial guess

$$w_0(\tau) = \alpha \exp(-\tau) + \gamma \left[\exp(-\tau) - \exp(-2\tau)\right], \tag{10.8}$$

where γ is an auxiliary parameter to be determined later. Under the *rule of solution expression* denoted by (10.7) and from Equation (10.4), it is straightforward to choose

$$\mathcal{L}f = \frac{df}{d\tau} + f \tag{10.9}$$

as the auxiliary linear operator, which has the property

$$\mathcal{L}[e^{-\tau}] = 0. \tag{10.10}$$

From Equation (10.4), we define the nonlinear integro-differential operator

$$\mathcal{N}\left[\Phi(\tau;q), \Lambda(q)\right] = \beta\Lambda^2(q)\frac{\partial\Phi(\tau;q)}{\partial\tau} - \Lambda(q)\left[\Phi(\tau;q) - \Phi^2(\tau;q)\right] + \Phi(\tau;q)\int_0^\tau \Phi(x;q)dx, \tag{10.11}$$

where $q \in [0, 1]$ is an embedding parameter, $\Phi(\tau; q)$ is a function of τ and q, and $\Lambda(q)$ is a function dependent of q. Let $\hbar \ne 0$ denote a nonzero auxiliary parameter and $H(\tau)$ a nonzero auxiliary function, respectively. We construct the zero-order deformation equation

$$(1-q)\,\mathcal{L}\left[\Phi(\tau;q) - w_0(\tau)\right] = q\,\hbar\,H(\tau)\,\mathcal{N}\left[\Phi(\tau;q), \Lambda(q)\right], \tag{10.12}$$

subject to the initial condition

$$\Phi(0;q) = \alpha, \tag{10.13}$$

where $q \in [0, 1]$ is an embedding parameter.

When $q = 0$, it is straightforward that

$$\Phi(\tau;0) = w_0(\tau). \tag{10.14}$$

When $q = 1$, since $\hbar \ne 0$ and $H(\tau) \ne 0$, the zero-order deformation equations (10.12) and (10.13) are equivalent to Equations (10.4) and (10.5), respectively, provided

$$\Phi(\tau;1) = w(\tau), \quad \Lambda(1) = \lambda. \tag{10.15}$$

Volterra's Population Model

Thus, as q increases from 0 to 1, $\Phi(\tau;q)$ varies from the initial guess $w_0(\tau)$ to the solution $w(\tau)$ of Equations (10.4) and (10.5), so does $\Lambda(q)$ from the initial guess

$$\Lambda(0) = \lambda_0 \tag{10.16}$$

to the exact time-scale parameter λ. Note that the zero-order deformation equation (10.12) contains the auxiliary parameter $\hbar$ and the auxiliary function $H(\tau)$. The initial guess $w_0(\tau)$ contains the auxiliary parameter γ. Assume that all of them are properly chosen so that in the whole region $q \in [0,1]$ there exist the solutions $\Phi(\tau;q), \Lambda(q)$ of the zero-order deformation equations (10.12) and (10.13), and also the terms

$$w_n(\tau) = \frac{1}{n!} \left.\frac{\partial^n \Phi(\tau;q)}{\partial q^n}\right|_{q=0}, \tag{10.17}$$

$$\lambda_n = \frac{1}{n!} \left.\frac{\partial^n \Lambda(q)}{\partial q^n}\right|_{q=0}. \tag{10.18}$$

Then, by Taylor's theorem and using (10.14) and (10.16), we expand $\Phi(\tau;q)$ and $\Lambda(q)$ in the series

$$\Phi(\tau;q) = w_0(\tau) + \sum_{n=1}^{+\infty} w_n(\tau) \, q^n, \tag{10.19}$$

$$\Lambda(q) = \lambda_0 + \sum_{n=1}^{+\infty} \lambda_n \, q^n. \tag{10.20}$$

Assuming that the auxiliary parameters $\hbar, \gamma$, and the auxiliary function $H(\tau)$ are properly chosen so that the above series converge at $q=1$, we have, using (10.15), the solution series

$$w(\tau) = w_0(\tau) + \sum_{n=1}^{+\infty} w_n(\tau), \tag{10.21}$$

$$\lambda = \lambda_0 + \sum_{n=1}^{+\infty} \lambda_n. \tag{10.22}$$

At the Mth order of approximation, we gain

$$w(\tau) \approx w_0(\tau) + \sum_{n=1}^{M} w_n(\tau), \tag{10.23}$$

$$\lambda \approx \lambda_0 + \sum_{n=1}^{M} \lambda_n. \tag{10.24}$$

10.1.2 High-order deformation equation

For brevity, define the vector

$$\vec{w}_n = \{w_0(\tau), w_1(\tau), \cdots, w_n(\tau)\}, \quad \vec{\lambda}_n = \{\lambda_0, \lambda_1, \cdots, \lambda_n\}.$$

Differentiating the zero-order deformation equations (10.12) and (10.13) n times with respect to the embedding parameter q and then dividing by $n!$ and finally setting $q = 0$, we have the high-order deformation equation

$$\mathcal{L}\left[w_n(\tau) - \chi_n \, w_{n-1}(\tau)\right] = \hbar \, H(\tau) \, R_n(\vec{w}_{n-1}, \vec{\lambda}_{n-1}), \tag{10.25}$$

subject to the initial condition

$$w_n(0) = 0, \tag{10.26}$$

where χ_n is defined by (2.42) and

$$\begin{aligned}
R_n(\vec{w}_{n-1}, \vec{\lambda}_{n-1}) \\
&= \frac{1}{(n-1)!} \left.\frac{\partial^{n-1} \mathcal{N}\left[\Phi(\tau;q), \Lambda(q)\right]}{\partial q^{n-1}}\right|_{q=0} \\
&= \beta \sum_{j=0}^{n-1} w'_{n-1-j}(\tau) \sum_{i=0}^{j} \lambda_i \lambda_{j-i} - \sum_{j=0}^{n-1} \lambda_j \, w_{n-1-j}(\tau) \\
&\quad + \sum_{j=0}^{n-1} \lambda_{n-1-j} \sum_{i=0}^{j} w_i(\tau) w_{j-i}(\tau) \\
&\quad + \sum_{j=0}^{n-1} w_{n-1-j}(\tau) \int_0^\tau w_j(x) dx.
\end{aligned} \tag{10.27}$$

There are two unknowns: λ_{n-1} and $w_n(\tau)$. However, we have only one differential equation (10.25) for $w_n(\tau)$. Thus, the problem is not closed and an additional algebraic equation is needed to determine λ_{n-1}. Using (10.8) and (10.27), it is straightforward to get

$$R_1(\vec{w}_0, \vec{\lambda}_0) = \sum_{m=1}^{4} a_{1,m} \exp(-m\,\tau) \tag{10.28}$$

where

$$a_{1,1} = (\alpha + \gamma)\left(\alpha + \frac{\gamma}{2} - \lambda_0 - \beta\,\lambda_0^2\right)$$

and $a_{1,j}$ ($j = 2, 3, 4$) are coefficients. Note that the auxiliary function $H(\tau)$ is unknown right now. According to the *rule of solution expression* denoted by (10.7) and from Equation (10.25), the auxiliary function should be in the form

$$H(\tau) = \exp(\kappa\,\tau),$$

where κ is an integer. It is found that, when $\kappa \geq 1$, the solution $w_n(\tau)$ of Equation (10.25) contains a constant term that does not vanish at infinity. This disobeys the *rule of solution expression* denoted by (10.7). When $k \leq -2$, the solution $w_n(\tau)$ of Equation (10.25) does not contain the term $\exp(-2\tau)$, and this disobeys the so-called *rule of coefficient ergodicity*. So, κ should be either 0 or 1. When $\kappa = 1$, we cannot give an additional algebraic equation for λ_0 so that the problem is still not closed, and this disobeys the *rule of solution existence*. So, only $\kappa = 0$ is possible, which uniquely determines the auxiliary function

$$H(\tau) = 1. \tag{10.29}$$

In this case, the right-hand side of the first-order deformation equation (10.25) contains the term $\exp(-\tau)$. Then, according to (10.10), $w_1(\tau)$ contains the term $\tau \exp(-\tau)$, which disobeys the *rule of solution expression* denoted by (10.7). To obey the *rule of solution expression*, we had to enforce $a_{1,1} = 0$, i.e.,

$$(\alpha + \gamma)\left(\alpha + \frac{\gamma}{2} - \lambda_0 - \beta \lambda_0^2\right) = 0, \tag{10.30}$$

which provides us with the additional equation for λ_0 with the positive solution

$$\lambda_0 = \frac{\sqrt{1 + 2\beta(\gamma + 2\alpha)} - 1}{2\beta}. \tag{10.31}$$

Thereafter, it is straightforward to get the solution

$$w_1(\tau) = \hbar \sum_{m=2}^{4} \left(\frac{a_{1,m}}{m-1}\right) \left(\exp^{-\tau} - \exp^{-m\tau}\right). \tag{10.32}$$

In general, the term $R_n(\vec{w}_{n-1}, \vec{\lambda}_{n-1})$ can be generally expressed by

$$R_n(\vec{w}_{n-1}, \vec{\lambda}_{n-1}) = \sum_{m=1}^{2(n+1)} a_{n,m} \exp(-m\tau), \tag{10.33}$$

where $a_{n,m}$ is a coefficient, and we can get λ_{n-1} by enforcing

$$a_{n,1} = 0. \tag{10.34}$$

In this way, we successively gain the solution

$$w_n(\tau) = \chi_{n-1} w_{n-1}(\tau) + \hbar \sum_{m=2}^{2(n+1)} \left(\frac{a_{n,m}}{m-1}\right)\left(e^{-\tau} - e^{-m\tau}\right) \tag{10.35}$$

of high-order deformation equations (10.25) and (10.26).

10.1.3 Recursive expression

Considering the importance of Volterra's population model, it is helpful to give an explicit analytic expression of the solution. It is found that $w_n(\tau)$ can be expressed by

$$w_n(\tau) = \sum_{m=1}^{2(n+1)} b_{n,m} \exp(-m\,\tau), \qquad (10.36)$$

where $b_{n,m}$ is a coefficient. Substituting it into Equations (10.25) and (10.26), we have the recursive formulae ($n \geq 2, i \geq 2$)

$$\lambda_{n-1} = \frac{\Delta_{n,1} - \sum_{j=0}^{n-2} (\lambda_j + \beta\delta_j) b_{n-1-j,1} - \beta b_{0,1} \sum_{i=1}^{n-2} \lambda_i \lambda_{n-1-i}}{(1+2\beta\lambda_0)b_{0,1}}, \qquad (10.37)$$

$$b_{n,i} = \chi_n \chi_{2n+2-i} b_{n-1,i} + \frac{\hbar \left(\Pi_{n,i} + \Delta_{n,i} - \chi_{2n+2-i} \Gamma_{n,i}\right)}{(1-i)}, \qquad (10.38)$$

$$b_{n,1} = -\sum_{i=2}^{2(n+1)} b_{n,i}, \qquad (10.39)$$

where

$$\Pi_{n,i} = \sum_{j=\max\{0,[(i+1)/2]-2\}}^{n-1} \lambda_{n-1-j}\, d_{j,i}, \qquad 2 \leq i \leq 2(n+1),$$

$$\Delta_{n,i} = \sum_{j=0}^{n-1} \sum_{s=\max\{1, i-2(j+1)\}}^{\min\{2(n-j),i\}} b_{n-1-j,s}\, c_{j,i-s}, \qquad 1 \leq i \leq 2(n+1),$$

$$\Gamma_{n,i} = \sum_{j=0}^{\min\{n-1, n-[(i+1)/2]\}} (i\beta\delta_j + \lambda_j) b_{n-1-j,i}, \qquad 1 \leq i \leq 2n,$$

under the definitions

$$d_{n,m} = \sum_{i=0}^{n} \sum_{j=\max\{1, m-2(n-i+1)\}}^{\min\{2(i+1), m-1\}} b_{i,j}\, b_{n-i,m-j}, \qquad 2 \leq m \leq 2(n+1)$$

and

$$c_{n,m} = -\frac{b_{n,m}}{m}, \qquad 1 \leq m \leq 2(n+1),$$

$$c_{n,0} = \sum_{m=1}^{2(n+1)} \frac{b_{n,m}}{m},$$

$$\delta_n = \sum_{i=0}^{n} \lambda_i\, \lambda_{n-i}.$$

Volterra's Population Model

in which the term $[x]$ denotes the integer part of x. From (10.8) we have the first two coefficients

$$b_{0,1} = \alpha + \gamma, \qquad b_{0,2} = -\gamma. \tag{10.40}$$

From these two coefficients and using the above recursive formulae, we can successively calculate all coefficients $b_{n,j}$. The Mth-order approximation of $u(t)$ is given by

$$u(t) \approx \sum_{n=0}^{M} \sum_{m=1}^{2(n+1)} b_{n,m} \exp(-m\,\lambda\,t), \tag{10.41}$$

where

$$\lambda \approx \sum_{n=0}^{M-1} \lambda_n. \tag{10.42}$$

When $M \to +\infty$ we gain the explicit analytic solution

$$u(t) = \sum_{n=0}^{+\infty} \sum_{m=1}^{2(n+1)} b_{n,m} \exp(-m\,\lambda\,t), \tag{10.43}$$

where

$$\lambda = \sum_{n=0}^{+\infty} \lambda_n. \tag{10.44}$$

10.1.4 Convergence theorem

THEOREM 10.1
If the solution series (10.21) and (10.22) converge, where $w_n(\tau)$ is governed by Equations (10.25) and (10.26) under the definitions (10.27) and (2.42), they must be the solution of Equations (10.4) and (10.5).

Proof: If the solution series (10.21) and (10.22) converge, it is necessary that

$$\lim_{m \to +\infty} w_m(\tau) = 0. \tag{10.45}$$

Then, using (10.9), (10.25), and (2.42), it holds

$$\hbar\, H(\tau) \sum_{n=1}^{+\infty} R_n(\vec{w}_{n-1}, \vec{\lambda}_{n-1})$$

$$= \lim_{m \to +\infty} \mathcal{L}\left[w_m(\tau)\right] = \mathcal{L}\left[\lim_{m \to +\infty} w_m(\tau)\right] = 0, \tag{10.46}$$

which gives, since $\hbar \neq 0$ and $H(\tau) = 1$,

$$\sum_{n=1}^{+\infty} R_n(\vec{w}_{n-1}, \vec{\lambda}_{n-1}) = 0. \tag{10.47}$$

Substituting (10.27) into the above expression and simplifying it, we have

$$\beta \left(\sum_{n=0}^{+\infty} \lambda_n\right)^2 \frac{d}{d\tau}\left[\sum_{n=0}^{+\infty} w_n(\tau)\right]$$
$$= \left(\sum_{n=0}^{+\infty} \lambda_n\right) \left\{\left[\sum_{n=0}^{+\infty} w_n(\tau)\right] - \left[\sum_{n=0}^{+\infty} w_n(\tau)\right]^2\right\}$$
$$- \left[\sum_{n=0}^{+\infty} w_n(\tau)\right] \int_0^\tau \left[\sum_{n=0}^{+\infty} w_n(x)\right] dx. \qquad (10.48)$$

From (10.8) and (10.26), we have

$$\sum_{n=0}^{+\infty} w_n(0) = \alpha. \qquad (10.49)$$

Comparing these two expressions with Equations (10.4) and (10.5), the solution series (10.21) and (10.22) must be the solution of the Volterra's population model, as long as they are convergent. This ends the proof.

10.2 Result analysis

According to Theorem 10.1 we should only focus on ensuring that the solution series (10.21) and (10.22) converge. Note that there exists the integral term

$$\int_0^t u(x)dx$$

in Equation (10.1) and the value

$$\mu = \int_0^{+\infty} u(x)dx \qquad (10.50)$$

denotes the total scaled population and thus has an important meaning. Under the transformation (10.3), the above expression becomes

$$\lambda \mu = \int_0^{+\infty} w(\xi)d\xi. \qquad (10.51)$$

10.2.1 Choosing a plain initial approximation

For the given α and β, we have freedom to choose the auxiliary parameters $\hbar$ and γ, which influence the convergence region and rate of the solution series

(10.21) and (10.22). Generally speaking, for any chosen value of γ, we can investigate the influence of $\hbar$ by plotting the so-called $\hbar$-curves (see page 26 and §3.5.1) of $\int_0^{+\infty} u(x)dx$. For example, consider the case $\alpha = 1/10$ and $\beta = 1/5$. The corresponding $\hbar$-curves of $\int_0^{+\infty} u(x)dx$ at the 10th order of approximation when $\gamma = 1, 2, 3, 4$ are as shown in Figure 10.1. Note that the corresponding valid region of $\hbar$ increases when γ decreases from 4 to 2, but there is no such valid region when $\gamma = 1$. From these $\hbar$-curves, it is clear that when $\alpha = 1/10$ and $\beta = 1/5$, the approximation series of $\int_0^{+\infty} u(x)dx$ converges if $2 \leq \gamma \leq 4$ and $\hbar$ is chosen in the corresponding valid region. For instance, when $\gamma = 3$ corresponding to

$$\lambda_0 = 1.27492$$

given by (10.31), the approximation series of $\int_0^{+\infty} u(x)dx$ is convergent by means of $\hbar = -1/2$, as shown in Table 10.1. It is found that, in general, as long as the series of $\int_0^{+\infty} u(x)dx$ converges, the corresponding series (10.22) of λ also converges, as shown in Table 10.1 for the special case. The homotopy-Padé technique (see page 38 and §3.5.2) greatly enhances the convergence rate of the series of λ and $\int_0^{+\infty} u(x)dx$, as shown in Table 10.2. It is found that the $[m, m]$ homotopy-Padé approximants are independent of $\hbar$. Also, as long as the series of $\int_0^{+\infty} u(x)dx$ converges, the corresponding series (10.21) converges in the whole region $0 \leq t < +\infty$ to the numerical results [100, 101, 102], as shown in Figure 10.2 for the special case of $\alpha = 1/10$ and $\beta = 1/5$. In this way we can gain the analytic solutions for any given values of α and β.

10.2.2 Choosing the best initial approximation

Note that the initial approximation $w_0(\tau)$ is determined by the auxiliary parameter γ. Clearly, a better choice of γ should give a better series which approximates the solution more efficiently. At the zero order of approximation we have

$$\lambda \mu \approx \int_0^{+\infty} w_0(x)dx. \tag{10.52}$$

At the first order of approximation it holds

$$\lambda \mu \approx \int_0^{+\infty} w_0(x)dx + \int_0^{+\infty} w_1(x)dx. \tag{10.53}$$

Obviously, we can choose γ in such a way that the zero-order approximation (10.52) is so accurate that its first-order approximation (10.53) cannot give a better result of $\lambda \mu$, i.e.,

$$\int_0^{+\infty} w_1(x)dx = 0. \tag{10.54}$$

This gives an algebraic equation

$$24\beta \gamma \lambda_0^2 + 2(6\alpha^2 + 6\gamma + 4\alpha \gamma + \gamma^2)\lambda_0 - 3(4\alpha^2 + 8\alpha\gamma + 3\gamma^2) = 0. \tag{10.55}$$

Solving the set of two algebraic equations (10.30) and (10.55), we can obtain the "best" positive values of λ_0 and γ for any given values of α and β. For example, when $\alpha = 1/10$ and $\beta = 1/5$, the best values of λ_0 and γ are

$$\lambda_0 = 1.02682, \quad \gamma = 2.27538. \tag{10.56}$$

The above "best" value of γ explains why the valid region of $\hbar$ corresponding to $\gamma = 2$ is longer than that of $\gamma = 3$ and $\gamma = 4$, as shown in Figure 10.1. The corresponding solution series are indeed convergent more quickly, as shown in Table 10.3. Also, the homotopy-Padé technique can greatly accelerate the convergence of the solution series, as shown in Table 10.4. It is interesting that, when $\alpha = 1/10$ and $\beta = 1/5$, the series of the time-scale parameter λ converge to the same value 0.986, although two different values of λ_0 are used, as shown in Tables 10.1 to 10.4. Like $\int_0^{+\infty} u(x)dx$, the time-scale parameter λ depends upon α and β. It decreases as the term $\int_0^{+\infty} u(x)dx$ increases, thus, the time-scale parameter λ might have some physical meanings.

Using the "best" values of γ and λ_0 given by (10.30) and (10.55), we can analytically solve Volterra's population model more efficiently. The convergent analytic results of $u(t)$ when $\alpha = 1/10$ and $\beta = 1/10, 1/5, 1/2, 1$ and 10 are as shown in Figure 10.3. The total scaled population $\int_0^{+\infty} u(x)dx$, the time-scale parameter λ, and the corresponding best values of γ and λ_0 for some α and β are listed in Table 10.5 .

This example illustrates that the homotopy analysis method is valid for nonlinear integro-differential equations.

TABLE 10.1
The analytic approximations of $\int_0^\infty u(x)dx$ and λ given by (10.23) and (10.24) when $\alpha = 1/10, \beta = 1/5$ by means of $\gamma = 3, \lambda_0 = 1.27492$, and $\hbar = -1/2$.

Order of approximation	$\int_0^{+\infty} u(x)dx$	λ
10	1.194	1.014
20	1.196	0.988
30	1.196	0.983
40	1.197	0.983
50	1.197	0.983
60	1.197	0.984
70	1.197	0.985
80	1.197	0.985

TABLE 10.2
The $[m,m]$ homotopy-Padé approximations of $\int_0^\infty u(x)dx$ and λ when $\alpha = 1/10, \beta = 1/5$ by means of $\gamma = 3$ and $\lambda_0 = 1.27492$.

$[m,m]$	$\int_0^{+\infty} u(x)dx$	λ
$[5,5]$	1.196	0.982
$[10,10]$	1.197	0.987
$[15,15]$	1.197	0.986
$[20,20]$	1.197	0.986
$[25,25]$	1.197	0.986
$[30,30]$	1.197	0.986
$[35,35]$	1.197	0.986
$[40,40]$	1.197	0.986

TABLE 10.3
The analytic approximations of $\int_0^\infty u(x)dx$ and λ when $\alpha = 1/10, \beta = 1/5$ by means of $\gamma = 2.27538, \lambda_0 = 1.02682$, and $\hbar = -1$.

Order of approximation	$\int_0^{+\infty} u(x)dx$	λ
10	1.195	0.997
20	1.197	0.985
30	1.197	0.985
40	1.197	0.986
50	1.197	0.986
60	1.197	0.986
70	1.197	0.986
80	1.197	0.986

TABLE 10.4
The $[m, m]$ homotopy-Padé approximations of $\int_0^\infty u(x)dx$ and λ when $\alpha = 1/10, \beta = 1/5$ by means of $\gamma = 2.27538$ and $\lambda_0 = 1.02682$.

$[m, m]$	$\int_0^{+\infty} u(x)dx$	λ
$[5, 5]$	1.197	0.986
$[10, 10]$	1.197	0.986
$[15, 15]$	1.197	0.986
$[20, 20]$	1.197	0.986
$[25, 25]$	1.197	0.986
$[30, 30]$	1.197	0.986
$[35, 35]$	1.197	0.986
$[40, 40]$	1.197	0.986

TABLE 10.5
The convergent analytic results and the corresponding "best" values of γ and λ_0 for $\alpha = 1/10$ and different values of β.

β	λ_0	γ	$\int_0^{+\infty} u(x)dx$	λ
1/10	1.19933	2.48633	1.100	1.000
1/5	1.02682	2.27538	1.197	0.986
1/2	0.75414	1.87701	1.418	0.836
1	0.55274	1.51653	1.627	0.626
10	0.15621	0.57101	2.572	0.157

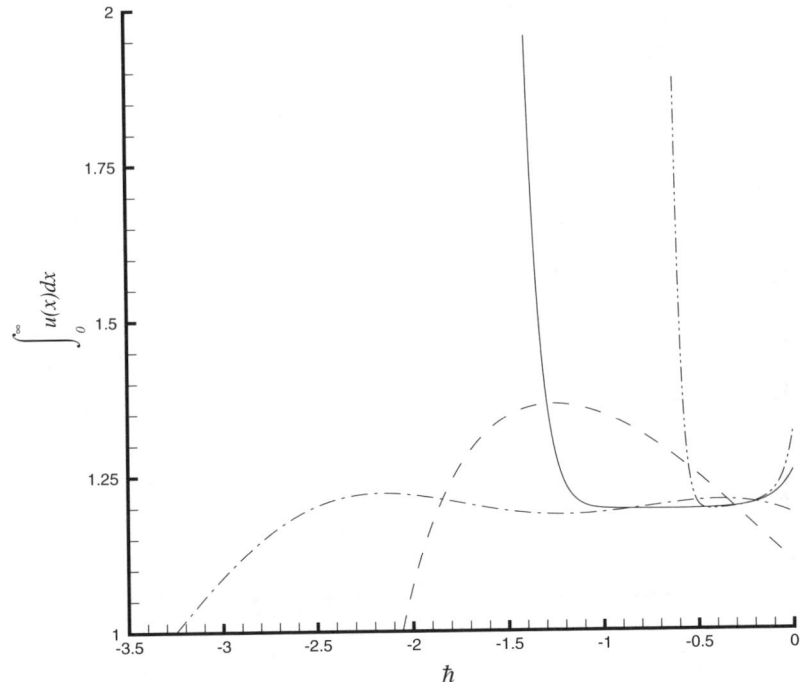

FIGURE 10.1
The $\hbar$-curves of $\int_0^{+\infty} u(x)dx$ at the 10th order of approximation when $\alpha = 1/10$ and $\beta = 1/5$ with different values of γ. Dashed line: $\gamma = 1$; dash-dotted line: $\gamma = 2$; solid line: $\gamma = 3$; dash-dot-dotted line: $\gamma = 4$.

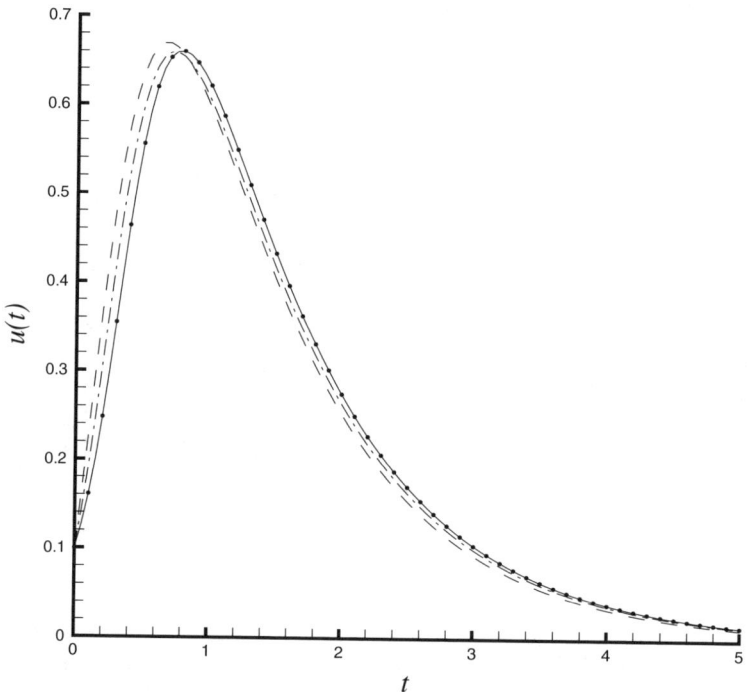

FIGURE 10.2
Comparison of the numerical result [100, 101, 102] with the analytic approximations of $u(t)$ when $\alpha = 1/10$ and $\beta = 1/5$ by means of $\gamma = 3$ and $\hbar = -1/2$. Symbol: numerical result; dashed line: 10th-order analytic approximation; dash-dotted line: 20th-order analytic approximation; solid line: 50th-order analytic approximation.

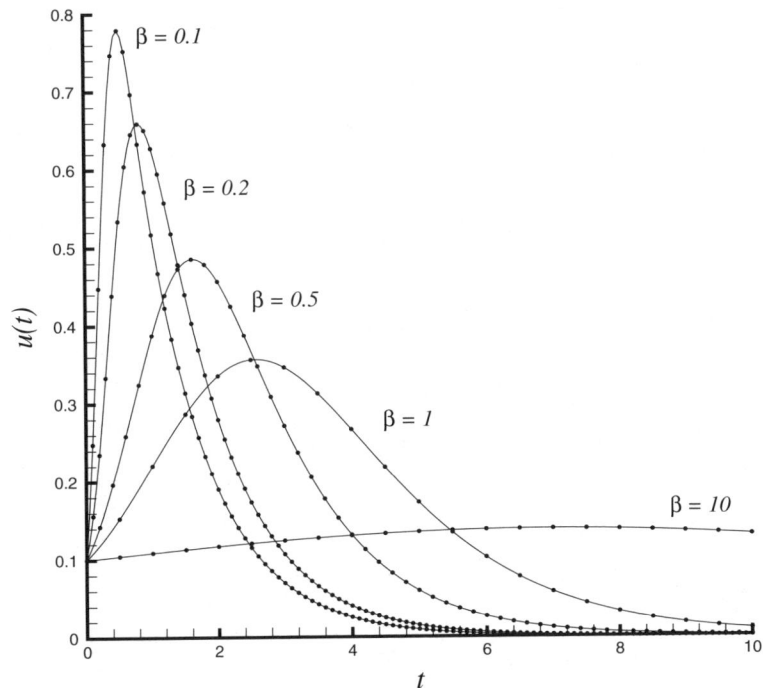

FIGURE 10.3
Comparison of the numerical results [100, 101, 102] with the analytic approximations of $u(t)$ when $\alpha = 1/10$ and $\beta = 1/10, 1/5, 1/2, 1, 10$ by means of $\hbar = -1$ and the "best" values of λ_0 and γ given by (10.30) and (10.55). Symbol: numerical result; solid line: analytic results at the 20th ($\beta = 1/10$), 30th ($\beta = 1/5$), 40th ($\beta = 1/2$), 50th ($\beta = 1$), and 50th order ($\beta = 10$) of approximation, respectively.

11

Free oscillation systems with odd nonlinearity

Consider free oscillations of a conservative system with odd nonlinearity governed by

$$\ddot{U}(t) = f[U(t), \dot{U}(t), \ddot{U}(t)], \tag{11.1}$$

where t denotes the time, the dot denotes derivative with respect to t, and $f[U(t), \dot{U}(t), \ddot{U}(t)]$ is a known function of $U(t), \dot{U}(t)$, and $\ddot{U}(t)$, respectively. Unlike perturbation techniques, it is unnecessary to assume the existence of any small/large parameters in Equation (11.1). This equation is very general and describes many problems in science and engineering.

Physically speaking, free oscillation of a conservative system is a periodic motion. Let ω and a denote the frequency and amplitude of the oscillation, respectively. Physically, the frequency ω can be regarded as a time scale. For a linear system, the frequency is independent of the amplitude. However, for a nonlinear system, it is important to know the relationship between the frequency and amplitude. In a nonlinear conservative system, the amplitude a is physically determined by initial conditions and is related to the total kinetic energy. Without the loss of any generality, we may consider free oscillations with amplitude a under the initial conditions

$$\dot{U}(0) = 0, U(0) = a. \tag{11.2}$$

11.1 Homotopy analysis solution

11.1.1 Zero-order deformation equation

Obviously, free oscillations of a conservative system with odd nonlinearity can be expressed by the base functions

$$\{\cos(m\omega t) \mid m = 1, 2, 3, \cdots\}. \tag{11.3}$$

Under the transformation $\tau = \omega t$ and $U(t) = u(\tau)$, Equation (11.1) becomes

$$\omega^2 u''(\tau) = f[u(\tau), \omega u'(\tau), \omega^2 u''(\tau)], \tag{11.4}$$

subject to the initial conditions

$$u(\tau) = a, \quad u'(\tau) = 0, \quad \text{when } \tau = 0, \tag{11.5}$$

where the prime denotes derivative with respect to τ. From (11.3), $u(\tau)$ can be expressed by the base functions

$$\{\cos(m\tau) \mid m = 1, 2, 3, \cdots\} \tag{11.6}$$

in the form

$$u(\tau) = \sum_{k=1}^{+\infty} c_k \cos(k\tau), \tag{11.7}$$

where c_k is a coefficient. This provides us with the so-called *rule of solution expression*.

Let ω_0 denote the initial guess of the frequency ω. Obviously, under the *rule of solution expression* denoted by (11.7) and using the initial condition (11.5), it is easy to choose

$$u_0(\tau) = a \cos \tau \tag{11.8}$$

as the initial guess of $u(\tau)$, where a is the amplitude of oscillation. Under the *rule of solution expression* denoted by (11.7), we choose the auxiliary linear operator

$$\mathcal{L}[\Phi(\tau; q)] = \omega_0^2 \left[\frac{\partial^2 \Phi(\tau; q)}{\partial \tau^2} + \Phi(\tau; q) \right], \tag{11.9}$$

with the property

$$\mathcal{L}(C_1 \sin \tau + C_2 \cos \tau) = 0. \tag{11.10}$$

From Equation (11.4), we define the nonlinear operator

$$\mathcal{N}[\Phi(\tau; q), \Omega(q)] = \Omega^2(q) \frac{\partial^2 \Phi(\tau; q)}{\partial \tau^2} \\ - f\left[\Phi(\tau; q), \Omega(q) \frac{\partial \Phi(\tau; q)}{\partial \tau}, \Omega^2(q) \frac{\partial^2 \Phi(\tau; q)}{\partial \tau^2}\right], \tag{11.11}$$

where $\Phi(\tau; q)$ is a function of τ and q, $\Omega(q)$ is a function of q. Let $\hbar$ denote a nonzero auxiliary parameter and $H(\tau)$ a nonzero auxiliary function, respectively. We then construct the zero-order deformation equation

$$(1 - q) \mathcal{L}[\Phi(\tau; q) - u_0(\tau)] = q \hbar H(\tau) \mathcal{N}[\Phi(\tau; q), \Omega(q)], \tag{11.12}$$

subject to the initial conditions

$$\Phi(0; q) = a, \quad \left.\frac{\partial \Phi(\tau; q)}{\partial \tau}\right|_{\tau=0} = 0. \tag{11.13}$$

When $q = 0$, it is clear that Equations (11.12) and (11.13) have the solution

$$\Phi(\tau; 0) = u_0(\tau), \quad \Omega(0) = \omega_0. \tag{11.14}$$

When $q = 1$, since $\hbar \neq 0$ and $H(\tau) \neq 0$, Equations (11.12) and (11.13) are equivalent to Equations (11.4) and (11.5), respectively, provided

$$\Phi(\tau; 1) = u(\tau), \quad \Omega(1) = \omega. \tag{11.15}$$

Free Oscillation Systems with Odd Nonlinearity

Therefore, as q increases from 0 to 1, $\Phi(\tau;q)$ deforms from the initial guess $u_0(\tau) = a\cos\tau$ to the exact solution $u(\tau)$, and $\Omega(q)$ varies from the initial guess ω_0 to the exact frequency ω, respectively.

Using (11.14) and Taylor's theorem, $\Phi(\tau;q)$ and $\Omega(q)$ can be expanded in the power series of q as follows:

$$\Phi(\tau;q) = u_0(\tau) + \sum_{m=1}^{+\infty} u_m(\tau)\, q^m, \tag{11.16}$$

$$\Omega(q) = \omega_0 + \sum_{m=1}^{+\infty} \omega_m\, q^m, \tag{11.17}$$

where

$$u_m(\tau) = \frac{1}{m!}\left.\frac{\partial^m \Phi(\tau;q)}{\partial q^m}\right|_{q=0}, \quad \omega_m = \frac{1}{m!}\left.\frac{\partial^m \Omega(q)}{\partial q^m}\right|_{q=0}. \tag{11.18}$$

Note that the zero-order deformation equation (11.12) contains the auxiliary parameter $\hbar$ and the auxiliary function $H(\tau)$. Thus, $\Phi(\tau;q)$ and $\Omega(q)$ are also dependent upon them. Assuming that $\hbar$ and $H(\tau)$ are properly chosen so that the above series converge at $q = 1$, we have, using (11.15), the solution series

$$u(\tau) = u_0(\tau) + \sum_{m=1}^{+\infty} u_m(\tau), \tag{11.19}$$

$$\omega = \omega_0 + \sum_{m=1}^{+\infty} \omega_m. \tag{11.20}$$

11.1.2 High-order deformation equation

For brevity, define the vectors

$$\vec{u}_n = \{u_0(\tau), u_1(\tau), \cdots, u_n(\tau)\}, \quad \vec{\omega}_n = \{\omega_0, \omega_1, \cdots, \omega_n\}.$$

Differentiating the zero-order deformation equations (11.12) and (11.13) m times with respect to q, then setting $q = 0$, and finally dividing it by $m!$, we gain the so-called high-order deformation equation

$$\mathcal{L}[u_m(\tau) - \chi_m u_{m-1}(\tau)] = \hbar\, H(\tau)\, R_m(\vec{u}_{m-1}, \vec{\omega}_{m-1}), \tag{11.21}$$

subject to the initial conditions

$$u_m(0) = u'_m(0) = 0, \tag{11.22}$$

where χ_m is defined by (2.42) and

$$R_m(\vec{u}_{m-1}, \vec{\omega}_{m-1}) = \frac{1}{(m-1)!}\left.\frac{d^{m-1}\mathcal{N}[\Phi(\tau;q), \Omega(q)]}{dq^{m-1}}\right|_{q=0}. \tag{11.23}$$

Note that there are two unknowns: $u_m(\tau)$ and ω_{m-1}. However, we have only Equations (11.21) and (11.22) for $u_m(\tau)$. Thus, the problem is not closed and an additional algebraic equation is needed to determine ω_{m-1}. Under the *rule of solution expression* denoted by (11.7) and due to the odd nonlinearity of the conservative system, $R_m(\vec{u}_{m-1}, \vec{\omega}_{m-1})$ can be expressed by

$$R_m(\vec{u}_{m-1}, \vec{\omega}_{m-1}) = \sum_{n=0}^{\varphi(m)} b_{m,n}(\vec{\omega}_{m-1}) \cos[(2n+1)\tau], \quad (11.24)$$

where $b_{m,n}(\vec{\omega}_{m-1})$ is a coefficient dependent of $\vec{\omega}_{m-1}$, and the integer $\varphi(m)$ depends upon m and the form of Equation (11.1). In order to comply with the *rule of solution expression* denoted by (11.7), $H(\tau)$ should be in the form

$$H(\tau) = \cos(2\kappa\tau), \quad \kappa = 0, 1, 2, 3, \cdots.$$

For simplicity, we choose $\kappa = 0$, corresponding to

$$H(\tau) = 1. \quad (11.25)$$

According to the property (11.10) of $\mathcal{L}$, the solution of Equation (11.21) involves the so-called secular term $\tau \cos \tau$ if

$$R_m(\vec{u}_{m-1}, \vec{\omega}_{m-1})$$

contains the term $\cos \tau$. This disobeys *the rule of solution expression* denoted by (11.7). Thus, the coefficient $b_{m,0}$ in (11.24) must be enforced to be zero. This provides us with the additional algebraic equation

$$b_{m,0}(\vec{\omega}_{m-1}) = 0, \quad (11.26)$$

which yields ω_{m-1}. This equation is often nonlinear when $m = 1$ for ω_0 but is always linear otherwise. Thereafter, it is easy to gain the solution of Equation (11.21):

$$u_m(\tau) = \chi_m u_{m-1}(\tau) + \frac{\hbar}{\omega_0^2} \sum_{n=2}^{\varphi(m)} \frac{b_{m,n}(\vec{\omega}_{m-1})}{(1-n^2)} \cos(n\tau)$$
$$+ C_1 \sin \tau + C_2 \cos \tau, \quad (11.27)$$

where C_1 and C_2 are two coefficients. Using (11.22), we obtain $C_1 = 0$. To ensure that the oscillation amplitude equals to a, we use

$$u_m(0) - u_m(\pi) = 0, \quad m = 1, 2, 3, \cdots, \quad (11.28)$$

which determines C_2, thus producing ω_{m-1} and $u_m(\tau)$ successively. At the Mth-order of approximation,

$$u(\tau) \approx \sum_{m=0}^{M} u_m(\tau), \quad (11.29)$$

$$\omega \approx \sum_{m=0}^{M} \omega_m. \quad (11.30)$$

Free Oscillation Systems with Odd Nonlinearity

The above approach is very general and valid even for free oscillations of conservative systems with odd nonlinearity in more general cases, governed by

$$F\left[U(t), \dot{U}(t), \ddot{U}(t), sign U(t), sign \dot{U}(t), sign \ddot{U}(t)\right] = 0, \quad (11.31)$$

where

$$sign(x) = \begin{cases} 1, & \text{when } x > 0, \\ -1, & \text{when } x < 0. \end{cases} \quad (11.32)$$

With transformation $\tau = \omega t$ and $U(t) = u(\tau)$,

$$F\left[u(\tau), \omega u'(\tau), \omega^2 u''(\tau), sign(u), sign(u'), sign(u'')\right] = 0. \quad (11.33)$$

Let a ($a > 0$) denote the amplitude and $u_0(\tau) = a\cos\tau$ the initial approximation of oscillation. For free oscillations of conservative systems with odd nonlinearity,

$$sign(u) = sign(u_0) = sign(\cos\tau), \quad (11.34)$$

and similarly,

$$sign(u') = -sign(\sin\tau), \; sign(u'') = -sign(\cos\tau). \quad (11.35)$$

Thus, Equation (11.33) is equivalent to

$$F\left[u(\tau), \omega u'(\tau), \omega^2 u''(\tau), sign(\cos\tau), -sign(\sin\tau), -sign(\cos\tau)\right] = 0.$$

Using

$$sign(\cos\tau) = \frac{4}{\pi}\sum_{k=0}^{+\infty}\frac{(-1)^k}{2k+1}\cos[(2k+1)\tau], \quad (11.36)$$

$$sign(\sin\tau) = \frac{4}{\pi}\sum_{k=0}^{+\infty}\frac{1}{2k+1}\sin[(2k+1)\tau], \quad (11.37)$$

we write

$$f[u(\tau), \omega u'(\tau), \omega^2 u''(\tau)]$$
$$= F\left[u(\tau), \omega u'(\tau), \omega^2 u''(\tau), sign(\cos\tau), -sign(\sin\tau), -sign(\cos\tau)\right].$$

Similarly, we are able to solve free oscillations of conservative systems with odd nonlinearity, governed by Equation (11.31).

Note that

$$|x| = x \, sign(x).$$

Equation (11.31) is therefore equivalent to the equation

$$G\left[U(t), \dot{U}(t), \ddot{U}(t), |U(t)|, |\dot{U}(t)|, |\ddot{U}(t)|\right] = 0, \quad (11.38)$$

where G is a function of $U(t), \dot{U}(t), \ddot{U}(t), |U(t)|, |\dot{U}(t)|,$ and $|\ddot{U}(t)|$.

11.2 Illustrative examples
11.2.1 Example 11.2.1

Consider free oscillations of a conservation system with odd nonlinearity governed by

$$\ddot{U}(t) + U(t) = \epsilon\, U(t)\, \dot{U}^2(t). \tag{11.39}$$

Under the transformation $\tau = \omega t$ and $U(t) = u(\tau)$,

$$\omega^2 u''(\tau) + u(\tau) = \epsilon\, \omega^2\, u(\tau) u'^2(\tau). \tag{11.40}$$

All other related formulae are the same as those given in §11.1. From (11.23) and (11.39),

$$\begin{aligned}
R_m(\vec{u}_{m-1}, \vec{\omega}_{m-1}) \\
= \sum_{n=0}^{m-1}\left(\sum_{j=0}^{n} \omega_j \omega_{n-j}\right) u''_{m-1-n} + u_{m-1} \\
- \epsilon \sum_{n=0}^{m-1}\left(\sum_{i=0}^{n} u_{n-i} \sum_{r=0}^{i} \omega_r \omega_{i-r}\right)\left(\sum_{j=0}^{m-1-n} u'_j u'_{m-1-n-j}\right).
\end{aligned} \tag{11.41}$$

When $m = 1$ we obtain from (11.26) the algebraic equation

$$a - a\omega_0^2 - \frac{1}{4}a^3 \epsilon \omega_0^2 = 0, \tag{11.42}$$

which gives

$$\omega_0 = \frac{1}{\sqrt{1 + \frac{1}{4}\epsilon a^2}}. \tag{11.43}$$

The frequency ω at the first and second order of approximation is given by

$$\omega \approx \omega_0 + \frac{\hbar\,(\epsilon a^2)\,[2 + (\epsilon a^2 - 2)\omega_0^2]}{32(4 + \epsilon a^2)\omega_0}$$

and

$$\omega \approx \omega_0 + \frac{\hbar\,(\epsilon a^2)\,[2 + (\epsilon a^2 - 2)\omega_0^2]}{16(4 + \epsilon a^2)\omega_0}$$
$$+ \frac{\hbar^2(\epsilon a^2)}{6144(4 + \epsilon a^2)^2 \omega_0^3}\left[39\omega_0^4(\epsilon a^2)^3 + 4\omega_0^2(43\omega_0^2 + 17)(\epsilon a^2)^2 \right.$$
$$\left. + 4(97\omega_0^4 + 98\omega_0^2 - 3)(\epsilon a^2) - 192\,(9\omega_0^4 - 10\omega_0^2 + 1)\right],$$

respectively. These approximations contain the auxiliary parameter $\hbar$. When $\hbar = -1$, the series of frequency is convergent only in the region $0 \leq \epsilon a^2 < 5$,

Free Oscillation Systems with Odd Nonlinearity

as shown in Figure 11.1. Note that the convergence region becomes larger when $\hbar$ is chosen closer to 0, as shown in Figure 11.1. Therefore, $\hbar$ should be defined as a function of ϵa^2, whose absolute value should decrease as ϵa^2 increases. When $\hbar = -\omega_0^2 = -(1 + \epsilon a^2/4)^{-1}$ the series for the frequency converges quickly in the whole region $0 \leq \epsilon a^2 < +\infty$, as shown in Figure 11.1. Selecting

$$\hbar = -(1 + \epsilon a^2/4)^{-1},$$

we gain the first-order approximation

$$\omega \approx \frac{256 + 128\epsilon a^2 + 13(\epsilon a^2)^2}{8(4 + \epsilon a^2)^{5/2}}, \tag{11.44}$$

and the second-order approximation

$$\omega \approx \frac{393216 + 393216\epsilon a^2 + 142848(\epsilon a^2)^2 + 21248(\epsilon a^2)^3 + 1181(\epsilon a^2)^4}{768(4 + \epsilon a^2)^{9/2}}. \tag{11.45}$$

These two approximations agree with the numerical results in the whole region:

$$0 \leq \epsilon a^2 < +\infty,$$

as shown in Figure 11.1. This example illustrates that the auxiliary parameter $\hbar$ provides a convenient way to adjust the convergence region and rate of solution series.

11.2.2 Example 11.2.2

Consider free oscillations governed by

$$\ddot{U}(t) + U(t) + \epsilon\, U^3(t) = 0. \tag{11.46}$$

The exact frequency is

$$\omega = \frac{\pi\sqrt{1 + \epsilon a^2/2}}{2K(\mu)}, \tag{11.47}$$

where

$$\mu = -\frac{\epsilon a^2}{2 + \epsilon a^2}$$

and $K(\mu)$ is the complete elliptic integral of the first kind.

Under the transformation $\tau = \omega t$ and $U(t) = u(\tau)$, Equation (11.46) becomes

$$\omega^2 u''(\tau) + u(\tau) + \epsilon\, u^3(\tau) = 0. \tag{11.48}$$

All related formulae are the same as those given in §11.1. From (11.23) and (11.46),

$$R_m = \sum_{n=0}^{m-1}\left(\sum_{j=0}^{n}\omega_j\omega_{n-j}\right)u''_{m-1-n} + u_{m-1}$$

$$+ \epsilon \sum_{n=0}^{m-1}\left(\sum_{j=0}^{n} u_j u_{n-j}\right) u_{m-1-n}. \qquad (11.49)$$

When $m = 1$ we obtain from (11.26) the algebraic equation

$$a + \frac{3}{4}\epsilon a^3 - a\omega_0^2 = 0, \qquad (11.50)$$

which yields

$$\omega_0 = \sqrt{1 + \frac{3}{4}\epsilon a^2}. \qquad (11.51)$$

The frequency ω at the first and second order of approximation is given by

$$\omega \approx \omega_0 + \frac{\hbar(\epsilon a^2)}{128\omega_0^3}\left[2(1-\omega_0^2) + 3\,\epsilon a^2\right], \qquad (11.52)$$

and

$$\omega \approx \omega_0 + \frac{\hbar(\epsilon a^2)}{32768\omega_0^7}\left\{1024(\omega_0^4 - \omega_0^6) + 1536\omega_0^4(\epsilon a^2)\right.$$
$$- \hbar\left[(576\omega_0^6 - 640\omega_0^4 + 64\omega_0^2) - (940\omega_0^4 - 168\omega_0^2 - 4)(\epsilon a^2)\right.$$
$$\left.\left. + (84\omega_0^2 + 12)(\epsilon a^2)^2 + 9(\epsilon a^2)^3\right]\right\}, \qquad (11.53)$$

respectively. Note that the series of frequency contains the auxiliary parameter $\hbar$. When $-1 \leq \hbar < 0$ the series of frequency converges in the whole region $0 \leq \epsilon a^2 < +\infty$. Choosing $\hbar = -1$, we gain the first-order approximation

$$\omega \approx \frac{256 + 384\epsilon a^2 + 141\epsilon^2 a^4}{32(4 + 3\epsilon a^2)^{3/2}} \qquad (11.54)$$

and the second-order approximation

$$\omega \approx \frac{131072 + 393216\epsilon a^2 + 440832\epsilon^2 a^4 + 218880\epsilon^3 a^6 + 40599\epsilon^4 a^8}{1024(4 + 3\epsilon a^2)^{7/2}}. \qquad (11.55)$$

The maximum error of the first- and second-order approximation is only 0.09% and 0.07% in the whole region $0 \leq \epsilon a^2 < +\infty$, respectively! The first-order approximation given by the proposed approach agrees with the exact result, as shown in Figure 11.2.

11.2.3 Example 11.2.3

Consider free oscillations of conservative system with odd nonlinearity, governed by

$$\ddot{U}(t) + U(t) + \epsilon U(t)|U(t)| = 0. \qquad (11.56)$$

Free Oscillation Systems with Odd Nonlinearity 173

With transformation $\tau = \omega t$ and $U(t) = u(\tau)$,

$$\omega^2 u''(\tau) + u(\tau) + \epsilon u^2(\tau)\, sign[u(\tau)] = 0, \tag{11.57}$$

which is equivalent to the equation

$$\omega^2 u''(\tau) + u(\tau) + \epsilon u^2(\tau)\, sign[\cos \tau] = 0. \tag{11.58}$$

All related formulae are the same as those given in §11.1. From (11.23) and (11.46),

$$R_m = \sum_{n=0}^{m-1} \left(\sum_{j=0}^{n} \omega_j \omega_{n-j} \right) u''_{m-1-n}(\tau) + u_{m-1}(\tau)$$

$$+ \epsilon\, sign(\cos \tau) \sum_{n=0}^{m-1} u_n(\tau) u_{m-1-n}(\tau). \tag{11.59}$$

When $m = 1$ we obtain from (11.26) the algebraic equation

$$a + \frac{8\epsilon a^2}{3\pi} - a\omega_0^2 = 0, \tag{11.60}$$

which gives

$$\omega_0 = \sqrt{1 + \frac{8\epsilon a}{3\pi}}. \tag{11.61}$$

Note that the solution series for frequency contains the auxiliary parameter $\hbar$. We gain the frequency at the first several orders of approximation and find that the series of frequency is convergent when $-2 \leq \hbar < 0$. When $\hbar = -1$, the first order of approximation

$$\omega \approx \sqrt{1 + \frac{8\epsilon a}{3\pi}} - \frac{20,1789,3901,1695}{406,4428,1993,5152} \left(\frac{\epsilon a}{\pi}\right)^2 \left(1 + \frac{8\epsilon a}{3\pi}\right)^{-3/2} \tag{11.62}$$

agrees with the numerical results in the whole region $0 \leq \epsilon a < +\infty$, as shown in Figure 11.3.

11.3 The control of convergence region

As mentioned before, $\hbar = -1$ corresponds to the traditional method of constructing a homotopy. In Example 11.2.2 and Example 11.2.3, using $\hbar = -1$ we obtain accurate approximations valid in the whole regions $0 < \epsilon a^2 < +\infty$ and $0 \leq \epsilon a < +\infty$, respectively. In Example 11.2.2 and Example 11.2.3, using

(11.9) as the auxiliary linear operator $\mathcal{L}$, we gain accurate approximations by setting $\hbar = -1$, as shown in Figures 11.2 and 11.3. However, in Example 11.2.1, the series of frequency converges in a fairly small region $0 \le \epsilon a^2 < 5$ when $\hbar = -1$. Thus, we had to choose $\hbar = -(1 + \epsilon a^2/4)^{-1}$ to adjust its convergence region to ensure that it is valid in the whole region $0 \le \epsilon a^2 < +\infty$, as shown in Figure 11.1.

Note that the auxiliary linear operator $\mathcal{L}$ defined by (11.9) contains the term ω_0^2. If we replace (11.9) with

$$\mathcal{L}[\Phi(\tau;q)] = \frac{\partial^2 \Phi(\tau;q)}{\partial \tau^2} + \Phi(\tau;q), \tag{11.63}$$

we gain the frequency of Example 11.2.2 at the first order of approximation:

$$\omega \approx \omega_0 + \frac{\hbar(\epsilon a^2)}{128\omega_0}\left[2(1-\omega_0^2) + 3\,\epsilon a^2\right] \tag{11.64}$$

and at the second order of approximation:

$$\omega \approx \omega_0 + \frac{\hbar(\epsilon a^2)}{32768\omega_0^5}\left\{1024(\omega_0^4 - \omega_0^6) + 1536\omega_0^4(\epsilon a^2)\right.$$
$$-\hbar\,\omega_0^2\left[(576\omega_0^6 - 640\omega_0^4 + 64\omega_0^2) - (940\omega_0^4 - 168\omega_0^2 - 4)(\epsilon a^2)\right.$$
$$\left.\left. + (84\omega_0^2 + 12)(\epsilon a^2)^2 + 9(\epsilon a^2)^3\right]\right\}, \tag{11.65}$$

respectively, where ω_0 is defined by (11.51). Unfortunately, when $\hbar = -1$, the above approximations are valid in restricted regions much smaller than those of (11.54) and (11.55) given by the auxiliary linear operator (11.9) and the same value of $\hbar$, as shown in Figure 11.2. In this case we must choose

$$\hbar = -\omega_0^{-2} = -\left(1 + \frac{3}{4}\epsilon a^2\right)^{-1}$$

to adjust the convergence region so that the series of frequency is convergent in the whole region $0 \le \epsilon a^2 < +\infty$. Using the above expression of $\hbar$, we gain the same results as in (11.54) and (11.55), respectively.

Similarly, when $\hbar = -1$ and using the auxiliary linear operator defined by (11.63), the corresponding series of frequency of Example 11.2.3 also converges in a restricted region much smaller than that of (11.62) given by the auxiliary linear operator (11.9) and the same value of $\hbar$, as shown in Figure 11.3. In this case we must select

$$\hbar = -\left(1 + \frac{8\epsilon a}{3\pi}\right)^{-1}$$

to gain the accurate approximations valid in the whole region $0 \le \epsilon a < +\infty$. Using the above expression of $\hbar$, we gain exactly the same result as (11.62).

However, using the auxiliary linear operator $\mathcal{L}$ defined by (11.63), we gain the frequency ω of Example 11.2.1 at the first order of approximation:

$$\omega \approx \omega_0 + \frac{\hbar\,\omega_0\,(\epsilon a^2)\,[2 + (\epsilon a^2 - 2)\omega_0^2]}{32(4 + \epsilon a^2)}$$

and at the second order of approximation:

$$\omega \approx \omega_0 + \frac{\hbar\,\omega_0\,(\epsilon a^2)\,[2 + (\epsilon a^2 - 2)\omega_0^2]}{16(4 + \epsilon a^2)}$$
$$+ \frac{\hbar^2 \omega_0 (\epsilon a^2)}{6144(4 + \epsilon a^2)^2} \left[39\omega_0^4 (\epsilon a^2)^3 + 4\omega_0^2(43\omega_0^2 + 17)(\epsilon a^2)^2 \right.$$
$$\left. + 4(97\omega_0^4 + 98\omega_0^2 - 3)(\epsilon a^2) - 192\,(9\omega_0^4 - 10\omega_0^2 + 1) \right],$$

respectively. They are exactly the same as (11.44) and (11.45) when $\hbar = -1$.

These examples illustrate that, for given auxiliary linear operator and auxiliary function, the auxiliary parameter $\hbar$ provides a convenient way to control the convergence region and rate of solution series. The auxiliary parameter $\hbar$ plays an important role in the homotopy analysis method.

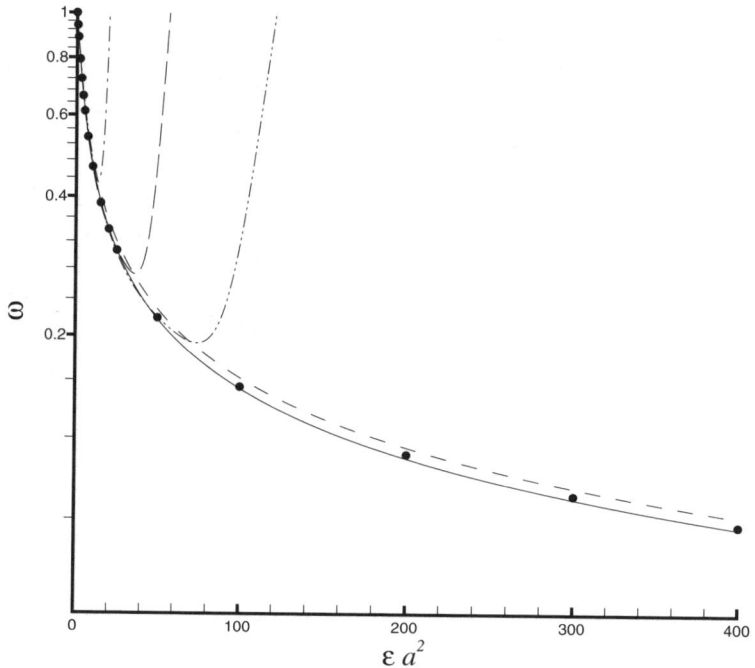

FIGURE 11.1
Comparison of the exact ω with the approximate results of Example 11.2.1 given by the auxiliary linear operator defined by (11.9). Symbols: exact result; dashed line: first-order approximation (11.44) when $\hbar = -(1+\epsilon a^2/4)^{-1}$; solid line: second-order approximation (11.45) when $\hbar = -(1 + \epsilon a^2/4)^{-1}$; dash-dotted line: sixth-order approximation when $\hbar = -1/2$; long-dashed line: sixth-order approximation when $\hbar = -1/5$; dash-dot-dotted line: sixth-order approximation when $\hbar = -1/10$.

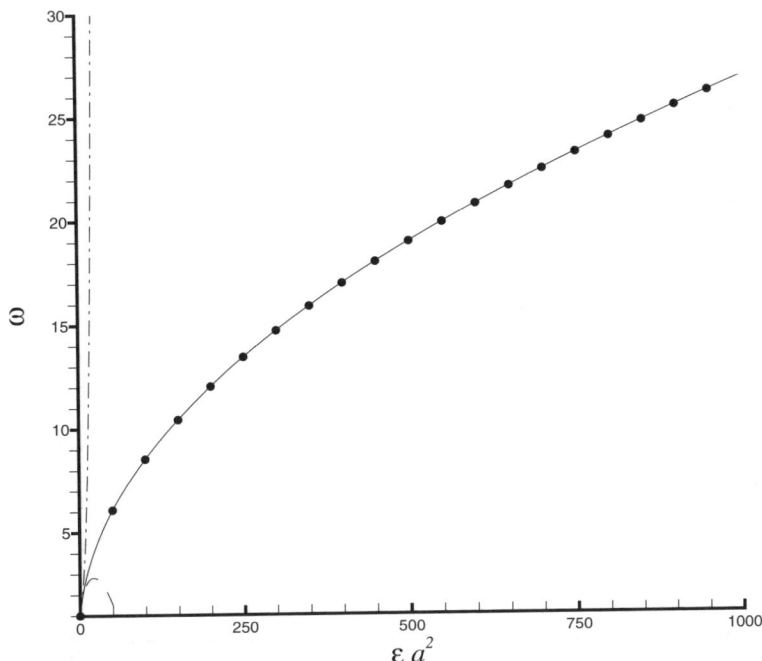

FIGURE 11.2
Comparison of the exact ω with the approximate results of Example 11.2.2 when $\hbar = -1$ by different auxiliary linear operators. Symbols: exact result; solid line: first-order approximation (11.54) given by the auxiliary operator defined by (11.9); dashed line: first-order approximation (11.64) given by the auxiliary linear operator defined by (11.63); dash-dotted line: second-order approximation (11.65) given by the auxiliary linear operator defined by (11.63).

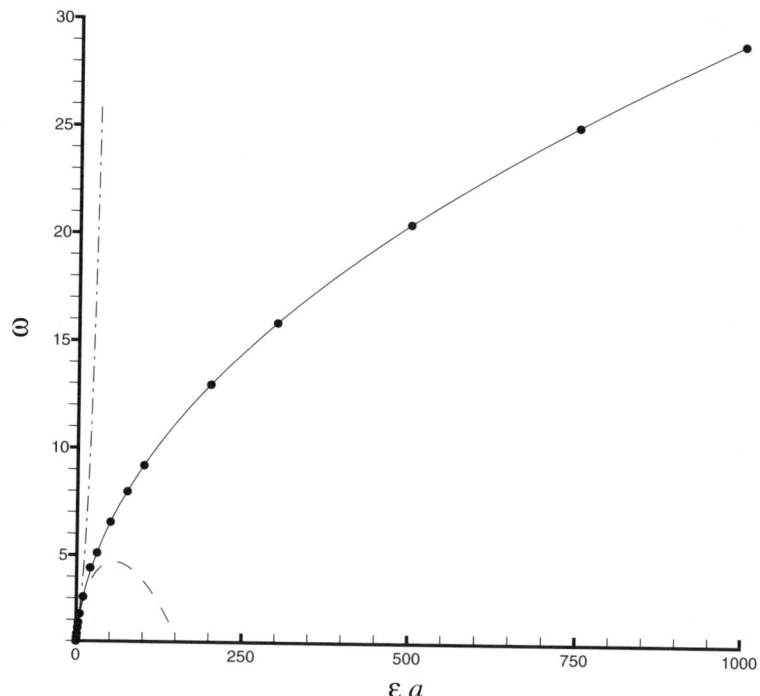

FIGURE 11.3
Comparison of the exact ω with the approximate results of Example 11.2.3 when $\hbar = -1$ by different auxiliary linear operators. Symbols: exact result; solid line: first-order approximation (11.62) given by the auxiliary linear operator defined by (11.9); dashed line: first-order approximation given by the auxiliary linear operator defined by (11.63) of $\mathcal{L}$; dash-dotted line: second-order approximation given by the auxiliary linear operator defined by (11.63).

12

Free oscillation systems with quadratic nonlinearity

Consider free oscillations of a conservative system with quadratic nonlinearity, governed by
$$\ddot{U}(t) = f[U(t), \dot{U}(t), \ddot{U}(t)], \qquad (12.1)$$
where t denotes the time, the dot denotes derivative with respect to t, and $f[U(t), \dot{U}(t), \ddot{U}(t)]$ is a known function of $U(t), \dot{U}(t)$, and $\ddot{U}(t)$. Physically, free oscillation of conservative systems is a periodic motion. Let ω and a denote the frequency and amplitude of oscillation, respectively. Define the mean of motion
$$\delta = \frac{1}{T} \int_0^T U(t) dt, \qquad (12.2)$$
where $T = 2\pi/\omega$ is the period of oscillation. For conservative systems with quadratic nonlinearity, the mean of motion δ is generally nonzero. This is the main difference between free oscillations of conservative system with odd nonlinearity and those with quadratic one. Obviously, both δ and ω have clear physical meanings. For conservative systems, the oscillation amplitude a is determined by initial conditions and is related to the total kinetic energy. Both ω and δ are dependent of a. Without the loss of generality, we consider oscillations with amplitude a under the initial conditions
$$\dot{U}(0) = 0 \qquad U(0) = a + \delta. \qquad (12.3)$$

Unlike perturbation techniques, we need not assume that Equation (12.1) contains any small/large parameters.

12.1 Homotopy analysis solution

12.1.1 Zero-order deformation equation

Obviously, the free oscillation can be described by the base functions
$$\{\cos(m\omega t) \mid m = 0, 1, 2, 3, \cdots\} \qquad (12.4)$$

in the form:
$$U(t) = \delta + \sum_{m=1}^{+\infty} c_m \cos(m\omega t), \qquad (12.5)$$
where c_m is a coefficient. Under the transformation
$$\tau = \omega t, \ U(t) = \delta + u(\tau), \qquad (12.6)$$
Equations (12.2) and (12.3) become
$$\omega^2 u''(\tau) = f[\delta + u(\tau), \omega u'(\tau), \omega^2 u''(\tau)], \qquad (12.7)$$
and
$$u(0) = a, \quad u'(0) = 0, \qquad (12.8)$$
respectively, where the prime denotes derivative with respect to τ. Obviously, $u(\tau)$ can be expressed by the base functions
$$\{\cos(m\tau) \mid m = 1, 2, 3, \cdots\} \qquad (12.9)$$
in the form:
$$u(\tau) = \sum_{m=1}^{+\infty} c_m \cos(m\tau). \qquad (12.10)$$
This provides us with the *rule of solution expression* for free oscillations of conservative systems with quadratic nonlinearity.

Note that the frequency ω and the mean of motion δ are unknown. Let ω_0, δ_0 denote the initial guesses of ω and δ, respectively. Under the *rule of solution expression* denoted by (12.10) and from (12.8), it is easy to choose
$$u_0(\tau) = a \cos \tau \qquad (12.11)$$
as the initial guess of $u(\tau)$, where a is the amplitude of oscillation. Moreover, under the *rule of solution expression* denoted by (12.10) and from Equation (12.7), we choose the auxiliary linear operator
$$\mathcal{L}[\Phi(\tau;q)] = \omega_0^2 \left[\frac{\partial^2 \Phi(\tau;q)}{\partial \tau^2} + \Phi(\tau;q) \right], \qquad (12.12)$$
with the property
$$\mathcal{L}(C_1 \sin \tau + C_2 \cos \tau) = 0. \qquad (12.13)$$
where q is an embedding parameter, $\Phi(\tau;q)$ is a function of τ and q, C_1 and C_2 are coefficients. From Equation (12.7), we define the nonlinear operator
$$\mathcal{N}[\Phi(\tau;q), \Omega(q), \Delta(q)]$$
$$= \Omega^2(q) \frac{\partial^2 \Phi(\tau;q)}{\partial \tau^2}$$
$$- f\left[\Delta(q) + \Phi(\tau;q), \Omega(q) \frac{\partial \Phi(\tau;q)}{\partial \tau}, \Omega^2(q) \frac{\partial^2 \Phi(\tau;q)}{\partial \tau^2}\right], \qquad (12.14)$$

where $\Omega(q)$ and $\Delta(q)$ are functions of the embedding parameter $q \in [0,1]$, corresponding to the frequency ω and the mean of motion δ, respectively.

The homotopy analysis method is based on such continuous variations $\Phi(\tau;q)$, $\Omega(q)$, and $\Delta(q)$ that, as the embedding parameter q increases from 0 to 1, $\Phi(\tau;q)$ varies from the initial guess $u_0(\tau)$ to the exact solution $u(\tau)$, so does $\Omega(q)$ from the initial guess ω_0 to the exact frequency ω, and $\Delta(q)$ from the initial guess δ_0 to the exact mean of motion δ, respectively. To ensure this, we construct such a homotopy in a more general form (see §3.6):

$$\mathcal{H}[\Phi(\tau;q), \Omega(q), \Delta(q), H(\tau), H_2(\tau), \hbar, \hbar_2, q]$$
$$= (1-q)\,\mathcal{L}\,[\Phi(\tau;q) - u_0(\tau)] - q\,\hbar\,H(\tau)\,\mathcal{N}[\Phi(\tau;q), \Omega(q), \Delta(q)]$$
$$- \hbar_2\,H_2(\tau)\,(1-q)\,\{(f[\Delta(q), 0, 0] - f[\delta_0, 0, 0]) + [\Omega^2(q) - \omega_0^2]\,u_0''(\tau)\}$$

where $q \in [0,1]$ is the embedding parameter, $\hbar$ and $\hbar_2$ are nonzero auxiliary parameters, $H(\tau)$ and $H_2(\tau)$ are nonzero auxiliary functions, respectively.

Writing

$$\mathcal{H}[\Phi(\tau;q), \Omega(q), \Delta(q), H(\tau), H_2(\tau), \hbar, \hbar_2, q] = 0,$$

we have the zero-order deformation equation

$$(1-q)\,\mathcal{L}\,[\Phi(\tau;q) - u_0(\tau)]$$
$$= q\,\hbar\,H(\tau)\,\mathcal{N}[\Phi(\tau;q), \Omega(q), \Delta(q)]$$
$$+ \hbar_2\,H_2(\tau)\,(1-q)\,(f[\Delta(q), 0, 0] - f[\delta_0, 0, 0])$$
$$+ \hbar_2\,H_2(\tau)\,(1-q)\,[\Omega^2(q) - \omega_0^2]\,u_0''(\tau), \qquad (12.15)$$

subject to the initial conditions

$$\Phi(0;q) = a, \qquad \left.\frac{\partial \Phi(\tau;q)}{\partial \tau}\right|_{\tau=0} = 0. \qquad (12.16)$$

When $q = 0$ it is easy using (12.11) and (12.15) to show that

$$\Phi(\tau;0) = u_0(\tau), \qquad \Omega(0) = \omega_0, \qquad \Delta(0) = \delta_0. \qquad (12.17)$$

When $q = 1$, since $\hbar \neq 0$ and $H(\tau) \neq 0$, Equations (12.15) and (12.16) are equivalent to the original ones (12.7) and (12.8), respectively, provided

$$\Phi(\tau;1) = u(\tau), \qquad \Omega(1) = \omega, \qquad \Delta(1) = \delta. \qquad (12.18)$$

Therefore, as q increases from 0 to 1, $\Phi(\tau;q)$ varies from the initial guess $u_0(\tau) = a \cos \tau$ to the exact solution $u(\tau)$, so does $\Omega(q)$ from the initial guess ω_0 to the exact frequency ω, and $\Delta(q)$ from the initial guess δ_0 to the exact mean of motion δ, respectively.

Note that the zero-order deformation equation (12.15) contains the auxiliary parameters $\hbar, \hbar_2$ and the auxiliary functions $H(\tau)$ and $H_2(\tau)$. Assume that all of them are properly chosen so that Equations (12.15) and (12.16) have solutions $\Phi(\tau;q)$, $\Omega(q)$, and $\Delta(q)$ for all $q \in [0,1]$ and, in addition, the so-called high-order deformation derivatives

$$u_0^{[m]}(\tau) = \left.\frac{\partial^m \Phi(\tau;q)}{\partial q^m}\right|_{q=0}, \quad \omega_0^{[m]} = \left.\frac{\partial^m \Omega(q)}{\partial q^m}\right|_{q=0}, \quad \delta_0^{[m]} = \left.\frac{\partial^m \Delta(q)}{\partial q^m}\right|_{q=0}$$

exist for $m \geq 1$. Then, by Taylor's theorem and using (12.17), we expand $\Phi(\tau;q)$, $\Omega(q)$, and $\Delta(q)$ in the power series of q as follows:

$$\Phi(\tau;q) = u_0(\tau) + \sum_{m=1}^{+\infty} u_m(\tau)\, q^m, \tag{12.19}$$

$$\Omega(q) = \omega_0 + \sum_{m=1}^{+\infty} \omega_m\, q^m, \tag{12.20}$$

$$\Delta(q) = \delta_0 + \sum_{m=1}^{+\infty} \delta_m\, q^m, \tag{12.21}$$

where

$$u_m(\tau) = \frac{u_0^{[m]}(\tau)}{m!}, \quad \omega_m = \frac{\omega_0^{[m]}}{m!}, \quad \delta_m = \frac{\delta_0^{[m]}}{m!}. \tag{12.22}$$

Assuming that the auxiliary parameters $\hbar$, $\hbar_2$ and the auxiliary functions $H(\tau)$ and $H_2(\tau)$ are properly chosen so that the above series converge at $q=1$, we have using (12.18) the solution series

$$u(\tau) = u_0(\tau) + \sum_{m=1}^{+\infty} u_m(\tau), \tag{12.23}$$

$$\omega = \omega_0 + \sum_{m=1}^{+\infty} \omega_m, \tag{12.24}$$

$$\delta = \delta_0 + \sum_{m=1}^{+\infty} \delta_m. \tag{12.25}$$

12.1.2 High-order deformation equation

For brevity, define the vectors

$$\vec{u}_n = \{u_0(\tau), u_1(\tau), \cdots, u_n(\tau)\}, \quad \vec{\omega}_n = \{\omega_0, \omega_1, \cdots, \omega_n\}$$

and

$$\vec{\delta}_n = \{\delta_0, \delta_1, \cdots, \delta_n\}.$$

Differentiating Equations (12.15) and (12.16) m times with respect to q, then setting $q=0$, and finally dividing it by $m!$, we have the so-called high-order

Free Oscillation Systems with Quadratic Nonlinearity

deformation equation

$$\mathcal{L}[u_m(\tau) - \chi_m u_{m-1}(\tau)] = \hbar\, H(\tau)\, R_m(\vec{u}_{m-1}, \vec{\omega}_{m-1}, \vec{\delta}_{m-1}) \\ + \hbar_2\, H_2(\tau)\, S_m(\tau, \vec{\omega}_m, \vec{\delta}_m), \quad (12.26)$$

subject to the initial conditions

$$u_m(0) = u'_m(0) = 0, \quad (12.27)$$

where χ_m is defined by (2.42),

$$R_m(\vec{u}_{m-1}, \vec{\omega}_{m-1}, \vec{\delta}_{m-1}) \\ = \frac{1}{(m-1)!} \left. \frac{d^{m-1}\mathcal{N}[\Phi(\tau;q), \Omega(q), \Delta(q)]}{dq^{m-1}} \right|_{q=0}, \quad (12.28)$$

$$S_m(\tau, \vec{\omega}_m, \vec{\delta}_m) \\ = -\left(\sum_{i=0}^{m} \omega_i \omega_{m-i} - \chi_m \sum_{i=0}^{m-1} \omega_i \omega_{m-1-i}\right) a\, \cos\tau \\ + \left[Q_m(\vec{\delta}_m) - \chi_m Q_{m-1}(\vec{\delta}_{m-1})\right], \quad (12.29)$$

and

$$Q_m(\vec{\delta}_m) = \frac{1}{m!} \left. \frac{d^m f[\Delta(q), 0, 0]}{dq^m} \right|_{q=0}. \quad (12.30)$$

Note that there exist three unknowns: $u_m(\tau), \omega_{m-1}$ and δ_{m-1} (when $\hbar_2 = 0$), or $u_m(\tau), \omega_m$ and δ_m (when $\hbar_2 \neq 0$). However, we have only Equations (12.26) and (12.27) for $u_m(\tau)$. So, the problem is not closed and two additional algebraic equations are needed to determined ω_{m-1} and δ_{m-1} (when $\hbar_2 = 0$), or ω_m and δ_m (when $\hbar_2 \neq 0$).

Under the *rule of solution expression* denoted by (12.10) and from Equation (12.26), the auxiliary functions $H(\tau)$ and $H_2(\tau)$ might appear as

$$H(\tau) = \cos(2\kappa_1 \tau), \quad H_2(\tau) = \cos(2\kappa_2 \tau),$$

where κ_1 and κ_2 are integers. For simplicity, we choose $\kappa_1 = \kappa_2 = 0$, corresponding to

$$H(\tau) = 1, \quad H_2(\tau) = 1. \quad (12.31)$$

Then, under *the rule of solution expression* denoted by (12.10) and due to the quadratic nonlinearity of conservative systems, the term on the right-hand side of Equation (12.26) may be expressed by

$$b_{m,0} + \sum_{n=1}^{\varphi(m)} b_{m,n} \cos(n\tau), \quad (12.32)$$

where the integer $\varphi(m)$ is dependent of m and the form of the original equation (12.1), and the coefficient $b_{m,n}$ becomes zero when $n > \varphi(m)$. From the property (12.13) of $\mathcal{L}$, the solution of the mth-order deformation equation (12.26) contains the so-called secular term $\tau \cos \tau$ if $b_{m,1} \neq 0$. Besides, when $b_{m,0} \neq 0$, its solution $u_m(\tau)$ contains a constant term $b_{m,0}/\omega_0^2$. However, these two terms do not conform to the *rule of solution expression* denoted by (12.10). Therefore, we enforce the coefficients $b_{m,0}$ and $b_{m,1}$ to be zero:

$$b_{m,0} = 0, \quad b_{m,1} = 0, \quad (m = 1, 2, 3, \cdots). \tag{12.33}$$

This provides a set of two algebraic equations for ω_{m-1} and δ_{m-1} (when $\hbar_2 = 0$), or for ω_m and δ_m (when $\hbar_2 \neq 0$). In this way, the problem is closed and the *rule of solution existence* is obeyed. Notice that Equations (12.33) are often nonlinear when $\hbar_2 = 0$ and $m = 1$, but are always linear otherwise. When $\hbar_2 = 0$ and $m = 1$, we must solve a set of nonlinear algebraic equations (12.33) to gain ω_0 and δ_0. However, when $\hbar_2 \neq 0$, we have the freedom to choose the initial guesses ω_0 and δ_0. It is advisable to first consider the case of $\hbar_2 = 0$, because this often gives accurate results even at low order of approximations, as shown by the illustrative examples.

Thereafter, it is easy to gain the solution of the mth-order deformation equation (12.26)

$$u_m(\tau) = \chi_m u_{m-1}(\tau) + \sum_{n=2}^{\varphi(m)} \frac{b_{m,n}}{\omega_0^2(1-n^2)} \cos(n\tau) + C_1 \sin \tau + C_2 \cos \tau, \tag{12.34}$$

where C_1 and C_2 are two integral constants. From (12.27), we have $C_1 = 0$. To ensure that the amplitude of oscillation equals to a, we use

$$u_m(0) - u_m(\pi) = 0, \quad m = 1, 2, 3, \cdots, \tag{12.35}$$

which determines the coefficient C_2. We gain $u_m(\tau)$ $(m = 1, 2, 3, \cdots)$ and $\omega_{m-1}, \delta_{m-1}$ (when $\hbar_2 = 0$) or ω_m, δ_m (when $\hbar_2 \neq 0$), successively. At the Mth-order approximation we have

$$u(\tau) \approx \sum_{m=0}^{M} u_m(\tau), \tag{12.36}$$

$$\omega \approx \sum_{m=0}^{M} \omega_m, \tag{12.37}$$

$$\delta \approx \sum_{m=0}^{M} \delta_m. \tag{12.38}$$

12.2 Illustrative examples

12.2.1 Example 12.2.1

Consider
$$\ddot{U}(t) + U(t) + \gamma \, U^2(t) = 0, \tag{12.39}$$
where γ is a constant. Under the transformation $\tau = \omega t$ and $U(t) = \delta + u(\tau)$,
$$\omega^2 u''(\tau) + \delta + u(\tau) + \gamma \, [\delta + u(\tau)]^2 = 0. \tag{12.40}$$

All related formulae are the same as those given in §12.1. From (12.28) and (12.30),
$$R_m = \sum_{n=0}^{m-1} \left(\sum_{j=0}^{n} \omega_j \, \omega_{n-j} \right) u''_{m-1-n}(\tau) + v_{m-1}(\tau)$$
$$+ \gamma \sum_{n=0}^{m-1} v_n(\tau) \, v_{m-1-n}(\tau) \tag{12.41}$$

and
$$Q_m = \delta_m + \gamma \sum_{n=0}^{m} \delta_n \, \delta_{m-n}, \tag{12.42}$$

where
$$v_k(\tau) = \delta_k + u_k(\tau). \tag{12.43}$$

Note that there exist two auxiliary parameters $\hbar$ and $\hbar_2$. First, let us consider the case of $\hbar_2 = 0$. In this case we gain from (12.33) the set of algebraic equations for ω_0 and δ_0,
$$a + 2a\gamma\delta_0 - a\omega_0^2 = 0, \tag{12.44}$$

and
$$\frac{\gamma a^2}{2} + \delta_0 + \gamma \delta_0^2 = 0, \tag{12.45}$$

which yield
$$\omega_0 = \left(1 - 2a^2\gamma^2\right)^{1/4}, \quad \delta_0 = \frac{\omega_0^2 - 1}{2\gamma}. \tag{12.46}$$

When $\hbar_2 = 0$, we have the first-order approximation
$$\omega \approx \omega_0 - \frac{\hbar(a\gamma)^2}{12\omega_0^3}, \quad \delta \approx \delta_0, \tag{12.47}$$

the second-order approximation
$$\omega \approx \omega_0 - \frac{\hbar(a\gamma)^2}{6\omega_0^3}\left(1 + \frac{\hbar}{2}\right) + \frac{\hbar^2(a\gamma)^4}{288\omega_0^7}, \quad \delta \approx \delta_0 + \frac{\hbar^2 a^4 \gamma^3}{144\omega_0^6}, \tag{12.48}$$

the third-order approximation

$$\omega \approx \omega_0 - \frac{\hbar(a\gamma)^2}{4\omega_0^3}\left(1 + \hbar + \frac{\hbar^2}{3}\right)$$
$$+ \frac{\hbar^2(a\gamma)^2}{1728\omega_0^7}(18 + 41\hbar) + \frac{\hbar^3(a\gamma)^6}{3456\omega_0^{11}}, \tag{12.49}$$

$$\delta \approx \delta_0 + \frac{\hbar^2 a^4 \gamma^3}{48\omega_0^6}\left(1 + \frac{2\hbar}{3}\right), \tag{12.50}$$

and so on. These results are dependent upon the auxiliary parameter $\hbar$. For any given a and γ we can investigate the influence of $\hbar$ on the convergence by plotting the so-called $\hbar$-curves (see page 26 or §3.5.1). The series for ω and δ are convergent when $-2 \leq \hbar < 0$. However, the convergence region depends upon the value of $\hbar$. We can adjust the convergence regions by choosing a proper value of $\hbar$. For example, when $\hbar = -4/5$ or $\hbar = -\omega_0^2$, the third-order approximation of ω agrees with the exact results in the region $|a\gamma| \leq 1/\sqrt{2}$ and is much better than the perturbation approximation, as shown in Figure 12.1. When $\hbar = -1/5$ or $\hbar = -\omega_0^2$, the third-order approximation of γ δ yields good agreement with the numerical results in the region $|a\gamma| \leq 1/\sqrt{2}$, as shown in Figure 12.2.

When $|a\gamma| > 1/\sqrt{2}$, the initial approximations ω_0 and δ_0 given by (12.46) have no physical meanings. By some simple calculations, we deduce from Equation (12.39) that solutions exist in the region $|a\gamma| \leq 3/4$. Using (12.46) as initial approximations, we certainly cannot gain results valid in the region $1/\sqrt{2} \leq |a\gamma| \leq 3/4$. To gain approximations valid in the region $0 \leq |a\gamma| \leq 3/4$, we must choose initial approximations that have physical meanings in the whole region. Fortunately, when $\hbar_2 \neq 0$, the proposed approach provides the freedom to choose such an initial approximation. Now, let us consider the case of $\hbar_2 \neq 0$. Notice that ω_0 defined by (12.46) gives good approximation for $|a\gamma| < 1/\sqrt{2}$. More importantly, it provides valuable information about the mathematical structure of the frequency. Therefore, it is reasonable to select the initial approximation

$$\tilde{\omega}_0 = \left(1 - \frac{16}{9}a^2\gamma^2\right)^{1/4}, \tag{12.51}$$

which is valid in the whole region $0 \leq |a\gamma| \leq 3/4$. From (12.46), we choose the initial approximation

$$\tilde{\delta}_0 = \frac{\tilde{\omega}_0^2 - 1}{2\gamma}, \tag{12.52}$$

which is also valid in the whole region $0 \leq |a\gamma| \leq 3/4$. Note that ω and δ now contain two auxiliary parameters: $\hbar$ and $\hbar_2$. For simplicity, let us consider the special case of $\hbar_2 = -1$. We successively gain the first-order approximation

$$\omega \approx \tilde{\omega}_0, \quad \delta \approx \tilde{\delta}_0 + \frac{\hbar}{4\gamma\tilde{\omega}_0^2}\left(\tilde{\omega}_0^4 - 1 + 2a^2\gamma^2\right), \tag{12.53}$$

the second-order approximation

$$\omega \approx \tilde{\omega}_0 - \frac{\hbar^2}{12\tilde{\omega}_0^3}\left(3\,\tilde{\omega}_0^4 - 3 + 5\,a^2\gamma^2\right),$$

$$\delta \approx \tilde{\delta}_0 + \frac{\hbar}{16\gamma\tilde{\omega}_0^6}\left(\tilde{\omega}_0^4 - 1 + 2a^2\gamma^2\right)\left[8\tilde{\omega}_0^4 + \hbar\left(3\,\tilde{\omega}_0^4 + 1 - 2\,a^2\gamma^2\right)\right],$$

the third-order approximation

$$\omega \approx \tilde{\omega}_0 - \frac{\hbar^2}{48\tilde{\omega}_0^7}\left\{12\tilde{\omega}_0^4\left(3\tilde{\omega}_0^4 - 3 + 5a^2\gamma^2\right)\right.$$
$$\left. + \hbar\left[(21\tilde{\omega}_0^8 - 18\tilde{\omega}_0^4 - 3) + 4\left(7\tilde{\omega}_0^4 + 3\right)a^2\gamma^2 - 12a^4\gamma^4\right]\right\}, \quad (12.54)$$

$$\delta \approx \tilde{\delta}_0 + \frac{\hbar}{288\gamma\tilde{\omega}_0^{10}}\left\{216\,\tilde{\omega}_0^8\left(\tilde{\omega}_0^4 - 1 + 2a^2\gamma^2\right)\right.$$
$$+ 54\,\tilde{\omega}_0^4\hbar\left[(3\,\tilde{\omega}_0^8 - 2\,\tilde{\omega}_0^4 - 1) + 4(\tilde{\omega}_0^4 + 1)a^2\gamma^2 - 4a^4\gamma^4\right]$$
$$+ \hbar^2\left[9(5\,\tilde{\omega}_0^{12} - 3\,\tilde{\omega}_0^8 - \tilde{\omega}_0^4 - 1) + 18(3\,\tilde{\omega}_0^8 + 2\,\tilde{\omega}_0^4 + 3)a^2\gamma^2\right.$$
$$\left.\left. -2(19\,\tilde{\omega}_0^4 + 54)a^4\gamma^4 + 72\,a^6\gamma^6\right]\right\}, \quad (12.55)$$

and so on. These results depend upon the auxiliary parameter $\hbar$. Its influence on the convergence regions can be investigated by plotting the so-called $\hbar$-curves (see page 26 and §3.5.1). We see that, at the third order of approximation, the frequency ω when $\hbar = -\tilde{\omega}_0$ and the mean of motion δ when $\hbar = -\tilde{\omega}_0/2$ agree with the numerical results in the whole region $0 \leq |a\gamma| \leq 3/4$, as shown in Figures 12.3 and 12.4, respectively. Using the better initial approximations (12.51) and (12.52) and properly choosing the two auxiliary parameters $\hbar$ and $\hbar_2$, we gain analytic results valid in the whole region $0 \leq |a\gamma| \leq 3/4$.

12.2.2 Example 12.2.2

Consider
$$\ddot{U}(t) - U(t) + U^4(t) = 0. \quad (12.56)$$

Under the transformation $U(t) = \delta + u(\tau)$ and $\tau = \omega t$, we see that

$$\omega^2 u''(\tau) - [u(\tau) + \delta] + [\delta + u(\tau)]^4 = 0. \quad (12.57)$$

All related formulae are the same as those given in §12.1. From (12.28) and (12.30),

$$R_m = \sum_{n=0}^{m-1}\left(\sum_{j=0}^{n}\omega_j\omega_{n-j}\right)u''_{m-1-n}(\tau) - v_{m-1}(\tau)$$
$$+ \sum_{n=0}^{m-1}\left[\sum_{i=0}^{n}v_i(\tau)v_{n-i}(\tau)\right]\left[\sum_{j=0}^{m-1-n}v_j(\tau)v_{m-1-n-j}(\tau)\right], \quad (12.58)$$

and

$$Q_m = -\delta_m + \sum_{n=0}^{m}\left(\sum_{i=0}^{n}\delta_i\delta_{n-i}\right)\left(\sum_{j=0}^{m-n}\delta_j\delta_{m-n-j}\right), \quad (12.59)$$

where $v_k(\tau)$ is defined by (12.43).

When $\hbar_2 = 0$, we have from (12.33) the algebraic equations for ω_0 and δ_0,

$$a - 3a^3\delta_0 - 4a\delta_0^3 + a\omega_0^2 = 0, \quad (12.60)$$

$$\frac{3}{8}a^4 - \delta_0 + 3a^2\delta_0^2 + \delta_0^4 = 0, \quad (12.61)$$

which give

$$\omega_0 = \sqrt{4\delta_0^3 + 3a^2\delta_0 - 1}, \quad (12.62)$$

and

$$\delta_0 = \frac{1}{2}\left(\sqrt{\mu_1} + \sqrt{\frac{2}{\sqrt{\mu_1}} - \mu_1 - 6a^2}\right), \quad (12.63)$$

where

$$\mu_1 = -2a^2 + \frac{3a^4}{\mu_0} + \frac{\mu_0}{2}, \quad (12.64)$$

$$\mu_0 = \left(4 - 4a^6 + 2\sqrt{4 - 8a^6 - 50a^{12}}\right)^{1/3}. \quad (12.65)$$

We have therefore the first-order approximation

$$\omega \approx \omega_0 + \frac{\hbar a^2}{(4\delta_0^3 + 6a^2\delta_0 - 1)\omega_0^3}\left[\frac{27}{160}a^4 + \left(\frac{1}{16} - \frac{9}{20}a^6\right)\delta_0\right.$$
$$+ \frac{3}{4}a^2\delta_0^2 - \frac{9}{5}a^4\delta_0^3 + \frac{5}{2}\delta_0^4 - \frac{15}{2}a^2\delta_0^5 - 11\delta_0^7$$
$$\left. + \left(\frac{1}{16}\delta_0 - \frac{3}{8}a^2\delta_0^2 - \frac{1}{4}\delta_0^4\right)\omega_0^2\right], \quad (12.66)$$

$$\delta \approx \delta_0 + \frac{\hbar a^4 \delta_0}{(4\delta_0^3 + 6a^2\delta_0 - 1)\omega_0^2}\left(\frac{3}{8}a^2 + \frac{9}{4}\delta_0^2\right), \quad (12.67)$$

and so on, where ω_0, δ_0 are given by (12.62) and (12.63), respectively. Similarly, the influence of $\hbar$ on the convergence region can be investigated by plotting the corresponding $\hbar$-curves (see page 26 and §3.5.1). It is found that the series of ω is convergent when $-2 < \hbar < 0$, so does the series of δ. Even at the first order of approximation, the frequency ω when $\hbar = -1$ and the mean of motion δ when $\hbar = -3/4$ agree with the numerical results, as shown in Figures 12.5 and 12.6. It is unnecessary to consider the case $\hbar_2 \neq 0$ for the second illustrative problem.

In this chapter we illustrate how to get better approximations by means of zero-order deformation equations in a more general form as mentioned in §3.6. The illustrative examples demonstrate the flexibility and potential of the homotopy analysis method.

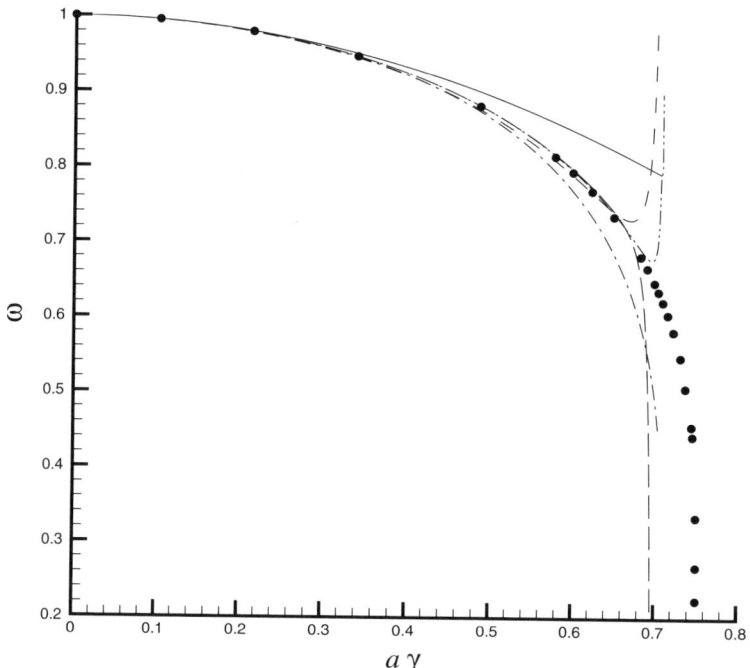

FIGURE 12.1
Comparison of the exact frequency ω of Example 12.2.1 with the approximate results when $\hbar_2 = 0$. Symbols: exact result; solid line: first-order perturbation approximation $\omega = 1 - 5a^2\gamma^2/12$; dashed line: first-order approximation (12.47) when $\hbar = -4/5$; long-dashed line: third-order approximation (12.49) when $\hbar = -4/5$; dash-dotted line: first-order approximation (12.47) when $\hbar = -\omega_0^2$; dash-dot-dotted line: third-order approximation (12.49) when $\hbar = -\omega_0^2$.

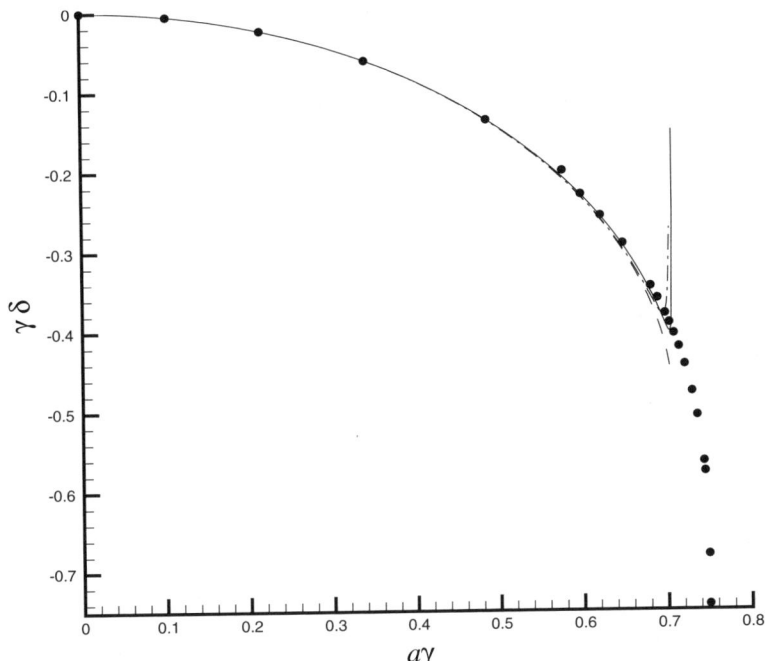

FIGURE 12.2
Comparison of the exact mean of motion δ of Example 12.2.1 with the approximate results when $\hbar_2 = 0$. Symbols: exact result; dashed line: first-order approximation (12.47); dash-dotted line: third-order approximation (12.50) when $\hbar = -1/5$; solid line: third-order approximation (12.50) when $\hbar = -\omega_0^2$.

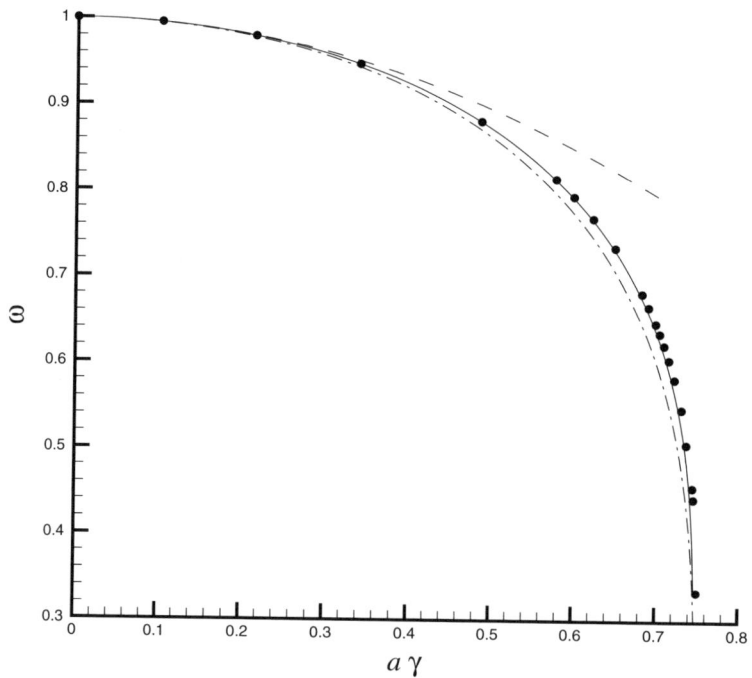

FIGURE 12.3
Comparison of the exact frequency ω of Example 12.2.1 with the approximate results when $\hbar_2 = -1$, $\tilde{\omega}_0 = (1 - 16a^2\gamma^2/9)^{1/4}$ and $\hbar = -\tilde{\omega}_0$. Symbols: exact result; dashed line: perturbation solution $\omega = 1 - 5a^2\gamma^2/12$; dash-dotted line: first-order approximation (12.53); solid line: third-order approximation (12.54).

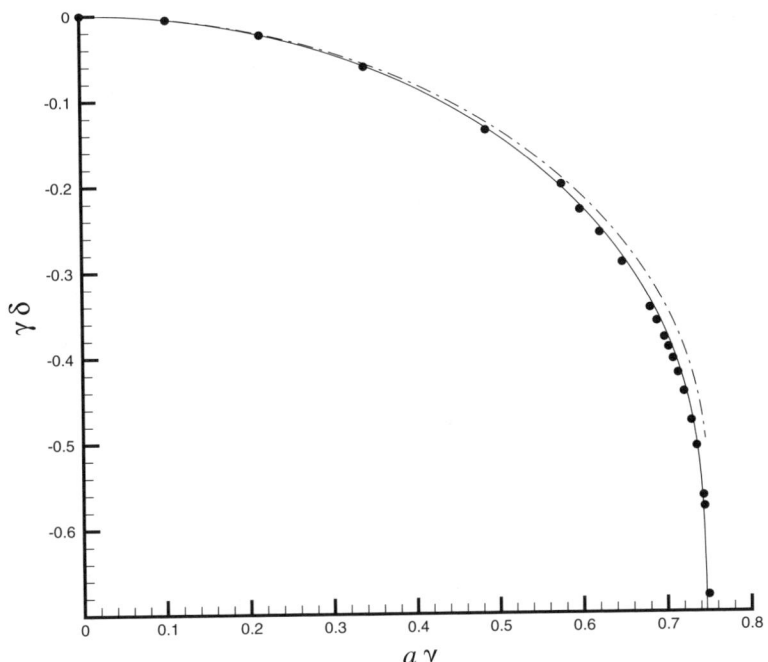

FIGURE 12.4
Comparison of the exact mean of motion δ of Example 12.2.1 with the approximate results when $\hbar_2 = -1$, $\tilde{\omega}_0 = (1 - 16a^2\gamma^2/9)^{1/4}$, and $\hbar = -\tilde{\omega}_0/2$. Symbols: exact result; dash-dotted line: first-order approximation (12.53); solid line: third-order approximation (12.55).

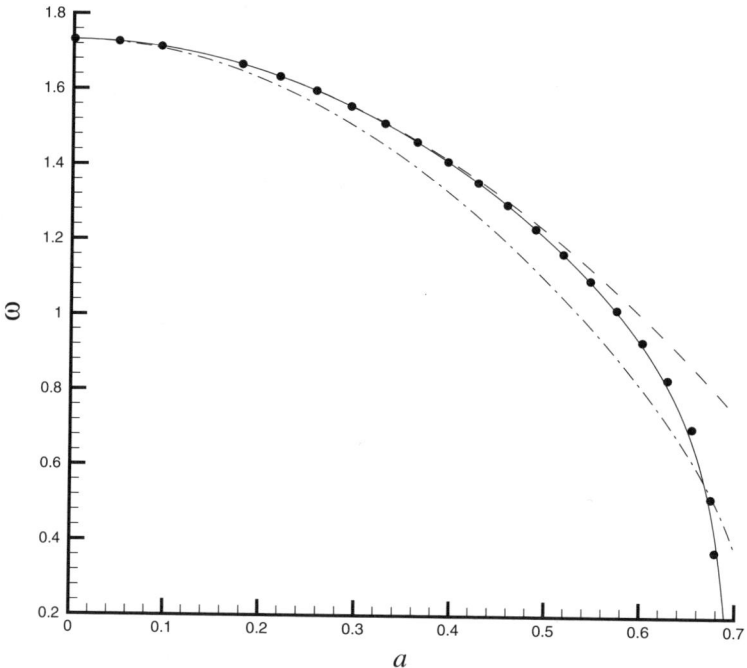

FIGURE 12.5
Comparison of the exact frequency of Example 12.2.2 with the first-order approximation when $\hbar_2 = 0$ and $\hbar = -1$. Symbols: exact result; dashed line: first-order perturbation approximation $\omega = \sqrt{3}(1 - 7a^2/6)$; dash-dotted line: initial approximation ω_0 given by (12.62); solid line: first-order approximation (12.66).

Free Oscillation Systems with Quadratic Nonlinearity 195

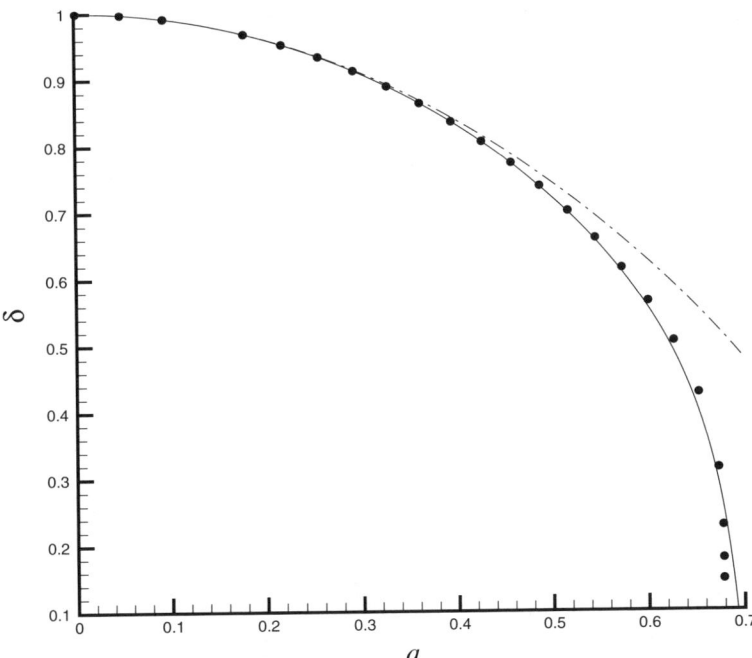

FIGURE 12.6
Comparison of the exact mean of motion δ of Example 12.2.2 with the first-order approximation when $\hbar_2 = 0$ and $\hbar = -3/4$. Symbols: exact result; dash-dotted line: initial approximation δ_0 given by (12.63); solid line: first-order approximation (12.67).

13

Limit cycle in a multidimensional system

The homotopy analysis method is successfully applied by Liao [39] to solve limit cycles of one-dimensional nonlinear dynamical systems, governed by

$$\ddot{u}(t) = f(u, \dot{u}, \ddot{u}), \tag{13.1}$$

where t denotes the time, the dot denotes derivative with respect to t, and $f(u, \dot{u}, \ddot{u})$ is a known function of $u, \dot{u}$, and $\ddot{u}$. Unlike perturbation techniques, it is unnecessary to assume the existence of any small/large quantities in the above equation. Here, we show that the homotopy analysis method can also be applied to gain limit cycles of multidimensional dynamical systems.

As an example, let us consider a two-dimensional nonlinear dynamical system governed by (see Kahn [103])

$$\ddot{x} + x = \epsilon \, \dot{x}(1 - x^2 \, w), \tag{13.2}$$

$$\dot{w} = -\epsilon \, (w^2 - \mu \, x^4), \tag{13.3}$$

where the dot denotes differentiation with respect to t, μ and ϵ are physical parameters, x and w are two unknown functions. Physically, a limit cycle is independent of initial conditions. Let T and $\alpha = \max[x(t)]$ denote the period and the maximum value of $x(t)$ of the limit cycle, respectively. Without loss of generality, we can define $t = 0$ so that

$$x(0) = \alpha, \quad \dot{x}(0) = 0. \tag{13.4}$$

Define

$$\delta = \frac{1}{T} \int_0^T w(t) dt \tag{13.5}$$

and let

$$\omega = T/2\pi$$

denote the frequency of $x(t)$ of the limit cycle. Under the transformations

$$\tau = \omega \, t, \quad x(t) = \alpha \, u(\tau), \quad w(t) = \delta + v(\tau), \tag{13.6}$$

Equations (13.2) and (13.3) become

$$\omega^2 \, u'' + u = \epsilon \, \omega \, u' \, (1 - \alpha^2 \delta \, u^2 - \alpha^2 \, u^2 \, v), \tag{13.7}$$

$$\omega \, v' = -\epsilon \, (\delta^2 + 2\delta \, v + v^2 - \mu \, \alpha^4 \, u^4), \tag{13.8}$$

subject to the initial conditions

$$u(0) = 1, \quad u'(0) = 0, \tag{13.9}$$

where the prime represents differentiation with respect to τ. Furthermore, from (13.5) and (13.6), it holds that

$$\int_0^{2\pi} v(\tau)d\tau = 0, \tag{13.10}$$

which provides us with the condition for $v(\tau)$. Note that $\alpha, \delta,$ and ω are unknown.

13.1 Homotopy analysis solution
13.1.1 Zero-order deformation equation

From a physical point of view, a limit cycle can be expressed by periodic functions. Clearly, $u(\tau)$ and $v(\tau)$ may be expressed in the forms:

$$u(\tau) = \sum_{n=1}^{+\infty} [a_n \cos(n\tau) + b_n \sin(n\tau)] \tag{13.11}$$

and

$$v(\tau) = \sum_{n=1}^{+\infty} [c_n \cos(n\tau) + d_n \sin(n\tau)], \tag{13.12}$$

where $a_n, b_n, c_n,$ and d_n are coefficients. The above expressions provide the so-called *rules of solution expression* for $u(\tau)$ and $v(\tau)$, respectively.

Under the above *rules of solution expressions* and from the initial conditions (13.9) and (13.10), it is convenient to choose

$$u_0(\tau) = \cos \tau, \quad v_0(\tau) = 0 \tag{13.13}$$

as the initial guesses of $u(\tau)$ and $v(\tau)$. Here, $v_0(\tau) = 0$ is chosen because of the lack of information about $v(\tau)$, especially the relationship between $u(\tau)$ and $v(\tau)$. Let $\alpha_0, \delta_0,$ and ω_0 denote the initial guesses of $\alpha, \delta,$ and ω, respectively. Under the *rules of solution expression* denoted by (13.11) and (13.12) and from Equations (13.7) and (13.8), we choose the auxiliary linear operators

$$\mathcal{L}_u f = \frac{\partial^2 f}{\partial \tau^2} + f \tag{13.14}$$

and

$$\mathcal{L}_v f = \frac{\partial f}{\partial \tau} \tag{13.15}$$

with the properties

$$\mathcal{L}_u (C_1 \cos \tau + C_2 \sin \tau) = 0, \quad \mathcal{L}_v(C_3) = 0, \qquad (13.16)$$

respectively, where C_1, C_2, and C_3 are coefficients and f is a real function. For simplicity, we define from Equations (13.7) and (13.8) the nonlinear operators

$$\mathcal{N}_u [U(\tau;q), V(\tau;q), A(q), \Delta(q), \Omega(q)]$$
$$= \Omega^2(q) \frac{\partial^2 U(\tau;q)}{\partial \tau^2} + U(\tau;q)$$
$$- \epsilon \, \Omega(q) \frac{\partial U(\tau;q)}{\partial \tau} \left[1 - A^2(q)\Delta(q)U^2(\tau;q) - A^2(q)U^2(\tau;q)V(\tau;q)\right] \qquad (13.17)$$

and

$$\mathcal{N}_v [U(\tau;q), V(\tau;q), A(q), \Delta(q), \Omega(q)]$$
$$= \Omega(q) \frac{\partial V(\tau;q)}{\partial \tau}$$
$$+ \epsilon \left[\Delta^2(q) + 2\Delta(q) \, V(\tau;q) + V^2(\tau;q) - \mu \, A^4(q) \, U^4(\tau;q)\right], \qquad (13.18)$$

where $q \in [0,1]$ is the imbedding parameter, $U(\tau;q)$ and $V(\tau;q)$ are real functions of τ and q, $A(q), \Delta(q)$, and $\Omega(q)$ are real functions of q.

Let $\hbar_u$ and $\hbar_v$ denote the nonzero auxiliary parameters, $H_u(\tau)$ and $H_v(\tau)$ the nonzero auxiliary functions, respectively. We construct the zero-order deformation equations

$$(1-q)\mathcal{L}_u [U(\tau;q) - u_0(\tau)]$$
$$= q \, \hbar_u \, H_u(\tau) \, \mathcal{N}_u [U(\tau;q), V(\tau;q), A(q), \Delta(q), \Omega(q)], \qquad (13.19)$$

$$(1-q)\mathcal{L}_v [V(\tau;q) - v_0(\tau)]$$
$$= q \, \hbar_v \, H_v(\tau) \, \mathcal{N}_v [U(\tau;q), V(\tau;q), A(q), \Delta(q), \Omega(q)], \qquad (13.20)$$

subject to the conditions

$$U(0;q) = 1, \quad \left.\frac{\partial U(\tau;q)}{\partial \tau}\right|_{\tau=0} = 0, \quad \int_0^{2\pi} V(\tau;q) d\tau = 0, \qquad (13.21)$$

where $\tau \geq 0$ and $q \in [0,1]$.

When $q = 0$, it is clear from (13.13) and the above zero-order deformation equations that
$$U(\tau;0) = u_0(\tau), \quad V(\tau;0) = v_0(\tau). \qquad (13.22)$$

When $q = 1$, Equations (13.19) to (13.21) are equivalent to Equations (13.7) to (13.10), respectively, provided

$$U(\tau;1) = u(\tau), \quad V(\tau;1) = v(\tau) \qquad (13.23)$$

and
$$A(1) = \alpha, \quad \Delta(1) = \delta, \quad \Omega(1) = \omega. \tag{13.24}$$

So, as the embedding parameter q increases from 0 to 1, $U(\tau; q)$, and $V(\tau; q)$ vary from the initial guesses $u_0(\tau)$ and $v_0(\tau)$ to the exact solutions $u(\tau)$ and $v(\tau)$, respectively, so do $A(q), \Delta(q)$, and $\Omega(q)$ from the initial guesses α_0, δ_0, and ω_0 to the corresponding exact values α, δ, and ω.

The zero-order deformation equations (13.19) and (13.20) contain the two auxiliary parameters $\hbar_u, \hbar_v$ and the two auxiliary functions $H_u(\tau), H_v(\tau)$. Assume that all of them are properly chosen so that the terms

$$u_n(\tau) = \left(\frac{1}{n!}\right) \left.\frac{d^n U(\tau; q)}{\partial q^n}\right|_{q=0}, \tag{13.25}$$

$$v_n(\tau) = \left(\frac{1}{n!}\right) \left.\frac{\partial^n V(\tau; q)}{\partial q^n}\right|_{q=0}, \tag{13.26}$$

and

$$\alpha_n = \left(\frac{1}{n!}\right) \left.\frac{d^n A(q)}{d\,q^n}\right|_{q=0}, \tag{13.27}$$

$$\delta_n = \left(\frac{1}{n!}\right) \left.\frac{d^n \Delta(q)}{d\,q^n}\right|_{q=0}, \tag{13.28}$$

$$\omega_n = \left(\frac{1}{n!}\right) \left.\frac{d^n \Omega(q)}{d\,q^n}\right|_{q=0} \tag{13.29}$$

exist for $n \geq 1$. Then, using Taylor's theorem and (13.22), we have the power series of q in the forms:

$$U(\tau; q) = u_0(\tau) + \sum_{n=1}^{+\infty} u_n(\tau)\, q^n, \tag{13.30}$$

$$V(\tau; q) = v_0(\tau) + \sum_{n=1}^{+\infty} v_n(\tau)\, q^n, \tag{13.31}$$

$$A(q) = \alpha_0 + \sum_{n=1}^{+\infty} \alpha_n\, q^n, \tag{13.32}$$

$$\Delta(q) = \delta_0 + \sum_{n=1}^{+\infty} \delta_n\, q^n, \tag{13.33}$$

$$\Omega(q) = \omega_0 + \sum_{n=1}^{+\infty} \omega_n\, q^n. \tag{13.34}$$

Assuming that $\hbar_u$, $\hbar_v$, $H_u(\tau)$, and $H_v(\tau)$ are properly chosen so that the above series are convergent at $q = 1$, we obtain, using (13.23) and (13.24),

Limit Cycle in a Multidimensional System

the solution series

$$u(\tau) = u_0(\tau) + \sum_{n=1}^{+\infty} u_n(\tau), \tag{13.35}$$

$$v(\tau) = v_0(\tau) + \sum_{n=1}^{+\infty} v_n(\tau), \tag{13.36}$$

$$\alpha = \alpha_0 + \sum_{n=1}^{+\infty} \alpha_n, \tag{13.37}$$

$$\delta = \delta_0 + \sum_{n=1}^{+\infty} \delta_n, \tag{13.38}$$

and

$$\omega = \omega_0 + \sum_{n=1}^{+\infty} \omega_n. \tag{13.39}$$

13.1.2 High-order deformation equation

For conciseness, define the vectors

$$\vec{u}_k = \{u_0(\tau), u_1(\tau), \cdots, u_k(\tau)\}, \quad \vec{v}_k = \{v_0(\tau), v_1(\tau), \cdots, v_k(\tau)\}, \tag{13.40}$$

$$\vec{\alpha}_k = \{\alpha_0, \alpha_1, \cdots, \alpha_k\}, \quad \vec{\delta}_k = \{\delta_0, \delta_1, \cdots, \delta_k\}, \tag{13.41}$$

and

$$\vec{\omega}_k = \{\omega_0, \omega_1, \cdots, \omega_k\}. \tag{13.42}$$

Differentiating the zero-order deformation equations (13.19) to (13.21) n times with respect to q, then dividing by $n!$, and finally setting $q = 0$, we have the high-order deformation equations

$$\mathcal{L}_u [u_n(\tau) - \chi_n \, u_{n-1}(\tau)]$$
$$= \hbar_u \, H_u(\tau) \, R_n^u(\vec{u}_{n-1}, \vec{v}_{n-1}, \vec{\alpha}_{n-1}, \vec{\delta}_{n-1}, \vec{\omega}_{n-1}), \tag{13.43}$$

$$\mathcal{L}_v [v_n(\tau) - \chi_n \, v_{n-1}(\tau)]$$
$$= \hbar_v \, H_v(\tau) \, R_n^v(\vec{u}_{n-1}, \vec{v}_{n-1}, \vec{\alpha}_{n-1}, \vec{\delta}_{n-1}, \vec{\omega}_{n-1}), \tag{13.44}$$

subject to the conditions

$$u_n(0) = 0, \quad u_n'(0) = 0, \quad \int_0^{2\pi} v_n(\tau) d\tau = 0, \tag{13.45}$$

where χ_n is defined by (2.42),

$$R_n^u(\vec{u}_{n-1}, \vec{v}_{n-1}, \vec{\alpha}_{n-1}, \vec{\delta}_{n-1}, \vec{\omega}_{n-1})$$

$$= \frac{1}{(n-1)!} \frac{d^{n-1} \mathcal{N}_u \left[U(\tau;q), V(\tau;q), A(q), \Delta(q), \Omega(q) \right]}{d\, q^{n-1}}$$

$$= \sum_{j=0}^{n-1} u''_{n-1-j}(\tau) \left(\sum_{i=0}^{j} \omega_i \omega_{j-i} \right) + u_{n-1}(\tau) - \epsilon\, F_{n-1}(\tau)$$

$$+ \epsilon \sum_{j=0}^{n-1} F_{n-1-j}(\tau) \sum_{i=0}^{j} [\delta_i + v_i(\tau)]\, W_{j-i}(\tau) \tag{13.46}$$

and

$$R_n^v(\vec{u}_{n-1}, \vec{v}_{n-1}, \vec{\alpha}_{n-1}, \vec{\delta}_{n-1}, \vec{\omega}_{n-1})$$

$$= \frac{1}{(n-1)!} \frac{d^{n-1} \mathcal{N}_v \left[U(\tau;q), V(\tau;q), A(q), \Delta(q), \Omega(q) \right]}{d\, q^{n-1}}$$

$$= \sum_{j=0}^{n-1} \omega_j\, v'_{n-1-j}(\tau) + \epsilon \sum_{j=0}^{n-1} [\delta_j\, \delta_{n-1-j} + 2\delta_j\, v_{n-1-j}(\tau)]$$

$$+ \epsilon \sum_{j=0}^{n-1} [v_j(\tau)\, v_{n-1-j}(\tau) - \mu\, W_j(\tau)\, W_{n-1-j}(\tau)], \tag{13.47}$$

under the definitions

$$F_k(\tau) = \sum_{j=0}^{k} \omega_{k-j}\, u'_j(\tau), \tag{13.48}$$

$$W_k(\tau) = \sum_{j=0}^{k} \left(\sum_{m=0}^{k-j} \alpha_m\, \alpha_{k-j-m} \right) \left[\sum_{n=0}^{j} u_n(\tau)\, u_{j-n}(\tau) \right]. \tag{13.49}$$

It should be emphasized that the linear high-order deformation equations (13.43) and (13.44) are uncoupled and can be easily solved.

There are five unknowns: $u_n(\tau), v_n(\tau), \alpha_{n-1}, \delta_{n-1}$, and ω_{n-1}, and we have only Equations (13.43), (13.44), and (13.45) for $u_n(\tau)$ and $v_n(\tau)$. Therefore, the problem is not closed and three additional algebraic equations are needed to determine $\alpha_{n-1}, \delta_{n-1}$, and ω_{n-1}. Under the *rules of solution expression* denoted by (13.11) and (13.12) and from Equations (13.43) and (13.44), $H_u(\tau)$ and $H_v(\tau)$ may be sine and cosine functions. For simplicity, we select

$$H_u(\tau) = H_v(\tau) = 1. \tag{13.50}$$

When $n = 1$, by substituting (13.13) into (13.46) and (13.47), we gain

$$R_1^u = a_{1,0}\, \cos \tau + b_{1,0} \sin \tau + b_{1,1} \sin(3\tau) \tag{13.51}$$

and

$$R_1^v = c_{1,0} + c_{1,1}\, \cos(2\tau) + c_{1,2} \cos(4\tau), \tag{13.52}$$

where $a_{1,0}$, $b_{1,0}$, $b_{1,1}$, $c_{1,0}$, $c_{1,1}$, and $c_{1,2}$ are coefficients independent of τ. If $a_{1,0} \neq 0$ and $b_{1,0} \neq 0$, from the property (13.16) of $\mathcal{L}_u$, the solution $u_1(\tau)$ of Equation (13.43) contains the so-called secular terms $\tau \sin \tau$ and $\tau \cos \tau$, which do not conform to the *rule of solution expression* denoted by (13.11). Moreover, if $c_{1,0} \neq 0$, from the property (13.16) of $\mathcal{L}_v$, the solution $v_1(\tau)$ of Equation (13.44) contains the secular term $c_{1,0} \tau$, which disregards the *rule of solution expression* denoted by (13.12). In order to conform to the *rules of solution expression* denoted by (13.43) and (13.44), we must enforce

$$a_{1,0} = 0, \quad b_{1,0} = 0, \quad c_{1,0} = 0,$$

which provides us with three additional algebraic equations

$$\omega_0 - \frac{\alpha_0^2 \delta_0}{4} = 0, \quad \omega_0^2 - 1 = 0, \quad \delta_0^2 - \frac{3\alpha_0^4 \mu}{8} = 0, \tag{13.53}$$

whose solutions are

$$\alpha_0 = \frac{2}{\sqrt[8]{6\mu}}, \quad \delta_0 = \sqrt[4]{6\mu}, \quad \omega_0 = 1. \tag{13.54}$$

Now the problem is solved in accordance with the *rules of solution expression* denoted by (13.11) and (13.12). We now have

$$R_1^u = b_{1,1} \sin(3\tau)$$

and

$$R_1^v = c_{1,1} \cos(2\tau) + c_{1,2} \cos(4\tau).$$

Solving the first-order deformation equations (13.43) and (13.44) under the conditions noted in (13.45), we have

$$u_1(\tau) = -\left(\frac{\epsilon}{8}\right) \hbar_u \left(3 \sin \tau - \sin 3\tau\right) \tag{13.55}$$

and

$$v_1(\tau) = -\left(4\epsilon \sqrt{\frac{\mu}{6}}\right) \hbar_v \left(\sin 2\tau + \frac{1}{8} \sin 4\tau\right). \tag{13.56}$$

Similarly, first solving a set of linear algebraic equations

$$\epsilon^2 \left(3\hbar_u + \sqrt[4]{176\mu} \, \hbar_v\right) - 48 \, \omega_1 = 0, \tag{13.57}$$

$$(246\mu^3)^{1/8} \alpha_1 + (46)^{3/4} \delta_1 = 0, \tag{13.58}$$

$$(126\mu)^{3/8} \alpha_1 - \delta_1 = 0 \tag{13.59}$$

to gain α_1, δ_1, and ω_1, and then the remaining second-order deformation equations (13.43) to (13.45), we obtain $u_2(\tau)$ and $v_2(\tau)$. In this way we successively obtain $\alpha_{n-1}, \delta_{n-1}, \omega_{n-1}, u_n(\tau)$, and $v_n(\tau)$.

At the nth-order of approximation, $u(\tau)$ and $v(\tau)$ can be expressed by

$$u(\tau) = \sum_{k=0}^{M_n^u} [a_{n,k} \cos(2k+1)\tau + b_{n,k} \sin(2k+1)\tau]$$

and

$$v(\tau) = \sum_{k=1}^{M_n^v} [c_{n,k} \cos(2k\tau) + d_{n,k} \sin(2k\tau)],$$

where M_n^u and M_n^v are integers dependent upon the order n of approximation. Thus, the frequency of the motion $w(t)$ is twice that of $x(t)$.

13.1.3 Convergence theorem

THEOREM 13.1
If the solution series (13.35) to (13.39) are convergent, where $u_n(\tau)$ and $v_n(\tau)$ are governed by Equations (13.43) to (13.45) under the definitions (13.46) to (13.49), and (2.42), they must be the solution of Equations (13.7) to (13.10).

Proof: If the solution series (13.35) and (13.36) are convergent, then

$$\lim_{m \to +\infty} u_m(\tau) = 0, \qquad \lim_{m \to +\infty} v_m(\tau) = 0.$$

From Equation (13.43) and using the definitions (2.42) and (13.14), we then have

$$\hbar_u H_u(\tau) \sum_{n=1}^{+\infty} R_n^u(\vec{u}_{n-1}, \vec{v}_{n-1}, \vec{\alpha}_{n-1}, \vec{\delta}_{n-1}, \vec{\omega}_{n-1})$$

$$= \sum_{n=1}^{+\infty} \mathcal{L}_u [u_n(\tau) - \chi_n u_{n-1}(\tau)]$$

$$= \lim_{m \to +\infty} \sum_{n=1}^{m} \mathcal{L}_u [u_n(\tau) - \chi_n u_{n-1}(\tau)]$$

$$= \lim_{m \to +\infty} \mathcal{L}_u [u_m(\tau)]$$

$$= \mathcal{L}_u \left[\lim_{m \to +\infty} u_m(\tau) \right]$$

$$= 0,$$

which yields, since $\hbar_u \ne 0$ and $H_u(\tau) \ne 0$,

$$\sum_{n=1}^{+\infty} R_n^u(\vec{u}_{n-1}, \vec{v}_{n-1}, \vec{\alpha}_{n-1}, \vec{\delta}_{n-1}, \vec{\omega}_{n-1}) = 0.$$

Similarly,
$$\sum_{n=1}^{+\infty} R_n^v(\vec{u}_{n-1}, \vec{v}_{n-1}, \vec{\alpha}_{n-1}, \vec{\delta}_{n-1}, \vec{\omega}_{n-1}) = 0.$$

Substituting (13.46) and (13.47) into the above expressions and simplifying them, we have, due to the convergence of the series (13.37) to (13.39), that

$$\left(\sum_{i=0}^{+\infty} \omega_i\right)^2 \frac{d^2}{d\tau^2}\left[\sum_{j=0}^{+\infty} u_j(\tau)\right] + \sum_{j=0}^{+\infty} u_j(\tau)$$

$$= \epsilon \left(\sum_{i=0}^{+\infty} \omega_i\right) \frac{d}{d\tau}\left[\sum_{j=0}^{+\infty} u_j(\tau)\right]$$

$$\times \left\{ 1 - \left(\sum_{i=0}^{+\infty} \alpha_i\right)^2 \left(\sum_{j=0}^{+\infty} u_j\right)^2 \sum_{k=0}^{+\infty} [\delta_k + v_k(\tau)] \right\}$$

and

$$\left(\sum_{i=0}^{+\infty} \omega_i\right) \frac{d}{d\tau}\left[\sum_{j=0}^{+\infty} v_j(\tau)\right]$$

$$= -\epsilon \left\{ \left[\sum_{k=0}^{+\infty} \delta_k + \sum_{k=0}^{+\infty} v_k(\tau)\right]^2 - \mu \left(\sum_{i=0}^{+\infty} \alpha_i\right)^4 \left(\sum_{j=0}^{+\infty} u_j\right)^4 \right\}.$$

From (13.13) and (13.45), we have

$$\sum_{i=0}^{+\infty} u_i(0) = 1, \quad \sum_{i=0}^{+\infty} u_i'(0) = 0, \quad \int_0^{2\pi} \left[\sum_{i=0}^{+\infty} v_i(\tau)\right] d\tau = 0.$$

Comparing the above equations with Equations (13.7) to (13.10), it is obvious that the series (13.35) to (13.39) are the solutions. This ends the proof.

13.2 Result analysis

According to Theorem 13.1 we should ensure that the solution series (13.35) to (13.39) converge. Note that these solution series contain two auxiliary parameters $\hbar_u$ and $\hbar_v$. For simplicity, let

$$\hbar_u = \hbar_v = \hbar$$

so that the approximations of $u(\tau), v(\tau), \omega, \alpha,$ and δ are dependent only on $\hbar$. Generally, for any given physical parameters ϵ and μ, we first investigate the influence of the auxiliary parameter $\hbar$ on the convergence of the series by plotting the so-called $\hbar$-curves (see page 26 and §3.5.1) of $\alpha, \delta,$ and ω. For example, when $\epsilon = 1/5$ and $\mu = 3$ the $\hbar$-curves are as shown in Figure 13.1, clearly indicating the valid regions of $\hbar$ for the corresponding series of $\alpha, \delta,$ and ω. Obviously, when $\epsilon = 1/5$ and $\mu = 3$, the solution series (13.37) to (13.39) converge if $-3/2 < \hbar < 0$. For instance, when $\hbar_u = \hbar_v = -3/4$, the solution series of $\omega, \alpha,$ and δ converge to 0.96968, 1.41399, and 2.07015, respectively, as shown in Table 13.1. We can employ the so-called homotopy-Padé technique (see page 38 and §3.5.2) to accelerate the convergence, as shown in Table 13.2. As long as the solution series of $\alpha, \delta,$ and ω are convergent, the corresponding series of $u(\tau)$ and $v(\tau)$ given by the same value of $\hbar$ also converge, as shown in Figures 13.2 to 13.4 when $\epsilon = 1/5$ and $\mu = 3$.

In this way, for any given physical parameters ϵ and μ, we can gain convergent analytic results of the limit cycle of the two-dimensional dynamical system. As ϵ increases, the nonlinearity becomes stronger so that a higher order of approximation is necessary. For example, when $\epsilon = 3/4$ and $\mu = 1$, the $\hbar$-curves of $\alpha, \delta,$ and ω clearly indicate that the solution series converge when $\hbar = -3/4$, as shown in Figure 13.5. However, higher-order approximations are needed to get accurate enough results, as shown in Figures 13.6 to 13.8.

The homotopy analysis method was applied to solve limit cycles of one-dimensional systems [39]. This example shows that the homotopy analysis method may be employed to gain limit-cycles of multidimensional systems.

TABLE 13.1
The mth-order approximations of ω, α, and δ when $\epsilon = 1/5, \mu = 3$ by means of $\hbar_u = \hbar_v = -3/4$ and $H_u(\tau) = H_v(\tau) = 1$.

m	ω	α	δ
1	1.00000	1.39354	2.05977
2	0.97063	1.40476	2.06318
3	0.96966	1.41020	2.06458
4	0.96963	1.41251	2.06668
5	0.96968	1.41346	2.06843
6	0.96969	1.41382	2.06944
7	0.96969	1.41395	2.06989
8	0.96968	1.41398	2.07006
9	0.96968	1.41399	2.07011
10	0.96968	1.41399	2.07013
11	0.96968	1.41399	2.07014
12	0.96968	1.41399	2.07015
13	0.96968	1.41399	2.07015
14	0.96968	1.41399	2.07015

TABLE 13.2
The $[m, m]$ homotopy-Padé approximations of ω, α, and δ when $\epsilon = 1/5, \mu = 3$ by means of $\hbar_u = \hbar_v = -3/4$ and $H_u(\tau) = H_v(\tau) = 1$.

$[m, m]$	ω	α	δ
[1, 1]	0.96889	1.39354	2.05977
[2, 2]	0.96977	1.41413	2.08345
[3, 3]	0.96968	1.41414	2.08735
[4, 4]	0.96968	1.41399	2.07015
[5, 5]	0.96968	1.41398	2.07016
[6, 6]	0.96968	1.41399	2.07015
[7, 7]	0.96968	1.41399	2.07015

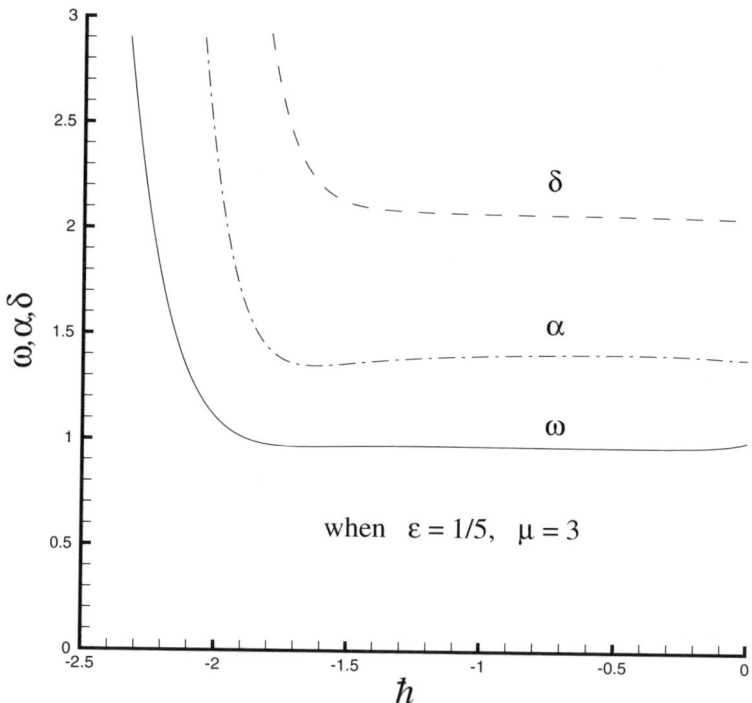

FIGURE 13.1
The $\hbar$-curves of ω, α, and δ at the 10th order of approximation when $\epsilon = 1/5$ and $\mu = 3$ by means of $H_u(\tau) = H_v(\tau) = 1$. Dashed line: δ; dash-dotted line: α; solid line: ω.

Limit Cycle in a Multidimensional System 209

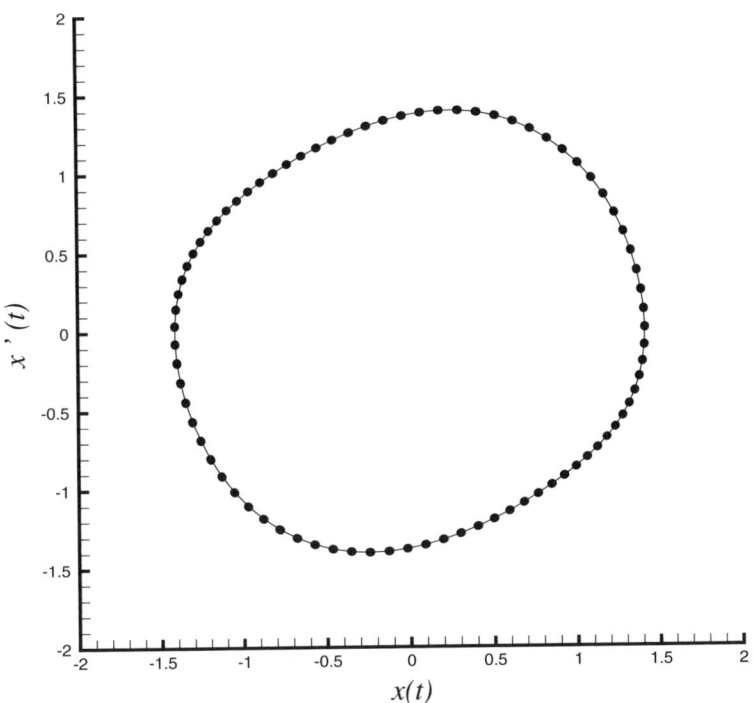

FIGURE 13.2
$x - \dot{x}$ plane projection of the limit cycle when $\epsilon = 1/5$ and $\mu = 3$. Solid line: fifth-order approximation by means of $\hbar_u = \hbar_v = -3/4$ and $H_u(\tau) = H_v(\tau) = 1$; symbols: numerical result.

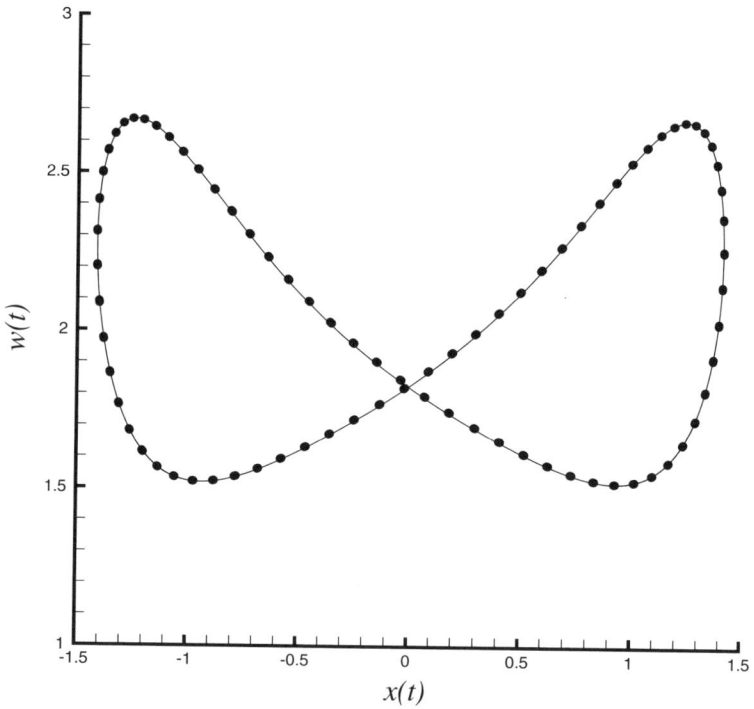

FIGURE 13.3
$x - w$ plane projection of the limit cycle when $\epsilon = 1/5$ and $\mu = 3$. Solid line: fifth-order approximation by means of $\hbar_u = \hbar_v = -3/4$ and $H_u(\tau) = H_v(\tau) = 1$; symbols: numerical result.

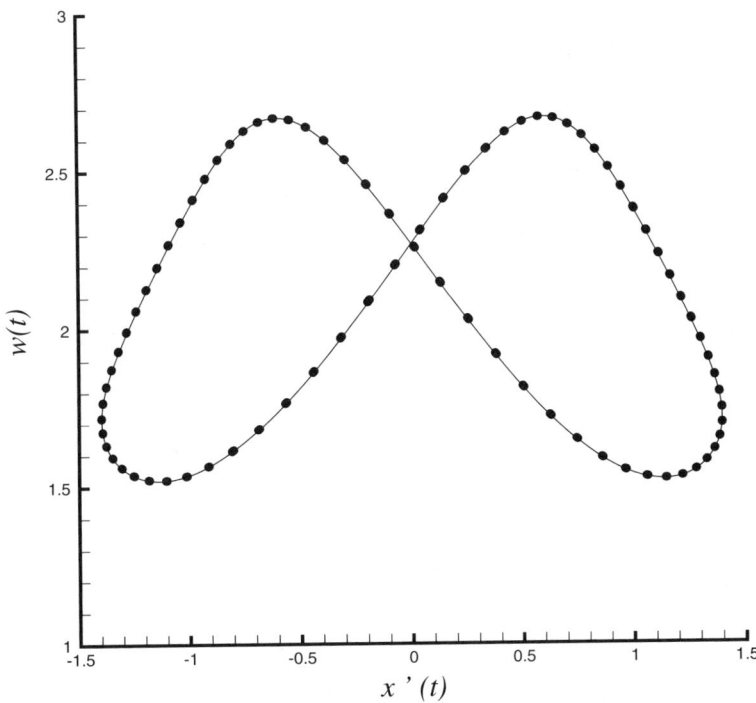

FIGURE 13.4
$\dot{x} - w$ plane projection of the limit cycle when $\epsilon = 1/5$ and $\mu = 3$. Solid line: fifth-order approximation by means of $\hbar_u = \hbar_v = -3/4$ and $H_u(\tau) = H_v(\tau) = 1$; symbols: numerical result.

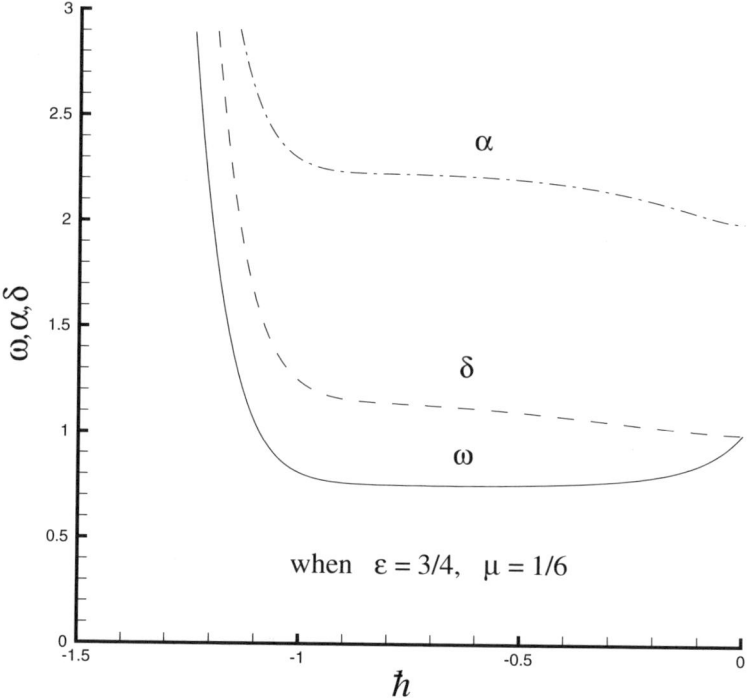

FIGURE 13.5
The $\hbar$-curves of ω, α, and δ at the 10th order of approximation when $\epsilon = 3/4$ and $\mu = 1/6$ by means of $H_u(\tau) = H_v(\tau) = 1$. Dashed line: δ; dash-dotted line: α; solid line: ω.

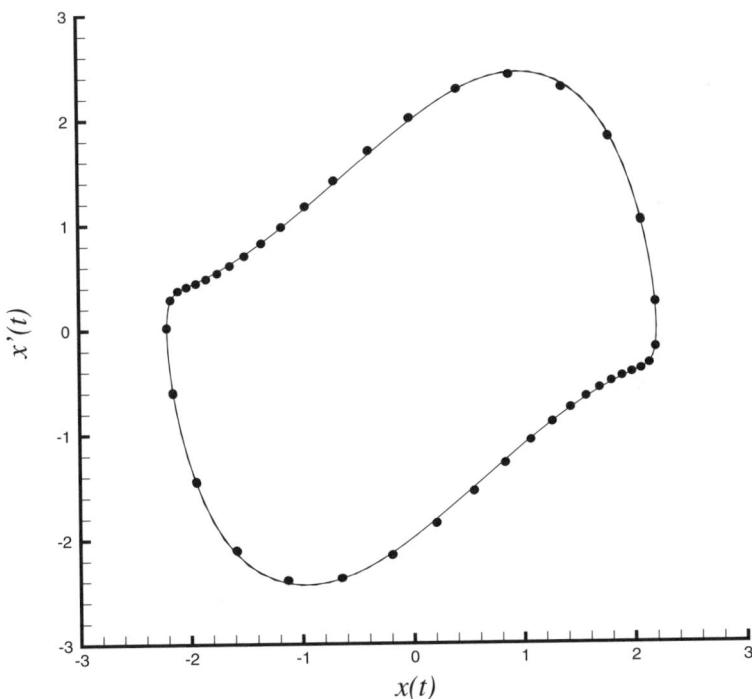

FIGURE 13.6
$x - \dot{x}$ plane projection of the limit cycle when $\epsilon = 3/4$ and $\mu = 1/6$. Solid line: 20th-order approximation by means of $\hbar_u = \hbar_v = -3/4$ and $H_u(\tau) = H_v(\tau) = 1$; Symbols: numerical result.

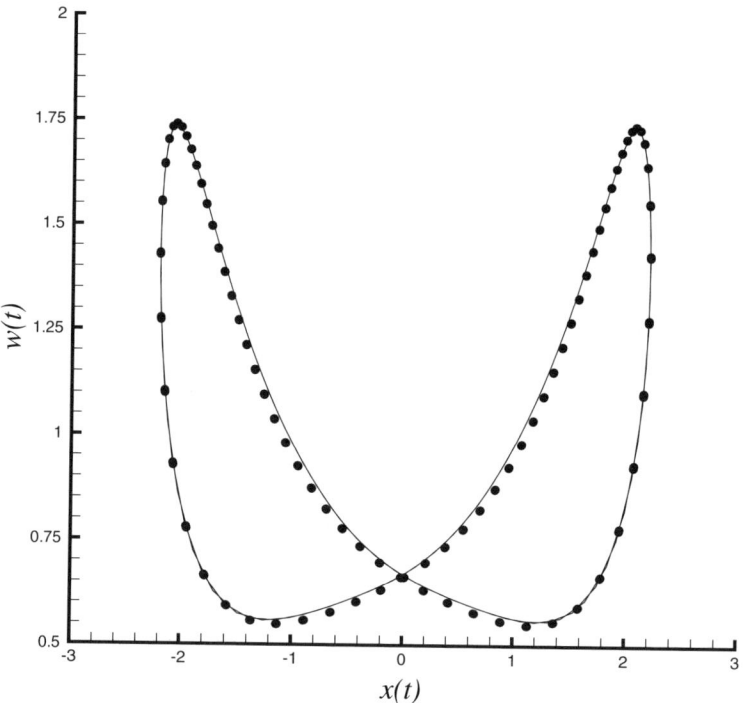

FIGURE 13.7
$x - w$ plane projection of the limit cycle when $\epsilon = 3/4$ and $\mu = 1/6$. Solid line: 20th-order approximation by means of $\hbar_u = \hbar_v = -3/4$ and $H_u(\tau) = H_v(\tau) = 1$; symbols: numerical result.

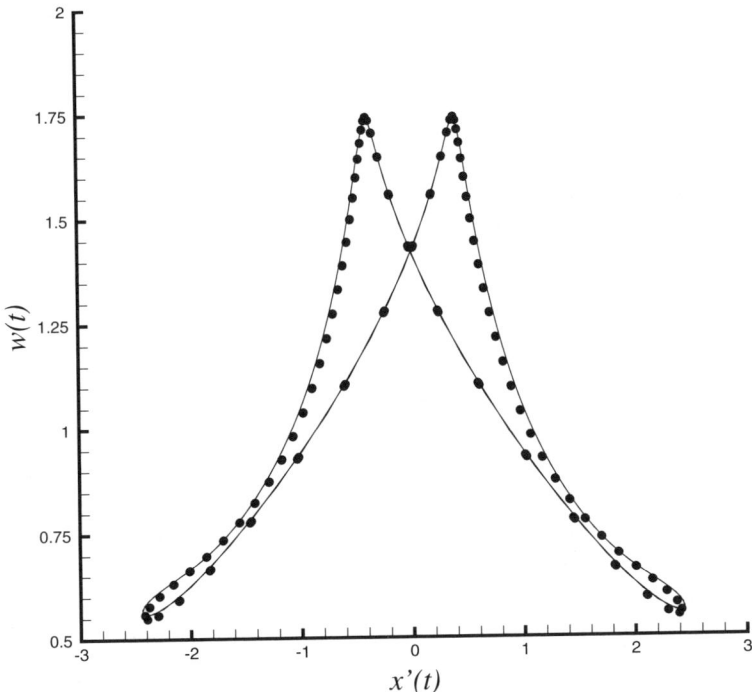

FIGURE 13.8
$\dot{x} - w$ plane projection of the limit cycle when $\epsilon = 3/4$ and $\mu = 1/6$. Solid line: 20th-order approximation by means of $\hbar_u = \hbar_v = -3/4$ and $H_u(\tau) = H_v(\tau) = 1$; symbols: numerical result.

14

Blasius' viscous flow

Consider the two-dimensional laminar viscous flow past a semi-infinite flat plate, governed by

$$f'''(\eta) + \frac{1}{2}f(\eta)f''(\eta) = 0, \tag{14.1}$$

subject to the boundary conditions

$$f(0) = f'(0) = 0, \quad f'(+\infty) = 1, \tag{14.2}$$

where the prime denotes the derivative with respect to the similarity variable $\eta = y\sqrt{U_\infty/(\nu x)}$, the dimensionless function $f(\eta)$ is related to the stream function $\psi(x,y)$ by $f(\eta) = \psi/\sqrt{\nu x U_\infty}$, U_∞ is the constant velocity of the mainstream at infinity, ν is the kinematic viscosity coefficient, and x and y are two independent variables. For details, the reader is referred to White [20].

In 1908, Blasius [104] provided a solution in power series

$$f(\eta) = \sum_{k=0}^{+\infty} \left(-\frac{1}{2}\right)^k \frac{A_k \sigma^{k+1}}{(3k+2)!} \eta^{3k+2}, \tag{14.3}$$

where $\sigma = f''(0)$ and

$$A_0 = A_1 = 1, \quad A_k = \sum_{r=0}^{k-1} \binom{3k-1}{3r} A_r A_{k-r-1} \quad (k \geq 2). \tag{14.4}$$

To get the unknown value of σ, Blasius [104] demonstrated another approximation of $f(\eta)$ for large η. Then, by means of matching two different approximations at a proper point, he obtained the numerical result $\sigma = 0.332$. In 1938, Howarth [105] gained a more accurate value $\sigma = 0.33206$ by means of a numerical technique. However, by means of $\sigma = 0.33206$, $f'(\eta)$ given by (14.3) is valid in a restricted region $0 \leq \eta < \rho_0$, where $\rho \approx 5.690$, as shown in Figure 14.1. Blasius' power series (14.3) is fundamentally an analytic-numerical solution, because the value of $\sigma = f''(0)$ is gained by numerical techniques.

In this chapter the homotopy analysis method is applied to yield the purely analytic solution expressions of Blasius' viscous flows by means of two different base functions.

14.1 Solution expressed by power functions
14.1.1 Zero-order deformation equation

Like Blasius [104], we express the solution of Equations (14.1) and (14.2) by the set of base functions

$$\{ \eta^{\alpha m+\beta} \mid m \geq 0 \} \tag{14.5}$$

in the form:

$$f(\eta) = \sum_{k=0}^{+\infty} a_k \, \eta^{\alpha k+\beta}, \tag{14.6}$$

where a_k is a coefficient, $\alpha > 0$ and $\beta \geq 0$ are constants. This provides us with the first *rule of solution expression* of Blasius' viscous flows.

Under the first *rule of solution expression* and using (14.2), it is easy to choose

$$f_0(\eta) = \frac{1}{2}\sigma\eta^2 \tag{14.7}$$

as the initial guess of $f(\eta)$, where $\sigma = f''(0)$. Then, under the first *rule of solution expression* denoted by (14.6) and from Equations (14.1) and (14.2), we choose the auxiliary linear operator

$$\mathcal{L}_0[\Phi(\eta;q)] = \frac{\partial^3 \Phi(\eta;q)}{\partial \eta^3} \tag{14.8}$$

with the property

$$\mathcal{L}_0 \left(C_0 + C_1 \eta + C_2 \eta^2 \right) = 0, \tag{14.9}$$

where C_0, C_1, and C_2 are coefficients, $\Phi(\eta;q)$ is a real function of η and q, $q \in [0,1]$ is an embedding parameter. For brevity, we define from Equation (14.1) the nonlinear operator

$$\mathcal{N}[\Phi(\eta;q)] = \frac{\partial^3 \Phi(\eta;q)}{\partial \eta^3} + \frac{1}{2}\Phi(\eta;q)\frac{\partial^2 \Phi(\eta;q)}{\partial \eta^2}. \tag{14.10}$$

Let $\hbar$ denote a nonzero auxiliary parameter and $H(\eta)$ a nonzero auxiliary function. We construct the so-called zero-order deformation equation

$$(1-q)\,\mathcal{L}_0\left[\Phi(\eta;q) - f_0(\eta)\right] = q\,\hbar\,H(\eta)\,\mathcal{N}[\Phi(\eta;q)], \tag{14.11}$$

subject to the boundary conditions

$$\Phi(0;q) = 0, \quad \left.\frac{\partial \Phi(\eta;q)}{\partial \eta}\right|_{\eta=0} = 0, \quad \left.\frac{\partial^2 \Phi(\eta;q)}{\partial \eta^2}\right|_{\eta=0} = \sigma, \tag{14.12}$$

where $q \in [0,1]$ is an embedding parameter.

When $q = 0$, it is easy to demonstrate from (14.7), (14.11), and (14.12) that
$$\Phi(\eta; 0) = f_0(\eta). \tag{14.13}$$

When $q = 1$, since $\hbar \neq 0$ and $H(\eta) \neq 0$, the zero-order deformation equations (14.11) and (14.12) are equivalent to Equations (14.1) and (14.2), respectively, provided
$$\Phi(\eta; 1) = f(\eta). \tag{14.14}$$

As the embedding parameter q increases from 0 to 1, $\Phi(\eta; q)$ varies from the initial guess $f_0(\eta)$ to the exact solution $f(\eta)$.

Using Taylor's theorem and Equation (14.13), we expand $\Phi(\eta; q)$ in the power series
$$\Phi(\eta; q) = f_0(\eta) + \sum_{k=1}^{+\infty} f_k(\eta)\, q^k, \tag{14.15}$$

where
$$f_k(\eta) = \frac{1}{k!} \left. \frac{\partial^k \Phi(\eta; q)}{\partial q^k} \right|_{q=0}. \tag{14.16}$$

Note that the zero-order deformation equation (14.11) contains the auxiliary parameter $\hbar$ and the auxiliary function $H(\eta)$. Assuming that both $\hbar$ and $H(\eta)$ are properly chosen so that the series (14.15) is convergent at $q = 1$, we have, using (14.14), that
$$f(\eta) = f_0(\eta) + \sum_{k=1}^{+\infty} f_k(\eta). \tag{14.17}$$

The mth-order approximation is given by
$$f(\eta) \approx f_0(\eta) + \sum_{k=1}^{m} f_k(\eta). \tag{14.18}$$

14.1.2 High-order deformation equation

For conciseness, define the vector
$$\vec{f}_n = \{f_0(\eta), f_1(\eta), f_2(\eta), \cdots, f_n(\eta)\}. \tag{14.19}$$

Differentiating the zeroth-order deformation equations (14.11) and (14.12) k times with respect to q, then setting $q = 0$, and finally dividing them by $k!$, we gain the high-order deformation equation
$$\mathcal{L}_0 \left[f_k(\eta) - \chi_k\, f_{k-1}(\eta) \right] = \hbar\, H(\eta)\, R_k(\vec{f}_{k-1}), \tag{14.20}$$

subject to the boundary conditions
$$f_k(0) = f_k'(0) = f_k''(0) = 0, \tag{14.21}$$

where χ_k is defined by (2.42) and

$$R_k(\vec{f}_{k-1}) = f'''_{k-1}(\eta) + \frac{1}{2}\sum_{n=0}^{k-1} f_n(\eta) f''_{k-1-n}(\eta). \tag{14.22}$$

Using (14.8), the solution of (14.20) is

$$f_k(\eta) = \chi_k f_{k-1}(\eta) + \hbar \int\int\int H(\eta)\, R_k(\vec{f}_{k-1})\, d\eta \\ + C_0 + C_1\eta + C_2\eta^2, \tag{14.23}$$

where the coefficients C_0, C_1, and C_2 are determined by the boundary conditions (14.21), which then yields $f_k(\eta)$.

14.1.3 Convergence theorem

THEOREM 14.1
If the solution series (14.17) converges, where $f_k(\eta)$ is governed by Equations (14.20) and (14.21) under the definitions (14.22) and (2.42), it must be the solution of Equations (14.1) and (14.2).

Proof: From (2.42) and (14.20), we have

$$\hbar\, H(\eta) \sum_{k=1}^{m} R_k(\vec{f}_{k-1}) = \mathcal{L}[f_m(\eta)].$$

If the series (14.17) converges, then

$$\lim_{m\to+\infty} f_m(\eta) = 0.$$

Using (14.8), we have

$$\hbar\, H(\eta) \sum_{k=1}^{+\infty} R_k(\vec{f}_{k-1}) = \lim_{m\to+\infty}\mathcal{L}[f_m(\eta)] = \mathcal{L}\left[\lim_{m\to+\infty} f_m(\eta)\right] = 0,$$

which yields, since $\hbar \neq 0$ and $H(\eta) \neq 0$,

$$\sum_{k=1}^{+\infty} R_k(\vec{f}_{k-1}) = 0.$$

Substituting (14.22) into the above expression and simplifying it, we obtain

$$\frac{d^3}{d\eta^3}\left[\sum_{k=0}^{+\infty} f_k(\eta)\right] + \frac{1}{2}\left[\sum_{k=0}^{+\infty} f_k(\eta)\right]\frac{d^2}{d\eta^2}\left[\sum_{k=0}^{+\infty} f_k(\eta)\right] = 0.$$

Blasius' Viscous Flow

From (14.7) and (14.21), we have

$$\sum_{k=0}^{+\infty} f_k(0) = \sum_{k=0}^{+\infty} f'_k(0) = 0, \quad \sum_{k=0}^{+\infty} f''_k(0) = \sigma.$$

Therefore, the solution series

$$f_0(\eta) + \sum_{k=1}^{+\infty} f_k(\eta)$$

must be the exact solution of Equations (14.1) and (14.2), as long as it is convergent. This ends the proof.

14.1.4 Result analysis

According to Theorem 14.1, we need only to focus on the convergence of the solution series (14.17) by properly choosing $\hbar$ and $H(\eta)$. Under the first *rule of solution expression* denoted by (14.6), the auxiliary function $H(\eta)$ takes the form

$$H(\eta) = \eta^\kappa,$$

where κ is a constant. When $\kappa < 0$, the solution of Equations (14.20) and (14.21) contains the term

$$\eta \ln \eta,$$

which, however, disobeys the first *rule of solution expression* denoted by (14.6). We have therefore

$$\kappa \geq 0. \tag{14.24}$$

Note that there exist two auxiliary parameters: $\hbar$ and κ. We gain therefore a two-parameter family of solution expressions.

14.1.4.1 Solution expression when $H(\eta) = 1$

Consider the case of $\kappa = 0$, corresponding to $H(\eta) = 1$. In this case, Liao [28] found that the mth-order approximation (14.18) may be expressed by

$$f(\eta) \approx \sum_{k=0}^{m} \left[\left(-\frac{1}{2}\right)^k \frac{A_k \sigma^{k+1}}{(3k+2)!} \eta^{3k+2} \right] \mu_0^{m,k}(\hbar), \tag{14.25}$$

where $\mu_0^{m,k}(\hbar)$ is exactly the same as the expression (2.58) defined in Chapter 2 on page 22. We have the exact solution

$$f(\eta) = \lim_{m \to +\infty} \sum_{k=0}^{m} \left[\left(-\frac{1}{2}\right)^k \frac{A_k \sigma^{k+1}}{(3k+2)!} \eta^{3k+2} \right] \mu_0^{m,k}(\hbar). \tag{14.26}$$

The above expression provides us with a one-parameter family of solution expressions, although the solution of Equations (14.1) and (14.2) is unique. Note that $\mu_0^{m,k}(\hbar)$ appears once again. As proved in Chapter 2 on page 22,

$$\mu_0^{m,k}(-1) = 1.$$

Thus, when $\hbar = -1$, the solution series (14.26) is exactly the same as that of Blasius' one (14.3). Therefore, Blasius' solution (14.3) is a special case of (14.26). As pointed out by Liao [28], the solution (14.26) is valid in the region

$$\rho_0 \leq \eta \leq \rho_0 \left[\frac{2}{|\hbar|} - 1 \right]^{1/3} \qquad (-2 < \hbar < 0), \qquad (14.27)$$

where $\rho_0 \approx 5.690$ is the convergence radius of Blasius' power series (14.3). Thus, as $\hbar$ varies from -1 to 0, the convergence region of the solution series (14.26) enlarges from $\eta \in [-\rho_0, \rho_0]$ to $\eta \in [-\rho_0, +\infty)$, as shown in Figure 14.1. We are able to adjust and control the convergence region of the solution series (14.26) through use of the auxiliary parameter $\hbar$.

The power series (14.26) can be theoretically valid in the whole region $\eta \in [0, +\infty)$. Unlike Blasius [104], we do not need an additional solution for large η any more. Using the condition $f'(+\infty) = 1$, we can gain the value of $f''(0)$ by numerically solving the algebraic equation

$$\sum_{k=0}^{m} \left[\left(-\frac{1}{2} \right)^k \frac{A_k \sigma^{k+1}}{(3k+1)!} \eta_0^{3k+1} \right] \mu_0^{m,k}(\hbar) = 1 \qquad (14.28)$$

for a proper value of $\hbar$ at a point $\eta = \eta_0$ far enough from $\eta = 0$. For a large enough order m of approximation and a small enough $\hbar$ ($-1 \leq \hbar < 0$), the above equation at two points $\eta = 8$ and $\eta = 9$ gives the same value $\sigma = f''(0) = 0.33206$ that agrees with Howarth's numerical result [105], as shown in Table 14.1.

14.1.4.2 Solution expression when $H(\eta) = \eta$

Consider the case of $\kappa = 1$, corresponding to $H(\eta) = \eta$. Here, the mth-order approximation can be expressed by

$$f(\eta) \approx \frac{\sigma}{2} \eta^2 + \sum_{k=6}^{4m+2} b_{m,k}(\hbar) \eta^k, \qquad (14.29)$$

where $b_{m,k}(\hbar)$ is a coefficient dependent of $\hbar$. Note that this solution does not contain the term η^5 and thus is different from the solution series (14.25). It also provides a new one-parameter family of solution expressions in $\hbar$. As $\hbar$ increases from -1 to 0, the convergence region of the solution series (14.29) enlarges, as shown in Figure 14.2. As $\hbar$ tends to 0 from below, the solution series (14.29) converges to the exact solution in the whole region $0 \leq \eta < +\infty$ as in the solution series (14.25).

Blasius' Viscous Flow

14.1.4.3 Solution expression when $H(\eta) = \sqrt{\eta}$

Consider the case of $\kappa = 1/2$, corresponding to $H(\eta) = \sqrt{\eta}$. Here, the mth-order approximation can be expressed by

$$f(\eta) \approx \frac{\sigma}{2}\eta^2 + \sum_{k=11}^{7m+4} c_{m,k}(\hbar)\,(\sqrt{\eta}\,)^k, \qquad (14.30)$$

where $c_{m,k}(\hbar)$ is a coefficient dependent of $\hbar$. Note that this solution expression contains the term $\eta^{11/2}$ and thus is different from the solution series (14.25) and (14.29). It provides another one-parameter family of solution expressions in $\hbar$. As $\hbar$ increases from -1 to 0, the convergence region of the solution series (14.30) also enlarges, as shown in Figure 14.3. As $\hbar$ tends to 0 from below, the solution series (14.30) converges to the exact solution in the whole region $0 \leq \eta < +\infty$, as in the series (14.25) and (14.29)

In general, for any given $\kappa \geq 0$, the corresponding solution series converges to the exact solution in the whole region $0 \leq \eta < +\infty$ as $\hbar$ tends to zero from below. And, for given value of $\hbar$, the convergence region of the solution series given by $H(\eta) = 1$ appears to be the largest.

It should be emphasized that the solution series (14.29) and (14.30) are quite different from Blasius solution (14.3), which is a Taylor series. Note that, both (14.29) and (14.30) can converge in the whole region $0 \leq \eta < +\infty$ and therefore are better than the Taylor series (14.3).

14.2 Solution expressed by exponentials and polynomials

14.2.1 Asymptotic property

Although the solution series (14.26), (14.29), and (14.30) given by power functions may be valid in the whole region $\eta \in [0, +\infty)$, as in Blasius' solution (14.3), it is still an analytic-numerical solution, because $\sigma = f''(0)$ had to be given by numerical techniques. Their convergence regions are dependent on $\hbar$, and when $|\hbar|$ is small, a large number of terms are needed to gain an accurate approximation for a large η. Therefore, they are not efficient solution expressions of Equations (14.1) and (14.2). This is mainly because the base functions defined by (14.5) do not automatically satisfy the boundary condition $f'(+\infty) = 1$ at infinity.

For very large η, Blasius [104] established from (14.1) and the boundary condition $f'(+\infty) = 1$ that

$$f'(\eta) \approx 1 + A\int \exp(-\eta^2/4)d\eta,$$

where A is an integral constant. Thus, $f(\eta) \to \eta$ exponentially as $\eta \to +\infty$. This is an important asymptotic property of $f(\eta)$. The velocity of the boundary layer physically tends to the mainstream velocity, exponentially.

To ensure that $f(\eta) \to \eta$ exponentially as $\eta \to +\infty$, we express $f(\eta)$ by the set of base functions

$$\{\eta, \eta^n \exp(-m\,\lambda\,\eta) \mid n \geq 0, m \geq 1, \lambda > 0\} \tag{14.31}$$

in the form:

$$f(\eta) = \eta + \sum_{m=1}^{+\infty} \sum_{n=0}^{+\infty} a_{m,n}\, \eta^n\, \exp(-m\,\lambda\,\eta), \tag{14.32}$$

where $\lambda > 0$ is the so-called spatial-scale parameter and $a_{m,n}$ is a coefficient. This provides us with the second *rule of solution expression* of Blasius viscous flow.

14.2.2 Zero-order deformation equation

According to the second *rule of solution expression* and using (14.2), it is easy to choose

$$\hat{f}_0(\eta) = \eta + \frac{1 - \exp(-\lambda\,\eta)}{\lambda} \tag{14.33}$$

as the initial guess of $f(\eta)$. Under the second *rule of solution expression* denoted by (14.32) and from Equations (14.1) and (14.2), we select the auxiliary linear operator

$$\hat{\mathcal{L}}[\Phi(\eta;q)] = \frac{\partial^3 \Phi(\eta;q)}{\partial \eta^3} + \lambda \frac{\partial^2 \Phi(\eta;q)}{\partial \eta^2} \tag{14.34}$$

with the property

$$\hat{\mathcal{L}}\left[C_0 + C_1 \eta + C_2 \exp(-\lambda\,\eta)\right] = 0, \tag{14.35}$$

where C_0, C_1, and C_2 are coefficients. Let $\hbar$ and $\hat{H}(\eta)$ denote a nonzero auxiliary parameter and a nonzero auxiliary function, respectively. Using the same definition $\mathcal{N}$ as (14.10), we construct the zero-order deformation equation

$$(1-q)\,\hat{\mathcal{L}}\left[\hat{\Phi}(\eta;q) - \hat{f}_0(\eta)\right] = q\,\hbar\,\hat{H}(\tau)\,\mathcal{N}[\hat{\Phi}(\eta;q)], \tag{14.36}$$

subject to the boundary conditions

$$\hat{\Phi}(0;q) = 0, \quad \left.\frac{\partial \hat{\Phi}(\eta;q)}{\partial \eta}\right|_{\eta=0} = 0, \quad \left.\frac{\partial \hat{\Phi}(\eta;q)}{\partial \eta}\right|_{\eta=+\infty} = 1, \tag{14.37}$$

where $q \in [0,1]$ is an embedding parameter, $\hat{\Phi}(\eta;q)$ is a real function of η and q.

Blasius' Viscous Flow

As shown in §14.1.1, we have the relationship

$$f(\eta) = \hat{f}_0(\eta) + \sum_{k=1}^{+\infty} \hat{f}_k(\eta), \qquad (14.38)$$

where

$$\hat{f}_k(\eta) = \frac{1}{k!} \left. \frac{\partial^k \hat{\Phi}(\eta; q)}{\partial q^k} \right|_{q=0}. \qquad (14.39)$$

14.2.3 High-order deformation equation

Define the vector

$$\mathbf{f}_n = \left\{ \hat{f}_0(\eta), \hat{f}_1(\eta), \hat{f}_2(\eta), \cdots, \hat{f}_n(\eta) \right\}.$$

Similarly, differentiating the zero-order deformation equations (14.36) and (14.37) k times with respect q, then setting $q = 0$, and finally dividing by $k!$, we have the high-order deformation equation

$$\hat{\mathcal{L}}\left[\hat{f}_k(\eta) - \chi_k \hat{f}_{k-1}(\eta)\right] = \hbar\, \hat{H}(\eta)\, \hat{R}_k(\mathbf{f}_{k-1}), \qquad (14.40)$$

subject to the boundary conditions

$$\hat{f}_k(0) = \hat{f}'_k(0) = \hat{f}'_k(+\infty) = 0, \qquad (14.41)$$

where χ_k is defined by (2.42) and

$$\hat{R}_k(\mathbf{f}_{k-1}) = \hat{f}'''_{k-1}(\eta) + \frac{1}{2}\sum_{n=0}^{k-1} \hat{f}_n(\eta)\, \hat{f}''_{k-1-n}(\eta). \qquad (14.42)$$

14.2.4 Recursive expressions

Considering the wide applications of Blasius' viscous flows, it is helpful to express its solution explicitly. Under the second *rule of solution expression* denoted by (14.32), the auxiliary function $\hat{H}(\eta)$ may be in the form

$$\hat{H}(\eta) = \eta^m \, \exp(-\lambda\, n\, \eta), \qquad m \geq 0, \quad n \geq 0.$$

For simplicity, we select

$$\hat{H}(\eta) = 1. \qquad (14.43)$$

Thereafter, by solving the first several high-order deformation equations (14.40) and (14.41), $\hat{f}_m(\eta)$ can be expressed by

$$\hat{f}_m(\eta) = b_0^{m,0} + \sum_{n=1}^{m+1} \exp(-n\lambda\, \eta) \sum_{k=0}^{2(m+1-n)} b_k^{m,n}\, \eta^k, \qquad (14.44)$$

where $b_k^{m,n}$ is a coefficient. Substituting this expression into Equations (14.40) and (14.41), we obtain the following recurrence formulae

$$b_0^{m,0} = \chi_m b_0^{m-1,0} - \lambda^{-1} \sum_{r=0}^{2m-1} \Gamma_r^{m,1} \Pi_r^{1,1} - \sum_{n=2}^{m+1} (n-1) \Gamma_0^{m,n} \Pi_0^{n,0}$$

$$+ \sum_{n=2}^{m+1} \sum_{r=1}^{2(m-n+1)} \Gamma_r^{m,n} \left(n \Pi_r^{n,0} - \Pi_r^{n,0} - \lambda^{-1} \Pi_r^{n,1} \right),$$

$$b_0^{m,1} = \chi_m b_0^{m-1,1} + \lambda^{-1} \sum_{r=0}^{2m-1} \Gamma_r^{m,1} \Pi_r^{1,1}$$

$$+ \sum_{n=2}^{m+1} \left[n \Gamma_0^{m,n} \Pi_0^{n,0} + \sum_{r=1}^{2(m-n+1)} \Gamma_r^{m,n} \left(n \Pi_r^{n,0} - \lambda^{-1} \Pi_r^{n,1} \right) \right],$$

$$b_k^{m,1} = \chi_m (1 - \chi_{k+3-2m}) b_k^{m-1,1} + \sum_{r=k-1}^{2m-1} \Gamma_r^{m,1} \Pi_r^{1,k} \quad (1 \le k \le 2m),$$

$$b_k^{m,n} = \chi_m (1 - \chi_{k+1-2m+2n}) b_k^{m-1,n} - \sum_{r=k}^{2(m-n+1)} \Gamma_r^{m,n} \Pi_r^{n,k}$$

$$(2 \le n \le m, 0 \le k \le 2m - 2n + 2)$$

and

$$b_0^{m,m+1} = -\Gamma_0^{m,m+1} \Pi_0^{m+1,0},$$

where

$$\Pi_r^{1,k} = \frac{r! \, (r-k+2)}{k! \, \lambda^{r-k+3}} (0 \le k \le r+1),$$

$$\Pi_r^{n,k} = \frac{r!}{k!(n-1)^{r-k+1} \lambda^{r-k+3}} \left[1 - \left(1 - \frac{1}{n}\right)^{r-k+1} \left(1 + \frac{r-k+1}{n}\right) \right]$$

$$(n \ge 2, 0 \le k \le r),$$

$$\Gamma_r^{m,n} = \hbar \left[(1 - \chi_{r+1-2m+2n}) \, d_r^{m-1,n} + \delta_r^{m,n} \right]$$

$$(1 \le n \le m, 0 \le r \le 2m - 2n + 2),$$

in which

$$\delta_r^{m,n} = \frac{1}{2} \sum_{k=0}^{m-1} \sum_{j=\max\{1,n+k-m\}}^{\min\{n,k+1\}} \sum_{i=\max\{0,r-2(m-k-n+j)\}}^{\min\{r,2(k-j+1)\}}$$

$$\times c_i^{k,j} \, b_{r-i}^{m-1-k,n-j} \, \Lambda_{r-i}^{m-1-k,n-j}$$

under the definitions

$$c_n^{m,k} = (n+1)(n+2)(1-\chi_{n+1-2m+2k})\, b_{n+2}^{m,k}$$
$$-2(k\lambda)(n+1)(1-\chi_{n-2m+2k})\, b_{n+1}^{m,k} + (k\lambda)^2\, b_n^{m,k},$$

$$d_n^{m,k} = (n+1)(1-\chi_{n-2m+2k})\, c_{n+1}^{m,k} - k\lambda c_n^{m,k}$$

and

$$\Lambda_k^{i,j} = \begin{cases} 0, & \text{when } i = j = 0, k \geq 2, \\ 0, & \text{when } i > 0, j = 0, k \geq 1, \\ 0, & \text{when } j > i+1, \\ 0, & \text{when } k > 2(i+1-j), \\ 1, & \text{otherwise.} \end{cases} \qquad (14.45)$$

Using (14.33), we gain the first three coefficients

$$b_0^{0,0} = -\lambda^{-1}, \quad b_1^{0,0} = 1, \quad b_0^{0,1} = \lambda^{-1}. \qquad (14.46)$$

From these three coefficients and using the above recurrence formulae, we can calculate all coefficients $b_k^{m,n}$. For details, the reader is referred to Liao [29].

Substituting (14.44) into (14.38), we obtain the solution

$$f(\eta) = \eta$$
$$+ \lim_{M \to +\infty} \left[\sum_{m=0}^{M} b_0^{m,0} + \sum_{n=1}^{M+1} \exp(-n\lambda\,\eta) \left(\sum_{m=n-1}^{M} \sum_{k=0}^{2(m-n+1)} b_k^{m,n}\, \eta^k \right) \right]. \qquad (14.47)$$

Note that the coefficient $b_k^{m,n}$ contains the auxiliary parameter $\hbar$ and the so-called spatial-scale parameter λ. It provides us with a two-parameter family of solution expressions. Note that the solution series (14.47) is explicit and has the asymptotic property $f'(\eta) \to 1$ exponentially as $\eta \to +\infty$.

14.2.5 Convergence theorem

THEOREM 14.2
If the solution series (14.38) converges, where $\hat{f}_k(\eta)$ is governed by Equations (14.40) and (14.41) under the definitions (14.42) and (2.42), it must be the solution of Equations (14.1) and (14.2).

Proof: From (2.42) and (14.40), it holds that

$$\hbar\, \hat{H}(\eta) \sum_{k=1}^{m} \hat{R}_k(\mathbf{f}_{k-1}) = \hat{\mathcal{L}}\,[\hat{f}_m(\eta)].$$

If the series (14.38) converges, it is necessary that
$$\lim_{m \to +\infty} \hat{f}_m(\eta) = 0.$$
Then, using (14.34), we have
$$\hbar \, \hat{H}(\eta) \sum_{k=1}^{+\infty} \hat{R}_k(\mathbf{f}_{k-1}) = \lim_{m \to +\infty} \hat{\mathcal{L}}[\hat{f}_m(\eta)] = \hat{\mathcal{L}}\left[\lim_{m \to +\infty} \hat{f}_m(\eta)\right] = 0,$$
which gives, since $\hbar \neq 0$ and $\hat{H}(\eta) \neq 0$,
$$\sum_{k=1}^{+\infty} \hat{R}_k(\mathbf{f}_{k-1}) = 0.$$
Substituting (14.42) into the above expression and simplifying it, we obtain
$$\frac{d^3}{d\eta^3}\left[\sum_{k=0}^{+\infty} \hat{f}_k(\eta)\right] + \frac{1}{2}\left[\sum_{k=0}^{+\infty} \hat{f}_k(\eta)\right] \frac{d^2}{d\eta^2}\left[\sum_{k=0}^{+\infty} \hat{f}_k(\eta)\right] = 0.$$
From (14.33) and (14.41),
$$\sum_{k=0}^{+\infty} \hat{f}_k(0) = \sum_{k=0}^{+\infty} \hat{f}'_k(0) = 0, \quad \sum_{k=0}^{+\infty} \hat{f}'_k(+\infty) = 1.$$
Therefore, the series
$$\hat{f}_0(\eta) + \sum_{k=1}^{+\infty} \hat{f}_k(\eta)$$
must be the exact solution of Equations (14.1) and (14.2) as long as it is convergent. This ends the proof.

14.2.6 Result analysis

According to Theorem 14.2, we need only to focus on correctly choosing the auxiliary parameter $\hbar$ and the spatial-scale parameter λ so that the solution series (14.47) is convergent. For simplicity, we consider first the convergence of $f''(0)$ that is dependent on $\hbar$ and λ. We first set $\hbar = -1$ and regard λ as an unknown variable. For large enough λ such as $\lambda \geq 4$, the approximation of $f''(0)$ converges to the same value, as shown in Figure 14.4. In general, for given $\lambda \geq 4$, we investigate the influence of $\hbar$ on the convergence of the solution series (14.47) using the so-called $\hbar$-curves (see page 26 and §3.5.1) of $f''(0)$. For example, the $\hbar$-curves of $f''(0)$ when $\lambda = 4$ clearly indicate that the valid region of $\hbar$ is $-3/2 \leq \hbar \leq -1/2$, as shown in Figure 14.5. For instance, when $\lambda = 4$ and $\hbar = -1$, the approximation sequence of $f''(0)$ given by

Blasius' Viscous Flow

(14.47) converges to 0.332057, which agrees with Howarth's [105] numerical result $f''(0) = 0.33206$, as shown in Table 14.2. The larger the absolute value of $\hbar$, the faster the sequence of $f''(0)$ converges. When $\hbar = -3/2$ and $\lambda = 4$, we obtain the accurate result $f''(0) = 0.332057$ at the 25th-order of approximation.

The homotopy-Padé technique (see page 38 and §3.5.2) can be applied to accelerate the convergence of $f''(0)$, as shown in Table 14.3. It is found that the $[m, m]$ homotopy-Padé approximants of $f''(0)$ do not depend upon the auxiliary parameter $\hbar$. Besides, the convergence rate of the $[m, m]$ homotopy-Padé approximants of $f''(0)$ is not sensitive to λ, as shown in Table 14.4. From Figure 14.4, it is clear that $f''(0)$ is divergent when $\hbar = -1$ and $\lambda \leq 2$. When $\lambda = 2$ and $\hbar = -1$, the 30th-order approximation of $f''(0)$ is equal to -3.7×10^9. However, even when $\lambda \leq 2$ such as $\lambda = 1$ and $\lambda = 2$, the sequence of the homotopy-Padé approximants of $f''(0)$ still converges to 0.332057, as shown in Table 14.4. We see that, the $[m, m]$ homotopy-Padé approximant of $f''(0)$ is not only independent of the auxiliary parameter $\hbar$ but also insensitive to the auxiliary parameter λ.

As long as the sequence of $f''(0)$ is convergent, the corresponding solution series of $f(\eta)$ and $f'(\eta)$ also converges to Howarth's [105] numerical result in the whole region $0 \leq \eta < +\infty$. For example, when $\lambda = 4$ and $\hbar = -1$, $f'(\eta)$ converges to Howarth's [105] numerical result in the whole region $0 \leq \eta < +\infty$, as shown in Table 14.5. Similarly, we can apply the homotopy-Padé technique to accelerate the convergence rate of the series (14.47); the $[m, m]$ homotopy-Padé approximants of $f(\eta)$ and $f'(\eta)$ are independent of $\hbar$. As mentioned before, the approximation sequence of $f''(0)$ diverges when $\lambda = 2$ and $\hbar = -1$. However, by means of the Homotopy-Padé technique, we obtain convergent results of $f(\eta)$ in the whole region $0 \leq \eta < +\infty$ even when $\lambda = 2$ and $\hbar = -1$, as shown in Table 14.6. When $\lambda = 2$ and $\hbar = -1$, even the $[5,5]$ homotopy-Padé approximant of $f'(\eta)$ is accurate enough, as shown in Figure 14.6.

Note that the solution (14.47) is explicit, purely analytic, and uniformly valid in the *whole* region $\eta \in [0, +\infty)$. Thus, it can be regarded as a definition of the solution of Blasius' viscous flow problems governed by Equations (14.1) and (14.2).

This example demonstrates once again that, using the homotopy analysis method, we can obtain many different solution expressions of a nonlinear problem, even if the solution is unique.

TABLE 14.1
Numerical values of $f''(0)$ given by (14.28) for different $\hbar, \eta_0$, and the order m of approximation.

m	$\hbar = -\frac{1}{10}$ and $\eta_0 = 8$	$\hbar = -\frac{1}{12}$ and $\eta_0 = 9$
20	0.32881	0.32743
40	0.33185	0.33149
60	0.33205	0.33201
80	0.33206	0.33205
90	0.33206	0.33206
100	0.33206	0.33206

Source: reprinted from *International Journal of Non-Linear Mechanics*, 32, Shi-Jun Liao, "A kind of approximate solution technique which does not depend upon small parameters (II): An application in fluid mechanics", 815-822, Copyright (1997), with permission from Elsevier.

TABLE 14.2
Analytic approximations of $f''(0)$ given by (14.47) when $\hbar = -1$ and $\lambda = 4$.

Order of approximation	$f''(0)$
10th	0.327756
20th	0.331851
30th	0.332040
40th	0.332055
45th	0.332057
50th	0.332057
55th	0.332057

Source: Shi-Jun Liao, "A uniformly valid analytic solution of two-dimensional viscous flow past a semi-infinite flat plate", *Journal of Fluid Mechanics* (1999), 385:101-128 Cambridge University Press Copyright ©1999 Cambridge University Press, reprinted with permission.

TABLE 14.3
The $[m, m]$ homotopy-Padé approximation of $f''(0)$ when $\hbar = -1$ and $\lambda = 4$.

$[m, m]$	homotopy-Padé approximant of $f''(0)$
[4, 4]	0.344675
[8, 8]	0.332055
[12, 12]	0.332056
[16, 16]	0.332057
[20, 20]	0.332057
[25, 25]	0.332057

TABLE 14.4
The $[m, m]$ homotopy-Padé approximation of $f''(0)$ for different values of λ.

$[m, m]$	$\lambda = 1$	$\lambda = 2$	$\lambda = 4$	$\lambda = 5$	$\lambda = 10$
[4, 4]	0.326857	0.331867	0.344675	0.362964	0.519751
[8, 8]	0.331808	0.331753	0.332055	0.332269	0.347726
[12, 12]	0.332008	0.332056	0.332056	0.332053	0.332908
[16, 16]	0.332043	0.332057	0.332057	0.332057	0.332084
[20, 20]	0.332054	0.332057	0.332057	0.332057	0.332057
[25, 25]	0.332057	0.332057	0.332057	0.332057	0.332057

TABLE 14.5
Comparison of Howarth's [105] numerical results of $f'(\eta)$ with the analytic approximations given by (14.47) when $\hbar = -1$ and $\lambda = 4$.

η	20th order	30th order	40th order	50th order	55th order	numerical result
0.4	0.132650	0.132756	0.132763	0.132764	0.132764	0.1328
0.8	0.264412	0.264488	0.264707	0.264709	0.264709	0.2647
1.2	0.393075	0.393755	0.393772	0.393776	0.393776	0.3938
1.6	0.514758	0.516680	0.516750	0.516756	0.516756	0.5168
2.0	0.626372	0.629553	0.629754	0.629764	0.629764	0.6298
2.4	0.727156	0.728494	0.728950	0.728980	0.728980	0.7290
2.8	0.814839	0.810980	0.811429	0.811503	0.811503	0.8115
3.2	0.885026	0.876124	0.875982	0.876066	0.876066	0.8761
3.6	0.935172	0.924321	0.923315	0.923312	0.923312	0.9233
4.0	0.966854	0.957245	0.955665	0.955518	0.955518	0.9555
4.4	0.984622	0.977780	0.976154	0.975900	0.975900	0.9759
5.0	0.995914	0.992920	0.991856	0.991599	0.991599	0.9916
6.0	0.999708	0.999317	0.999092	0.999006	0.999006	0.9990
7.0	0.999987	0.999961	0.999939	0.999926	0.999926	1.0000
8.0	1.000000	0.999999	0.999998	0.999997	0.999997	1.0000

Source: Shi-Jun Liao, "A uniformly valid analytic solution of two-dimensional viscous flow past a semi-infinite flat plate", *Journal of Fluid Mechanics* (1999), 385:101-128 Cambridge University Press Copyright ©1999 Cambridge University Press, reprinted with permission.

TABLE 14.6
Comparison of Howarth's [105] numerical results with the $[m,m]$ homotopy-Padé approximation of $f'(\eta)$ when $\lambda = 2$ and $\hbar = -1$.

η	[5, 5]	[10, 10]	[15, 15]	[20, 20]	[25, 25]	numerical result
0.4	0.133023	0.132814	0.132764	0.132764	0.132764	0.1328
0.8	0.264655	0.264688	0.264709	0.264709	0.264709	0.2647
1.2	0.393380	0.393774	0.393775	0.393776	0.393776	0.3938
1.6	0.516251	0.516751	0.516755	0.516757	0.516757	0.5168
2.0	0.629577	0.629759	0.629765	0.629766	0.629766	0.6298
2.4	0.729388	0.728968	0.728981	0.728982	0.728982	0.7290
2.8	0.812514	0.811489	0.811508	0.811509	0.811510	0.8115
3.2	0.877471	0.876059	0.876079	0.876081	0.876081	0.8761
3.6	0.924790	0.923298	0.923325	0.923329	0.923330	0.9233
4.0	0.956803	0.955468	0.955523	0.955518	0.955518	0.9555
4.4	0.976987	0.975798	0.975872	0.975871	0.975870	0.9759
5.0	0.992848	0.991460	0.991542	0.991542	0.991542	0.9916
6.0	1.000400	0.998920	0.998972	0.998974	0.998973	0.9990
7.0	0.999989	0.999920	0.999920	0.999922	0.999921	1.0000
8.0	0.999998	0.999995	0.999996	0.999997	0.999996	1.0000

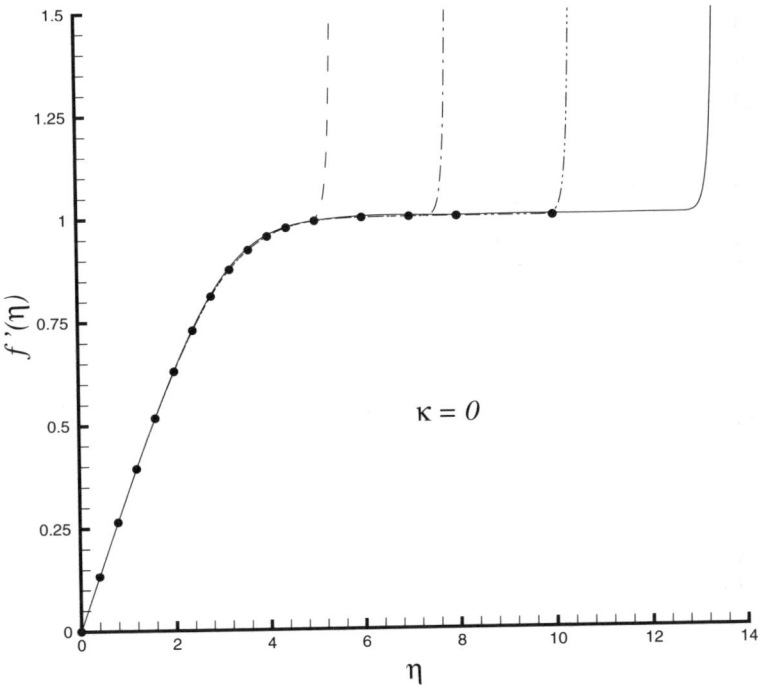

FIGURE 14.1
Comparison of Howarth's [105] numerical result of $f''(0)$ with the solution series (14.26) when $H(\eta) = 1$ by means of different values of $\hbar$. Symbols: numerical result; dashed line: $\hbar = -1$ (Blasius' power series); dash-dotted line: $\hbar = -1/2$; dash-dot-dotted line: $\hbar = -1/4$; solid line: $\hbar = -1/8$.

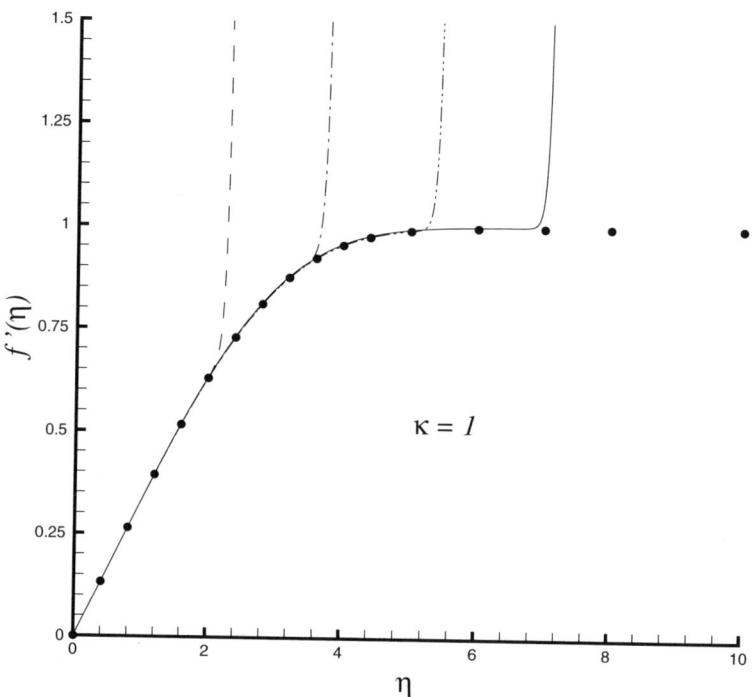

FIGURE 14.2
Comparison of Howarth's [105] numerical result of $f''(0)$ with the solution series (14.29) when $H(\eta) = \eta$ by means of different values of $\hbar$. Symbols: numerical result; dashed line: $\hbar = -1$ (Blasius' power series); dash-dotted line: $\hbar = -1/2$; dash-dot-dotted line: $\hbar = -1/4$; solid line: $\hbar = -1/8$.

Blasius' Viscous Flow

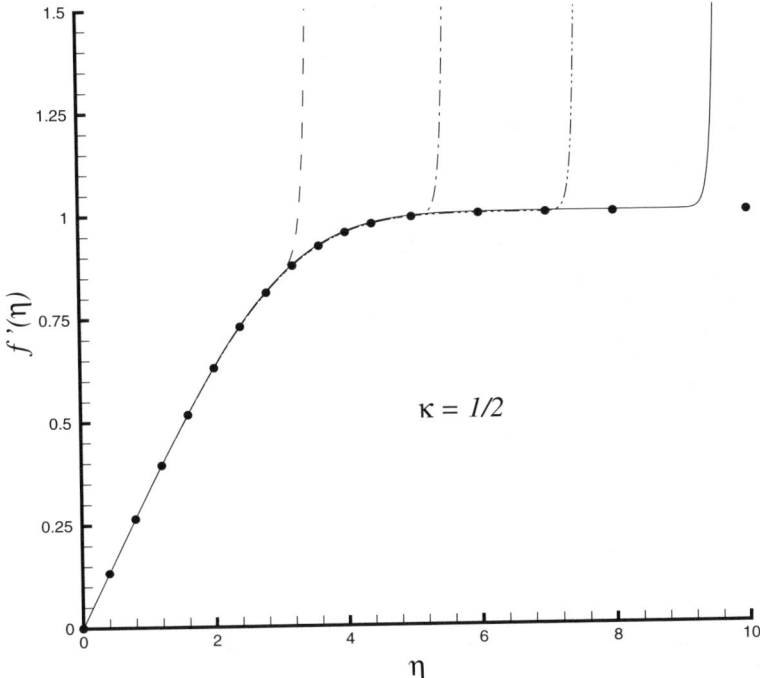

FIGURE 14.3
Comparison of Howarth's [105] numerical result of $f''(0)$ with the solution series (14.30) when $H(\eta) = \sqrt{\eta}$ by means of different values of $\hbar$. Symbols: numerical result; dashed line: $\hbar = -1$ (Blasius' power series); dash-dotted line: $\hbar = -1/2$; dash-dot-dotted line: $\hbar = -1/4$; solid line: $\hbar = -1/8$.

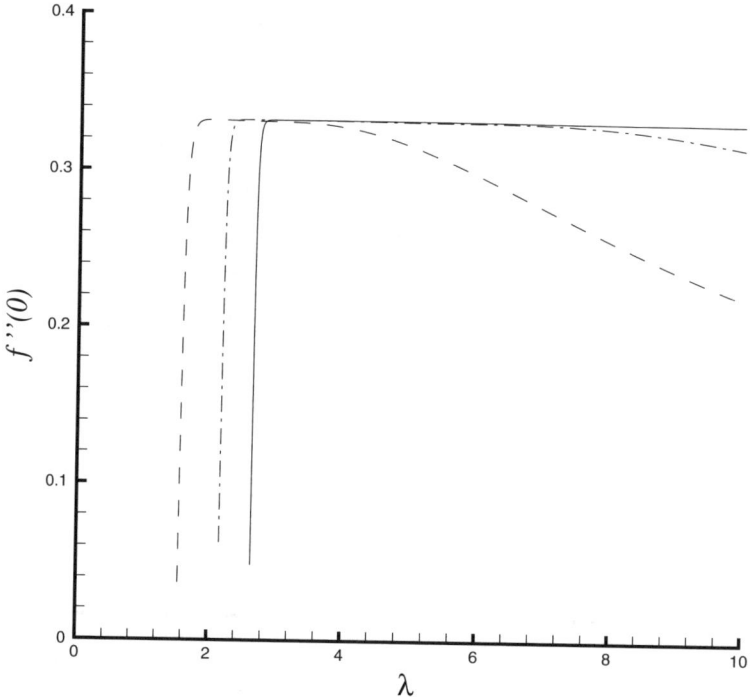

FIGURE 14.4
$f''(0)$ given by (14.47) for different λ when $\hbar = -1$. Dashed line: 10th-order approximation; dash-dotted line: 20th-order approximation; solid line: 30th-order approximation.

Blasius' Viscous Flow

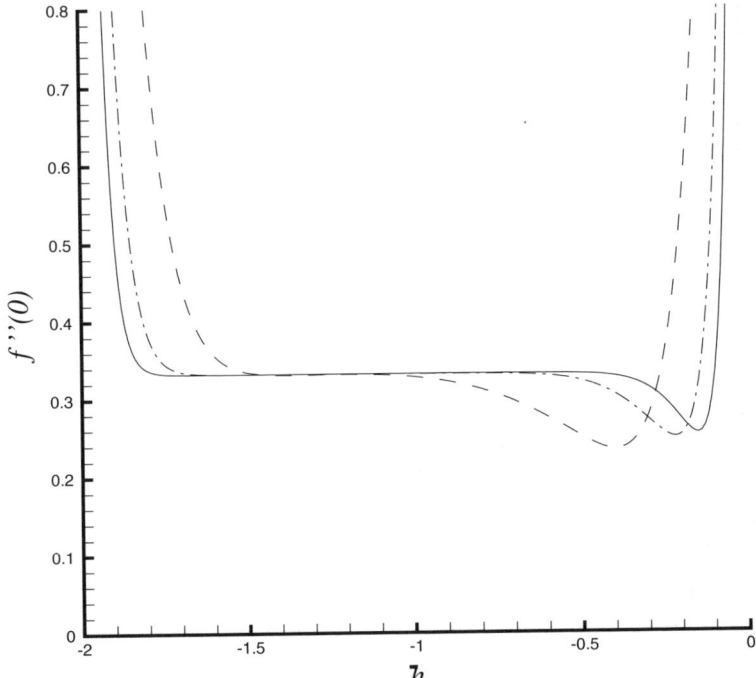

FIGURE 14.5
The $\hbar$-curves of $f''(0)$ given by (14.47) when $\lambda = 4$. Dashed line: 10th-order approximation; dash-dotted line: 20th-order approximation; solid line: 30th-order approximation.

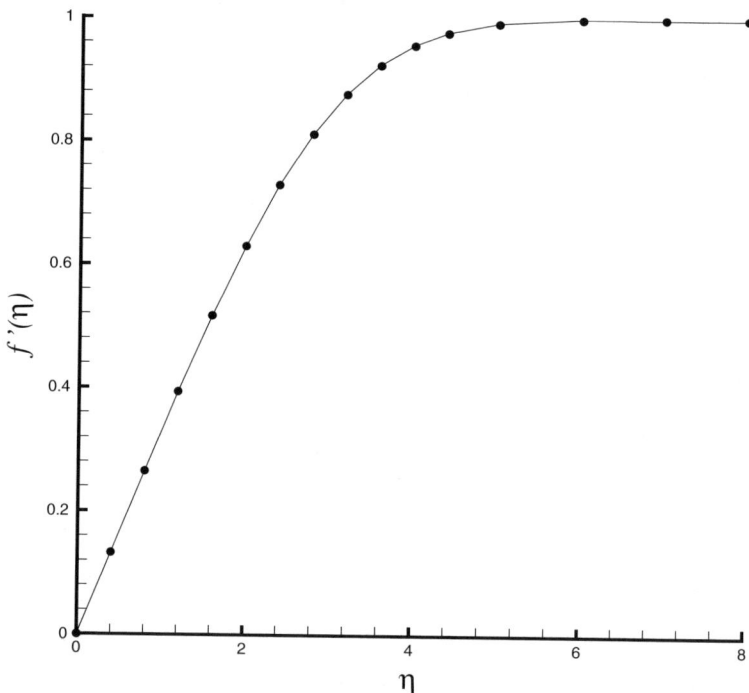

FIGURE 14.6
Comparison of Howarth's [105] numerical result of $f'(\eta)$ with the [5,5] homotopy-Padé approximation when $\lambda = 2$ and $\hbar = -1$. Symbols: numerical result; solid line: [5,5] homotopy-Padé approximation.

15

Boundary-layer flows with exponential property

Consider the two-dimensional laminar viscous flow over a semi-infinite flat plate. The family of similar solutions of the incompressible boundary layers was first obtained by Falkner and Skan [106] in 1931. Let x denote distance from the leading edge of a semi-infinite flat plate and y distance normal to the plate, U the velocity of the fluid in the mainstream, ν the kinematic viscosity, and u and v the components of the velocity of the fluid in the directions of x, y respectively. Falkner and Skan [106] demonstrated that, if $U \propto x^\kappa$, where κ is a constant, there exist solutions of the boundary layer equation

$$f'''(\eta) + f(\eta)f''(\eta) + \beta[1 - f'^2(\eta)] = 0, \qquad (15.1)$$

subject to the boundary conditions

$$f(0) = f'(0) = 0, \qquad f'(+\infty) = 1, \qquad (15.2)$$

where

$$\beta = \frac{2\kappa}{\kappa+1}, \qquad \eta = y\sqrt{\frac{(1+\kappa)U}{2\nu x}} \qquad (15.3)$$

and the prime denotes differentiation with respect to the similarity variable η. The components u, v of the fluid velocity are given by

$$u = Uf'(\eta), \qquad v = [f(\eta) - (\kappa-1)\eta f'(\eta)]\sqrt{\frac{\nu U}{2(\kappa+1)x}}. \qquad (15.4)$$

Note that $f(\eta)$ depends on the physical parameter β only. When $\kappa \geq 0$, from (15.3), it is easy to see that

$$0 \leq \beta \leq 2.$$

When $\kappa < 0$, the mainstream velocity $U \propto x^\kappa$ is singular at $x = 0$ so that Falkner-Skan's solution $f(\eta)$ cannot be taken right back to $x = 0$. This is a general difference between the solutions with a positive and negative β. In 1937 Hartree [107] numerically solved the Falkner-Skan's equations. For large η, Hartree [107] provided the asymptotic expression

$$1 - f'(\eta) \approx A\, \exp(-f^2/2)\, f^{-(2\beta+1)} + B\, f^{2\beta}, \qquad (15.5)$$

where A and B are coefficients. From the boundary condition $f'(+\infty) = 1$, it is clear that $f \sim \eta$ as $\eta \to +\infty$ so that $f \to +\infty$ as $\eta \to +\infty$.

Hartree [107] showed that, when β is positive, only one of the solutions, namely that with $B = 0$ in (15.5), would satisfy the conditions at infinity. For positive β, there exists a unique solution $f(\eta)$ so that $f'(\eta) \to 1$ exponentially as $\eta \to +\infty$. Hartree [107] numerically obtained the family of the unique solutions for $0 \le \beta \le 2$. However, when β is negative, any expression of the form (15.5) tends to 0 as η (and so f) tends to ∞, so that any value of $f''(0)$ yields a solution satisfying the condition at infinity. Thus, when β is negative, the boundary conditions (15.2) do not specify a unique solution. To make the solution for $\beta < 0$ unique, Hartree [107] replaced the condition at infinity by

$$f'(\eta) \to 1 \text{ from below as } \eta \to +\infty, \text{ and } f''(0) \text{ as large as possible}$$

and produced a family of numerical results for $\beta_0 \le \beta \le 2$, where $\beta_0 = -0.198$ corresponds to $f''(0) = 0$. The family of Hartree's solutions for $\beta_0 \le \beta \le 2$ has such properties that $f''(0) \ge 0$ and $f'(\eta) \to 1$ *exponentially* as $\eta \to +\infty$, and indicates neither reversed flow nor velocity overshoot.

Stewartson [108] proved that the Falkner-Skan equation (15.1) with the boundary conditions (15.2) has a unique solution when $\beta \ge 0$. To make the solution for $\beta < 0$ unique, Stewartson [108] replaced $f(\eta)$ by $F_\alpha(\eta)$ that satisfies the equation

$$F_\alpha'''(\eta) + F_\alpha(\eta) F_\alpha''(\eta) + \beta[1 - F_\alpha'^2(\eta)] = 0 \tag{15.6}$$

and the boundary conditions

$$F_\alpha(0) = F_\alpha'(0) = 0, F_\alpha'(\alpha) = 1. \tag{15.7}$$

Obviously,

$$f(\eta) = \lim_{\alpha \to +\infty} F_\alpha(\eta).$$

In this explanation Stewartson [108] found in the region $\beta_0 \le \beta < 0$ another new family of numerical solutions exhibiting the property $f''(0) < 0$, demonstrating reversed flow.

Stewartson [108] proved a theorem that, if $\beta < \beta_0 = -0.1988$, then in all the solutions of the Falkner-Skan equation with $f(0) = f'(0) = 0$, there is a range of values of η for which $f'(\eta) > 1$, expressing velocity overshoot in some regions. Unlike Hartree [107] and Stewartson [108], Libby and Liu [109] believed that the overshoot velocity profile might have physical definitions; therefore, they defined other breaches of numerical solutions for $\beta < \beta_0$. Their numerical calculations showed that when $\beta < \beta_0$ multiple (probably an infinite) number of solutions to (15.1) and (15.2) exist for any given values of $f''(0)$.

Note that all of above-mentioned solutions are either numerical or analytic-numerical. In this chapter the homotopy analysis method is applied to present an explicit, purely analytic solution of the Falkner-Skan boundary layer flow.

15.1 Homotopy analysis solution

15.1.1 Zero-order deformation equation

From (15.5), $f'(\eta) \to 1$ exponentially as $\eta \to +\infty$ if $\beta \geq 0$ or $B = 0$ in the case of $\beta < 0$. Thus, it is natural to express $f(\eta)$ by the set of base functions

$$\{\eta^m \exp(-n\,\lambda\,\eta) \mid m \geq 0, n \geq 0, \lambda > 0\} \qquad (15.8)$$

in the form:

$$f(\eta) = \sum_{m=0}^{+\infty} \sum_{n=0}^{+\infty} a_{m,n}\, \eta^m \exp(-n\,\lambda\,\eta), \qquad (15.9)$$

where $a_{m,n}$ is a coefficient and λ is the so-called spatial-scale parameter. This provides us with the *rule of solution expression* for the Falkner-Skan boundary layer flows.

Under the *rule of solution expression* and using (15.2), it is obvious to select

$$f_0(\eta) = \eta - \frac{1 - \exp(-\lambda\,\eta)}{\lambda} + \frac{\gamma[1 - (1 + \lambda\,\eta)\exp(-\lambda\,\eta)]}{\lambda^2} \qquad (15.10)$$

as the initial guess of $f(\eta)$, where γ is an auxiliary parameter. Note that

$$f_0''(0) = \lambda + \gamma. \qquad (15.11)$$

Under the *rule of solution expression* denoted by (15.9) and from Equations (15.1) and (15.2), we choose the auxiliary linear operator

$$\mathcal{L}[\Phi(\eta;q)] = \frac{\partial^3 \Phi(\eta;q)}{\partial \eta^3} + \lambda\, \frac{\partial^2 \Phi(\eta;q)}{\partial \eta^2} \qquad (15.12)$$

with the property

$$\mathcal{L}\left[C_0 + C_1 \eta + C_2 \exp(-\lambda\,\eta)\right] = 0, \qquad (15.13)$$

where C_0, C_1, and C_2 are coefficients, $\Phi(\eta;q)$ is a real function of η and q. From Equation (15.1), we define the nonlinear operator

$$\mathcal{N}[\Phi(\eta;q)] = \frac{\partial^3 \Phi(\eta;q)}{\partial \eta^3} + \Phi(\eta;q)\, \frac{\partial^2 \Phi(\eta;q)}{\partial \eta^2} + \beta\left\{1 - \left[\frac{\partial \Phi(\eta;q)}{\partial \eta}\right]^2\right\}. \qquad (15.14)$$

Let $\hbar$ denote a nonzero auxiliary parameter and $H(\eta)$ a nonzero auxiliary function. We construct the so-called zero-order deformation equation

$$(1-q)\, \mathcal{L}\left[\Phi(\eta;q) - f_0(\eta)\right] = q\, \hbar\, H(\eta)\, \mathcal{N}[\Phi(\eta;q)], \qquad (15.15)$$

subject to the boundary conditions

$$\Phi(0;q) = 0, \quad \left.\frac{\partial \Phi(\eta;q)}{\partial \eta}\right|_{\eta=0} = 0, \quad \left.\frac{\partial \Phi(\eta;q)}{\partial \eta}\right|_{\eta=+\infty} = 1, \qquad (15.16)$$

where $q \in [0,1]$ is an embedding parameter.

When $q = 0$, it is easy to demonstrate that

$$\Phi(\eta;0) = f_0(\eta). \qquad (15.17)$$

When $q = 1$, since $q \neq 0$ and $H(\eta) \neq 0$, Equations (15.15) and (15.16) are equivalent to Equations (15.1) and (15.2), respectively, provided

$$\Phi(\eta;1) = f(\eta). \qquad (15.18)$$

Thus, as the embedding parameter q increases from 0 to 1, $\Phi(\eta;q)$ varies from the initial guess $f_0(\eta)$ to the exact solution $f(\eta)$ of Equations (15.1) and (15.2). Then, by Taylor's theorem and using (15.17), we expand $\Phi(\eta;q)$ in the power series

$$\Phi(\eta;q) = f_0(\eta) + \sum_{k=1}^{+\infty} f_k(\eta) \, q^k, \qquad (15.19)$$

where

$$f_k(\eta) = \frac{1}{k!} \left.\frac{\partial^k \Phi(\eta;q)}{\partial q^k}\right|_{q=0}. \qquad (15.20)$$

Note that the zero-order deformation equation (15.15) contains the auxiliary parameter $\hbar$ and the auxiliary functions $H(\eta)$. The initial guess $f_0(\eta)$ contains the auxiliary parameter γ. Assuming that all of them are correctly chosen so that the series (15.19) converges when $q = 1$, we have, using (15.18),

$$f(\eta) = f_0(\eta) + \sum_{k=1}^{+\infty} f_k(\eta). \qquad (15.21)$$

This provides us with a relationship between the initial approximation $f_0(\eta)$ and the exact solution $f(\eta)$.

15.1.2 High-order deformation equation

For conciseness, define the vector

$$\vec{f}_n = \{f_0(\eta), f_1(\eta), f_2(\eta), \cdots, f_n(\eta)\}.$$

Differentiating the zero-order deformation equations (15.15) and (15.16) k times with respect to q, then setting $q = 0$, and finally dividing them by $k!$, we obtain the high-order deformation equation

$$\mathcal{L}[f_k(\eta) - \chi_k f_{k-1}(\eta)] = \hbar \, H(\eta) \, R_k(\vec{f}_{k-1}), \qquad (15.22)$$

Boundary-layer Flows with Exponential Property

subject to the boundary conditions

$$f_k(0) = f'_k(0) = f'_k(+\infty) = 0, \qquad (15.23)$$

where χ_k is defined by (2.42) and

$$R_k(\vec{f}_{k-1}) = f'''_{k-1}(\eta) + \sum_{n=0}^{k-1}\left[f_n(\eta)f''_{k-1-n}(\eta) - \beta f'_n(\eta)f'_{k-1-n}(\eta)\right]$$
$$+ \beta\,(1-\chi_k). \qquad (15.24)$$

It is easy to solve the linear differential equations (15.22) and (15.23), using symbolic calculation software.

Under the *rule of solution expression* denoted by (15.9) and from Equation (15.22), the auxiliary function $H(\eta)$ may be in the form

$$H(\eta) = \eta^{\kappa_1}\,\exp(-\lambda\,\kappa_2\,\eta),$$

where $\kappa_1 \geq 0$ and $\kappa_2 \geq 0$ are integers. For simplicity, we choose $\kappa_1 = \kappa_2 = 0$, corresponding to

$$H(\eta) = 1. \qquad (15.25)$$

Then, let $f_k^*(\eta)$ denote a special solution of the equation

$$\mathcal{L}[f_k^*(\eta)] = \hbar\, R_k(\vec{f}_{k-1}).$$

Then, from (15.13), we gain the solution

$$f_k(\eta) = \chi_k\,f_{k-1}(\eta) + f_k^*(\eta) + C_0 + C_1\eta + C_2\exp(-\lambda\,\eta), \qquad (15.26)$$

where $C_0, C_1,$ and C_2 are determined by the boundary conditions (15.23).

15.1.3 Recursive formulae

By solving the first several high-order deformation equations (15.22) and (15.23), $f_m(\eta)$ can be expressed by

$$f_m(\eta) = \sum_{k=0}^{m+1}\Psi_{m,k}(\eta)\exp(-k\lambda\,\eta), \qquad m \geq 0, \qquad (15.27)$$

where

$$\Psi_{0,0}(\eta) = b_0^{0,0} + b_1^{0,0}\,\eta, \qquad (15.28)$$
$$\Psi_{0,1}(\eta) = b_0^{0,1} + b_1^{0,1}\,\eta, \qquad (15.29)$$
$$\Psi_{m,0}(\eta) = b_0^{m,0}, \qquad m \geq 1 \qquad (15.30)$$

and

$$\Psi_{m,k}(\eta) = \sum_{n=0}^{2(m+1)-k} b_n^{m,k} \eta^n, \quad m \geq 1, 1 \leq k \leq m+1. \quad (15.31)$$

Substituting the above expressions into Equations (15.22) and (15.23), Liao [40] gained the recursive expressions of each coefficient $b_k^{m,n}$, where $m \geq 1$, $0 \leq n \leq m+1$ and $0 \leq k \leq 2(m+1)-n$, as follows:

$$b_0^{m,0} = \chi_m b_0^{m-1,0} - \lambda^{-1} \sum_{r=0}^{2m} \Gamma_r^{m,1} \Pi_r^{1,1} - \sum_{n=2}^{m+1} (n-1)\Gamma_0^{m,n} \Pi_0^{n,0}$$
$$+ \sum_{n=2}^{m+1} \sum_{r=1}^{2(m+1)-n} \Gamma_r^{m,n} \left(n\Pi_r^{n,0} - \Pi_r^{n,0} - \lambda^{-1}\Pi_r^{n,1} \right), \quad (15.32)$$

$$b_1^{m,0} = 0, \quad (15.33)$$

$$b_0^{m,1} = \chi_m b_0^{m-1,1} + \lambda^{-1} \sum_{r=0}^{2m} \Gamma_r^{m,1} \Pi_r^{1,1} + \sum_{n=2}^{m+1} n\Gamma_0^{m,n} \Pi_0^{n,0}$$
$$+ \sum_{n=2}^{m+1} \sum_{r=1}^{2(m+1)-n} \Gamma_r^{m,n} \left(n\Pi_r^{n,0} - \lambda^{-1}\Pi_r^{n,1} \right), \quad (15.34)$$

$$b_k^{m,1} = \chi_m (1 - \chi_{k+2-2m}) b_k^{m-1,1} + \sum_{r=k-1}^{2m} \Gamma_r^{m,1} \Pi_r^{1,k},$$
$$1 \leq k \leq 2m+1, \quad (15.35)$$

$$b_k^{m,n} = \chi_m (1 - \chi_{k+1-2m+n}) b_k^{m-1,n} - \sum_{r=k}^{2(m+1)-n} \Gamma_r^{m,n} \Pi_r^{n,k},$$
$$2 \leq n \leq m, 0 \leq k \leq 2(m+1)-n \quad (15.36)$$

and

$$b_k^{m,m+1} = - \sum_{r=k}^{m+1} \Gamma_r^{m,m+1} \Pi_r^{m+1,k}, \quad 1 \leq k \leq m+1, \quad (15.37)$$

where

$$\Pi_r^{1,k} = \frac{r!\,(r-k+2)}{k!\,\lambda^{r-k+3}}, \quad 0 \leq k \leq r+1,$$

$$\Pi_r^{n,k} = \frac{r!}{k!(n-1)^{r-k+1}\lambda^{r-k+3}} \left[1 - \left(1 - \frac{1}{n}\right)^{r-k+1} \left(1 + \frac{r-k+1}{n}\right) \right]$$

$$n \geq 2,\ 0 \leq k \leq r,$$

$$\Gamma_r^{m,n} = \hbar\left[(1 - \chi_{r+1-2m+n})\, d_r^{m-1,n} + \delta_r^{m,n} + \Delta_r^{m,n}\right]$$
$$1 \leq n \leq m, 0 \leq r \leq 2(m+1) - n,$$

$$\Gamma_r^{m,m+1} = \hbar(\delta_r^{m,m+1} + \Delta_r^{m,m+1}),$$

in which

$$\Delta_r^{m,n} = -\beta \sum_{k=0}^{m-1} \sum_{j=\max\{0,n+k-m\}}^{\min\{n,k+1\}} \sum_{i=\max\{0,r-2(m-k)+n-j\}}^{\min\{r,2(k+1)-j\}} a_i^{k,j}\, a_{r-i}^{m-1-k,n-j},$$

$$\delta_r^{m,n} = \sum_{k=0}^{m-1} \sum_{j=\max\{1,n+k-m\}}^{\min\{n,k+1\}} \sum_{i=\max\{0,r-2(m-k)+n-j\}}^{\min\{r,2(k+1)-j\}} c_i^{k,j}$$
$$\times b_{r-i}^{m-1-k,n-j}\, \Lambda_{r-i}^{m-1-k,n-j},$$

$$m \geq 1, 0 \leq n \leq m+1, 0 \leq r \leq 2(m+1) - n$$

under the definitions

$$a_n^{m,k} = (n+1)b_{n+1}^{m,k}\Lambda_{n+1}^{m,k} - (k\lambda)b_n^{m,k}\Lambda_n^{m,k}, \tag{15.38}$$

$$c_n^{m,k} = (n+1)(n+2)\, b_{n+2}^{m,k}\Lambda_{n+2}^{m,k} - 2(k\lambda)(n+1)b_{n+1}^{m,k}\Lambda_{n+1}^{m,k}$$
$$+ (k\lambda)^2\, b_n^{m,k}\Lambda_n^{m,k}, \tag{15.39}$$

$$d_n^{m,k} = (n+1)\, c_{n+1}^{m,k}\Lambda_{n+1}^{m,k} - (k\lambda)c_n^{m,k}\Lambda_n^{m,k} \tag{15.40}$$

and

$$\Lambda_k^{i,j} = \begin{cases} 0, & \text{when } i = j = 0, k \geq 2, \\ 0, & \text{when } i > 0, j = 0, k \geq 1, \\ 0, & \text{when } j > i+1, \\ 0, & \text{when } k > 2(i+1) - j, \\ 1, & \text{otherwise.} \end{cases} \tag{15.41}$$

From (15.10), we obtain the first four coefficients

$$b_0^{0,0} = \frac{\gamma - \lambda}{\lambda^2}, \quad b_1^{0,0} = 1, \quad b_0^{0,1} = \frac{\lambda - \gamma}{\lambda^2}, \quad b_1^{0,1} = -\frac{\gamma}{\lambda} \tag{15.42}$$

from which, we can calculate all other coefficients $b_k^{m,n}$ using the above recursive expressions.

The Mth-order approximation of Equations (15.1) and (15.2) is given by

$$f(\eta) \approx \eta + \left(\sum_{m=0}^{M} b_0^{m,0} \right)$$
$$+ \sum_{n=1}^{M+1} \exp(-n\lambda\,\eta) \left(\sum_{m=n-1}^{M} \sum_{k=0}^{2(m+1)-n} b_k^{m,n} \eta^k \right). \qquad (15.43)$$

Therefore, we have the explicit, purely analytic solution of Falkner-Skan laminar viscous flow over a semi-infinite flat plate

$$f(\eta) = \eta + \left(\sum_{m=0}^{+\infty} b_0^{m,0} \right)$$
$$+ \lim_{M \to +\infty} \sum_{n=1}^{M+1} \exp(-n\lambda\,\eta) \left(\sum_{m=n-1}^{M} \sum_{k=0}^{2(m+1)-n} b_k^{m,n} \eta^k \right). \qquad (15.44)$$

This exact solution obviously has the asymptotic property $f' \to 1$ exponentially as $\eta \to +\infty$.

15.1.4 Convergence theorem

THEOREM 15.1
The series

$$f_0(\eta) + \sum_{k=1}^{+\infty} f_k(\eta)$$

must be the exact solution of Equations (15.1) and (15.2) as long as it is convergent, where $f_k(\eta)$ is governed by Equations (15.22) and (15.23) under the definitions (15.10), (15.12), (15.24), and (2.42).

Proof: If the series is convergent, we have

$$\lim_{m \to +\infty} f_m(\eta) = 0.$$

From (15.22) and (2.42),

$$\hbar\, H(\eta) \sum_{k=1}^{m} R_k(\vec{f}_{k-1}) = \mathcal{L}[f_m(\eta)].$$

Using (15.12), we gain

$$\hbar\, H(\eta) \sum_{k=1}^{+\infty} R_k(\vec{f}_{k-1}) = \lim_{m \to +\infty} \mathcal{L}[f_m(\eta)] = \mathcal{L}\left[\lim_{m \to +\infty} f_m(\eta) \right] = 0,$$

Boundary-layer Flows with Exponential Property

which gives, since $\hbar \neq 0$ and $H(\eta) \neq 0$,

$$\sum_{k=1}^{+\infty} R_k(\vec{f}_{k-1}) = 0.$$

Substituting (15.24) into the above expression and simplifying it, we obtain

$$\frac{d^3}{d\eta^3}\left[\sum_{k=0}^{+\infty} f_k(\eta)\right] + \left[\sum_{k=0}^{+\infty} f_k(\eta)\right] \frac{d^2}{d\eta^2}\left[\sum_{k=0}^{+\infty} f_k(\eta)\right]$$
$$+ \beta\left\{1 - \left[\frac{d}{d\eta}\sum_{k=0}^{+\infty} f_k(\eta)\right]^2\right\} = 0.$$

Furthermore, from (15.10) and (15.23),

$$\sum_{k=0}^{+\infty} f_k(0) = \sum_{k=0}^{+\infty} f'_k(0) = 0, \quad \sum_{k=0}^{+\infty} f'_k(+\infty) = 1.$$

Therefore, if the series

$$f_0(\eta) + \sum_{k=1}^{+\infty} f_k(\eta)$$

is convergent, it must be an exact solution of Equations (15.1) and (15.2). This ends the proof.

15.2 Result analysis

According to Theorem 15.1, we need only ensure that the solution series (15.21) converges. Note that the solution (15.44) contains three auxiliary parameters $\hbar$, λ, and γ. We have therefore a three-parameter family of solution expressions. The spatial-scale parameter λ affects the rate of $f'(\eta) \to 0$ as $\eta \to +\infty$. From (15.11), the auxiliary parameter γ affects $f''_0(0)$, and we can investigate the relationship between the exact solution and different initial guesses $f_0(\eta)$, which becomes interesting when there exist multiple solutions in the case of $\beta < 0$. It should be emphasized that, for any given values of λ and γ, we still have the freedom to choose a proper value of the auxiliary parameter $\hbar$ to control and adjust the convergence region and rate of the solution (15.44), when necessary.

Physically, $f''(0)$ is related to the friction of the fluid on the plate and therefore has important physical meanings. From (15.21), we have

$$f''(0) = f''_0(0) + \sum_{k=1}^{+\infty} f''_k(0), \qquad (15.45)$$

which is dependent of the physical parameter β and the three auxiliary parameters $\hbar$, λ, and γ. First, consider the case of $\hbar = -1, \gamma = 0$ and regard λ as an unknown variable. For given values of β, $f''(0)$ converges to the same corresponding value, provided λ is large enough, as shown in Figure 15.1. Note that, when $\lambda \geq 5$, we can obtain convergent results of $f''(0)$ for $0 \leq \beta \leq 2$, corresponding to $0 \leq \kappa < +\infty$. It is therefore reasonable to choose $\lambda \geq 5$. Then, we consider the case of $\lambda = 5$ and $\gamma = 0$. The influence of $\hbar$ on the convergence of $f''(0)$ can be investigated by plotting the so-called $\hbar$-curves (see page 26 and §3.5.1) of $f''(0)$, as shown in Figure 15.2. Obviously, when $-5/4 \leq \hbar \leq -3/4$, we can obtain convergent results of $f''(0)$ for $0 \leq \beta \leq 2$. Furthermore, when $\lambda = 5, \gamma = 0$, and $\hbar = -1$, $f''(0)$ converges for $\beta_0 \leq \beta \leq 2$, where $\beta_0 = -0.1988$. The related 10th-, 20th- and 30th-order approximations of $f''(0)$ are given by

$$f''(0)$$
$$\approx 0.466892061269575 + 1.270377798259161\ \beta$$
$$-0.9366061372519299\ \beta^2 + 0.6565444804810052\ \beta^3$$
$$-0.2989667156611743\ \beta^4 + 8.714746301295173 \times 10^{-2}\ \beta^5$$
$$-1.646263530984164 \times 10^{-2}\ \beta^6 + 2.009360736004046 \times 10^{-3}\ \beta^7$$
$$-1.532383017041316 \times 10^{-4}\ \beta^8 + 6.654735449025599 \times 10^{-6}\ \beta^9$$
$$-1.259647041638817 \times 10^{-7}\ \beta^{10}, \qquad (15.46)$$

$$f''(0)$$
$$\approx 0.469470560483573 + 1.295165031248947\ \beta$$
$$-1.37974417506381\ \beta^2 + 2.191127183953301\ \beta^3$$
$$-3.010696768394167\ \beta^4 + 3.217599178710972\ \beta^5$$
$$-2.637727245237923\ \beta^6 + 1.672089788089693\ \beta^7$$
$$-0.8288927042391463\ \beta^8 + 0.3244321617350418\ \beta^9$$
$$-0.1009610729534239\ \beta^{10} + 2.508145750314333 \times 10^{-2}\ \beta^{11}$$
$$-4.979128795292073 \times 10^{-3}\ \beta^{12} + 7.879252632693684 \times 10^{-4}\ \beta^{13}$$
$$-9.872764016512919 \times 10^{-5}\ \beta^{14} + 9.675273712687701 \times 10^{-6}\ \beta^{15}$$
$$-7.265429168115138 \times 10^{-7}\ \beta^{16} + 4.042349583833827 \times 10^{-8}\ \beta^{17}$$
$$-1.573031139104484 \times 10^{-9}\ \beta^{18} + 3.831177670499221 \times 10^{-11}\ \beta^{19}$$
$$-4.409794935993077 \times 10^{-13}\ \beta^{20}, \qquad (15.47)$$

and

$$f''(0)$$
$$\approx 0.4695903615312177 + 1.298441994559965\ \beta$$
$$-1.491321283547855\ \beta^2 + 3.075663557218445\ \beta^3$$

Boundary-layer Flows with Exponential Property

$$\begin{aligned}
&-6.529797437132239\ \beta^4 + 12.02830971699564\ \beta^5\\
&-18.1643591437081\ \beta^6 + 22.24132274854839\ \beta^7\\
&-22.20184453035751\ \beta^8 + 18.25679986443915\ \beta^9\\
&-12.5044109522257\ \beta^{10} + 7.205799232502405\ \beta^{11}\\
&-3.523684854407225\ \beta^{12} + 1.472301275851469\ \beta^{13}\\
&-0.5283860769496572\ \beta^{14} + 0.1634753390904293\ \beta^{15}\\
&-4.369783644567797 \times 10^{-2}\ \beta^{16} + 1.010062677099088 \times 10^{-2}\ \beta^{17}\\
&-2.018052269391153 \times 10^{-3}\ \beta^{18} + 3.479126853101193 \times 10^{-4}\ \beta^{19}\\
&-5.159762018077551 \times 10^{-5}\ \beta^{20} + 6.552517694061283 \times 10^{-6}\ \beta^{21}\\
&-7.079292502238022 \times 10^{-7}\ \beta^{22} + 6.449249702165323 \times 10^{-8}\ \beta^{23}\\
&-4.894052058789618 \times 10^{-9}\ \beta^{24} + 3.041591067712079 \times 10^{-10}\ \beta^{25}\\
&-1.510937645005482 \times 10^{-11}\ \beta^{26} + 5.783596085142942 \times 10^{-13}\ \beta^{27}\\
&-1.606885758239585 \times 10^{-14}\ \beta^{28} + 2.896755640650334 \times 10^{-16}\ \beta^{29}\\
&-2.560020533395366 \times 10^{-18}\ \beta^{30}, \tag{15.48}
\end{aligned}$$

respectively. When $0 \leq \beta \leq 2$, the series of $f''(0)$ converges quickly and even the 10th-order approximation agrees with numerical results given by Hartree [107] and White [20], as shown in Table 15.1 and Figure 15.3, respectively. However, when $\beta_0 \leq \beta < 0$, the series of $f''(0)$ converges slowly. We fail to accelerate the convergence of $f''(0)$ by means of the Padé technique in the traditional way. But, using the homotopy-Padé technique (see page 38 and §3.5.2), the convergence of the series (15.45) is greatly accelerated, especially when $\beta_0 \leq \beta < 0$, as shown in Table 15.2 and Figure 15.3. Also, it is found that the $[m,m]$ homotopy-Padé approximant of $f''(0)$ does not depend upon the auxiliary parameter $\hbar$.

It is found that, as long as the series (15.45) of $f''(0)$ is convergent, the corresponding series of $f(\eta)$ and $f'(\eta)$ also converge in the whole region $0 \leq \eta < +\infty$. When $0 \leq \beta \leq 2$, the series of $f'(\eta)$ converges quickly and the 20th-order approximations agree with Hartree's numerical results [107], as shown in Figure 15.4. However, when $\beta_0 \leq \beta < 0$, the closer the value of β is to β_0, the more slowly the series of $f'(\eta)$ converges; thus higher-order approximations are necessary to get accurate enough results, as shown in Figure 15.4.

These verify Hartree's [107] conclusions that there exists a unique solution when $0 \leq \beta \leq 2$ and a solution when $\beta_0 \leq \beta < 0$, with the properties $f''(0) \geq 0$ and $f'(\eta) \to 1$ exponentially as $\eta \to +\infty$. When $\beta_0 \leq \beta < 0$, Stewartson [108] numerically found a kind of solution with the property $f''(0) < 0$ but still $f'(\eta) \to 1$ exponentially as $\eta \to +\infty$, showing reversed flow. To check Stewartson's [108] numerical results, we choose a negative value of γ so that

$$f_0''(0) = \gamma + \lambda < 0.$$

For given λ and γ, by means of plotting the corresponding $\hbar$-curves of $f''(0)$, a negative $\hbar$ with a small enough value of $|\hbar|$ may be chosen to ensure that the corresponding solution series converge. Several dozen cases are investigated. However, it is found that, as long as a solution series is convergent, it always converges to Hartree's family of solution. For example, when $\gamma = -5.5, \beta = -15/100, \lambda = 2$, and $\hbar = -1/10$, the approximate series converges to Hartree's family of solution. Thus, it seems that the solution (15.44) might not give Stewartson's [108] family of solutions with reversed flows. Similarly, it seems that the solution (15.44) might not give Libby and Liu's [109] family of solution with velocity overshoot. Note that Stewartson [108], and Libby and Liu [109] found their solutions by numerical techniques, and it is difficult to rigorously check if $f'(\eta) \to 1$ exponentially as $\eta \to +\infty$ by means of numerical methods, because all numerical methods are restricted in a finite domain and thus cannot correctly treat the quantity of infinity. It may therefore be doubtful that solutions given by Stewartson [108] and Libby and Liu [109] have the property $f'(\eta) \to 1$ exponentially as $\eta \to +\infty$ in rigorous mathematical meaning. More evidence is necessary to support this point of view. It is still an open question whether there exist multiple solutions of the Falkner-Skan viscous flow with the exponential property at infinity when $\beta < 0$. If the multiple solutions exist, it would be a challenge to apply the homotopy analysis method to discover all these solutions.

TABLE 15.1
Comparison of the analytic approximations of $f''(0)$ given by (15.45) when $\lambda = 5, \gamma = 0$, and $\hbar = -1$ with White's [20] and Hartree's [107] numerical results.

β	10th-order approx.	20th-order approx.	25th-order approx.	30th-order approx.	numerical result
2.0	1.68647	1.68719	1.68721	1.68722	1.6872
1.6	1.51709	1.52148	1.52152	1.52152	1.5215
1.2	1.33147	1.33578	1.33572	1.33572	1.3357
1.0	1.23079	1.23266	1.23258	1.23259	1.2326
0.8	1.12210	1.12027	1.12028	1.12027	1.1203
0.6	1.00107	0.99572	0.99585	0.99584	0.9958
0.5	0.93379	0.92755	0.92767	0.92768	0.9277
0.4	0.86038	0.85435	0.85440	0.85442	0.8544
0.3	0.77922	0.77483	0.77474	0.77475	0.7748
0.2	0.68830	0.68691	0.68674	0.68671	0.6867
0.1	0.59519	0.58711	0.58707	0.58705	0.5870
0.0	0.46689	0.46947	0.46956	0.46959	0.4696
-0.1	0.32980	0.32363	0.32197	0.32096	0.319
-0.14	0.26876	0.25374	0.24960	0.24682	0.239
-0.16	0.23676	0.21559	0.20947	0.20515	0.190
-0.18	0.20372	0.17499	0.16622	0.15971	0.128
-0.19	0.18679	0.15369	0.14329	0.13537	0.086
-0.198	0.17306	0.13614	0.12426	0.11504	0

Source: Shi-Jun Liao, "A uniformly valid analytic solution of two-dimensional viscous flow past a semi-infinite flat plate", *Journal of Fluid Mechanics* (1999), 385:101-128 Cambridge University Press Copyright ©1999 Cambridge University Press, reprinted with permission.

TABLE 15.2
Comparison of the $[m,m]$ homotopy-Padé approximations of $f''(0)$ when $\lambda = 5, \gamma = 0$, and $\hbar = -1$ with White's [20] and Hartree's [107] numerical results.

β	[5,5]	[10,10]	[15,15]	Numerical result
2.0	1.68636	1.68722	1.68722	1.6872
1.6	1.52026	1.52151	1.52151	1.5215
1.2	1.33399	1.33571	1.33572	1.3357
1.0	1.23063	1.23260	1.23259	1.2326
0.8	1.11816	1.12027	1.12027	1.1203
0.6	0.99372	0.99584	0.99584	0.9958
0.5	0.92563	0.92769	0.92768	0.9277
0.4	0.85246	0.85443	0.85442	0.8544
0.3	0.77287	0.77474	0.77476	0.7748
0.2	0.68478	0.68670	0.68671	0.6867
0.1	0.58484	0.58697	0.58704	0.5870
0.0	0.46736	0.46960	0.46960	0.4696
-0.1	0.32291	0.31935	0.31927	0.319
-0.14	0.25476	0.24074	0.23984	0.239
-0.16	0.21813	0.19380	0.19128	0.190
-0.18	0.17980	0.13870	0.13160	0.128
-0.19	0.16004	0.10677	0.09455	0.086
-0.198	0.14398	0.07833	0.05912	0

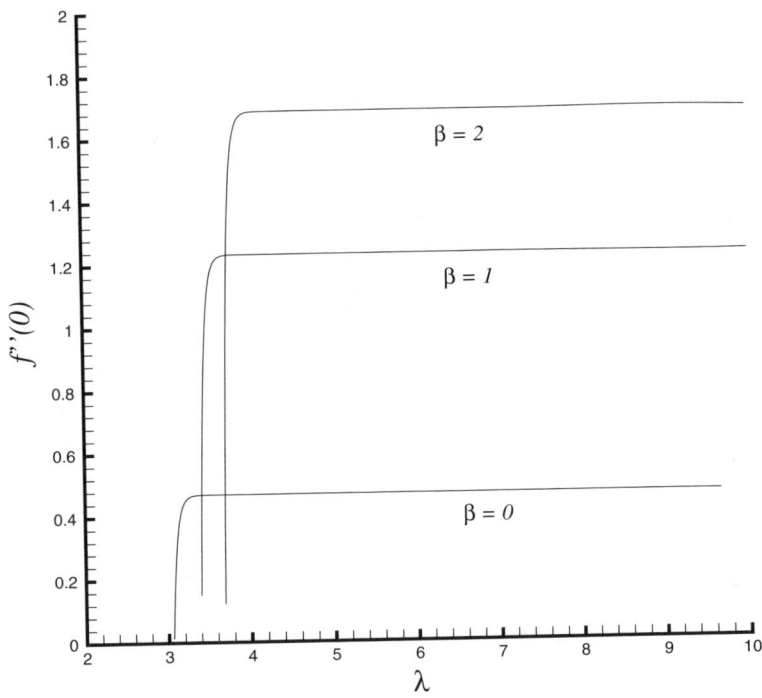

FIGURE 15.1
The 20th-order approximation of $f''(0)$ versus λ when $\hbar = -1$, $\gamma = 0$, and $\beta = 0, 1, 2$.

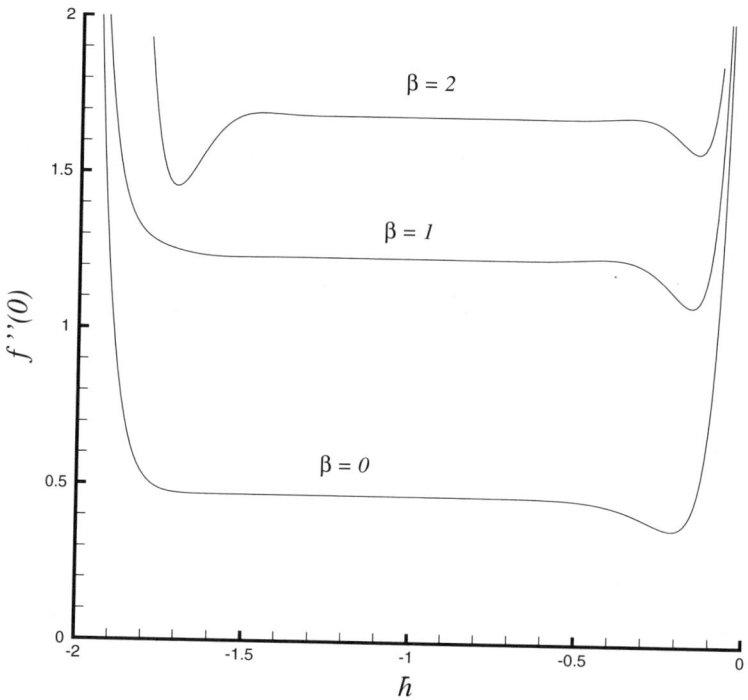

FIGURE 15.2
The $\hbar$-curves of $f''(0)$ at the 20th order of approximation when $\lambda = 5$, $\gamma = 0$, and $\beta = 0, 1, 2$.

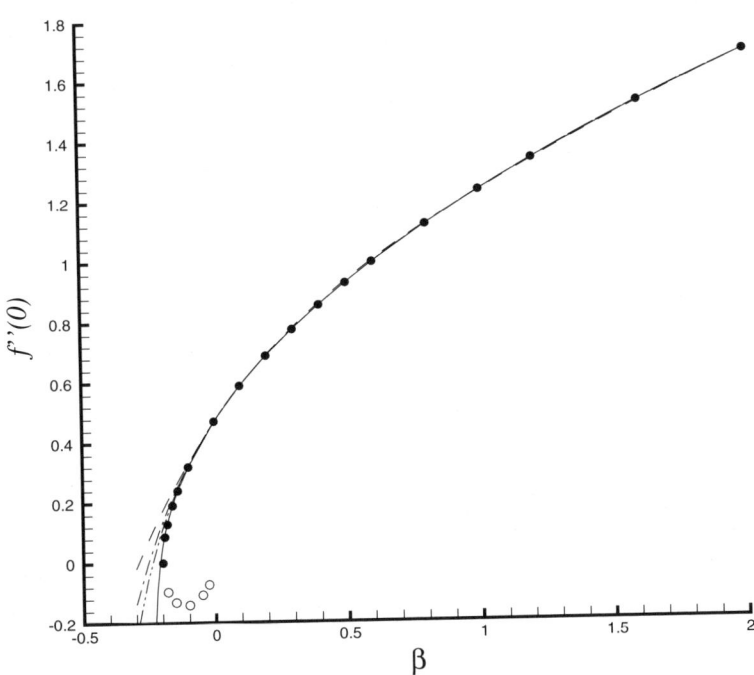

FIGURE 15.3
Comparison of the analytic approximations $f''(0)$ when $\lambda = 5, \gamma = 0$, and $\hbar = -1$ with the numerical results. Dashed line: 10th-order analytic approximations (15.46); dash-dotted line: 20th-order analytic approximations (15.47); dash-dot-dotted line: 30th-order analytic approximations (15.48); solid line: [15,15] homotopy-Padé approximation; filled circle: numerical results given by Hartree [107]; open circle: numerical results given by Stewartson [108].

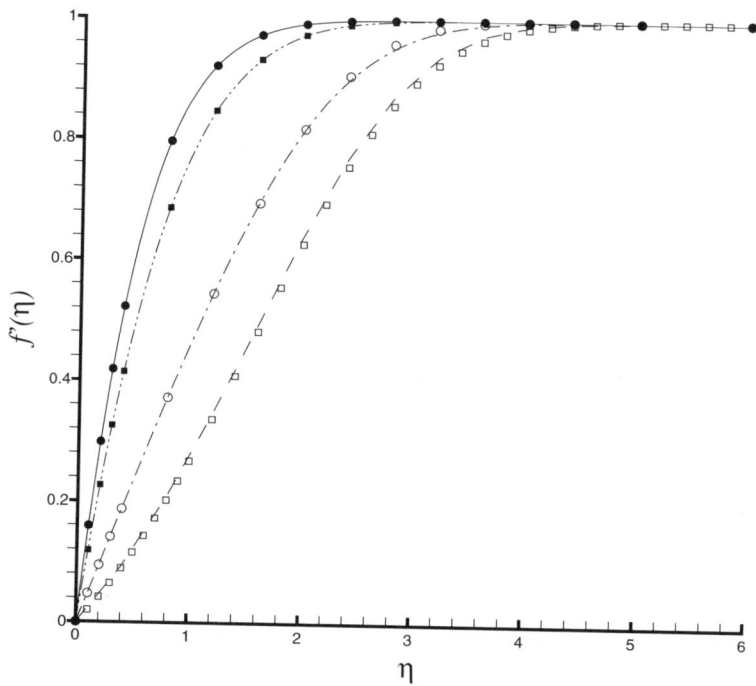

FIGURE 15.4
Comparison of the analytic approximation of $f'(\eta)$ given by (15.43) when $\lambda = 5, \gamma = 0$, and $\hbar = -1$ with Hartree's [107] numerical results. Solid line: numerical result when $\beta = 2$; dash-dot-dotted line: numerical result when $\beta = 1$; dash-dotted line: numerical result when $\beta = 0$; dashed line: numerical result when $\beta = -0.16$; filled circle: 20th-order analytic approximations when $\beta = 2$; filled square: 20th-order analytic approximations when $\beta = 1$; open circle: 20th-order analytic approximation when $\beta = 0$; open square: 50th-order analytic approximations when $\beta = -0.16$.

16

Boundary-layer flows with algebraic property

In most cases the velocity profiles and temperature of boundary layer flows decay exponentially. However, as reported by Kuiken [110, 111], solutions of some boundary layer flows behave algebraically at infinity.

For instance Kuiken [111] analyzed a glass-fiber production process and obtained, by means of various similarity transformations, a set of two coupled nonlinear differential equations

$$f'''(\eta) + \theta(\eta) - f'^2(\eta) = 0, \tag{16.1}$$
$$\theta''(\eta) = 3\,\sigma\,f'(\eta)\,\theta(\eta), \tag{16.2}$$

subject to the boundary conditions

$$f(0) = f'(0) = 0, \theta(0) = 1, \tag{16.3}$$
$$f'(+\infty) = \theta(+\infty) = 0, \tag{16.4}$$

where the prime denotes differentiation with respect to the similarity variable η, σ is the Prandtl number, $f(\eta)$ and $\theta(\eta)$ relate to the velocity profile and temperature distribution of the boundary layer, respectively. For details, the reader is referred to Kuiken [111].

Kuiken [111] presented a solution that contains a parameter that had to be determined by numerical methods. Kuiken's solution [111] is fundamentally analytic-numerical. To the best of the author's knowledge, no one has reported an explicit, fully analytic solution of the coupled nonlinear equations (16.1) and (16.2). In this chapter the homotopy analysis method is employed to yield this type of solution.

16.1 Homotopy analysis solution

16.1.1 Asymptotic property

From (16.4), both $f'(\eta)$ and $\theta(\eta)$ tend to zero as $\eta \to +\infty$. So, it is important to know the behavior of the solution at infinity. As pointed out by Kuiken [111], both $f(\eta)$ and $\theta(\eta)$ decay algebraically as $\eta \to +\infty$.

Under the transformation

$$\xi = 1 + \lambda\,\eta, \quad F(\xi) = f'(\eta), \quad S(\xi) = \theta(\eta), \qquad (16.5)$$

where $\lambda > 0$ is the so-called spatial-scale parameter, Equations (16.1) and (16.2) become

$$\lambda^2\,F''(\xi) + S(\xi) - F^2(\xi) = 0, \qquad (16.6)$$
$$\lambda^2\,S''(\xi) = 3\,\sigma\,F(\xi)\,S(\xi), \qquad (16.7)$$

subject to the boundary conditions

$$F(1) = 0,\, S(1) = 1, \qquad (16.8)$$
$$F(+\infty) = S(+\infty) = 0. \qquad (16.9)$$

Define

$$F \sim \xi^{\alpha_1}, \quad S \sim \xi^{\alpha_2} \qquad (16.10)$$

as the asymptotic expressions of $F(\xi)$ and $S(\xi)$ as $\xi \to +\infty$. Substituting them into Equations (16.6) and (16.7) and balancing the main terms of each equation, we have

$$\alpha_1 = -2, \quad \alpha_2 = -4. \qquad (16.11)$$

Thus, considering their algebraic property at infinity, $F(\xi)$ and $S(\xi)$ can be expressed by the base functions

$$\{\xi^{-n} \mid n \geq 2\} \qquad (16.12)$$

in the forms:

$$F(\xi) = \sum_{n=2}^{+\infty} \frac{a_n}{\xi^n}, \qquad (16.13)$$

$$S(\xi) = \sum_{n=4}^{+\infty} \frac{b_n}{\xi^n}, \qquad (16.14)$$

respectively, where a_n, b_n are coefficients. The above expressions provide the so-called *rules of solution expression* of $F(\xi)$ and $S(\xi)$, respectively.

16.1.2 Zero-order deformation equation

Under the *rules of solution expression* denoted by (16.13) and (16.14) and from Equations (16.8) and (16.9), it is easy to choose

$$F_0(\xi) = \gamma\left(\xi^{-2} - \xi^{-3}\right), \quad S_0(\xi) = \xi^{-4} \qquad (16.15)$$

as the initial guesses of $F(\xi)$ and $S(\xi)$, respectively, where γ is an auxiliary parameter. Furthermore, under the *rules of solution expression* denoted by

Boundary-layer Flows with Algebraic Property

(16.13) and (16.14) and from Equations (16.6) and (16.7), we select the auxiliary linear operators

$$\mathcal{L}_F \Phi = \left(\frac{\xi}{3}\right) \frac{\partial^2 \Phi}{\partial \xi^2} + \frac{\partial \Phi}{\partial \xi}, \qquad (16.16)$$

$$\mathcal{L}_S \Phi = \left(\frac{\xi}{5}\right) \frac{\partial^2 \Phi}{\partial \xi^2} + \frac{\partial \Phi}{\partial \xi} \qquad (16.17)$$

with the properties

$$\mathcal{L}_F \left(C_1 + C_2\, \xi^{-2}\right) = 0, \qquad (16.18)$$

$$\mathcal{L}_S \left(C_3 + C_4\, \xi^{-4}\right) = 0, \qquad (16.19)$$

where C_1, C_2, C_3, and C_4 are coefficients. For simplicity, from Equations (16.6) and (16.7), we define the nonlinear operators

$$\mathcal{N}_F \left[\Phi(\xi; q), \Theta(\xi; q)\right] = \lambda^2\, \frac{\partial \Phi(\xi; q)}{\partial \xi^2} + \Theta(\xi; q) - \Phi^2(\xi; q), \qquad (16.20)$$

$$\mathcal{N}_S \left[\Phi(\xi; q), \Theta(\xi; q)\right] = \lambda^2\, \frac{\partial^2 \Theta(\xi; q)}{\partial \xi^2} - 3\, \sigma\, \Phi(\xi; q)\, \Theta(\xi; q), \qquad (16.21)$$

where $q \in [0, 1]$ is an embedding parameter, $\Phi(\xi; q)$ and $\Theta(\xi; q)$ are real functions of ξ and q. Let $\hbar_F$ and $\hbar_S$ denote the nonzero auxiliary parameters, $H_F(\xi)$ and $H_S(\xi)$ the nonzero auxiliary functions, respectively. We construct the zero-order deformation equations

$$(1 - q)\, \mathcal{L}_F \left[\Phi(\xi; q) - F_0(\xi)\right]$$
$$= q\, \hbar_F\, H_F(\xi)\, \mathcal{N}_F \left[\Phi(\xi; q), \Theta(\xi; q)\right], \qquad (16.22)$$

$$(1 - q)\, \mathcal{L}_S \left[\Theta(\xi; q) - S_0(\xi)\right]$$
$$= q\, \hbar_S\, H_S(\xi)\, \mathcal{N}_S \left[\Phi(\xi; q), \Theta(\xi; q)\right], \qquad (16.23)$$

subject to the boundary conditions

$$\Phi(1; q) = \Phi(+\infty; q) = \Theta(+\infty; q) = 0, \quad \Theta(1; q) = 1, \qquad (16.24)$$

where $q \in [0, 1]$ is the embedding parameter.

When $q = 0$, it is easy to demonstrate that

$$\Phi(\xi; 0) = F_0(\xi), \quad \Theta(\xi; 0) = S_0(\xi), \qquad (16.25)$$

where $F_0(\xi)$ and $S_0(\xi)$ are the initial guesses defined by (16.15). When $q = 1$, since

$$\hbar_F \neq 0, \quad \hbar_S \neq 0, \quad H_F(\xi) \neq 0, \quad H_S(\xi) \neq 0,$$

the zero-order deformation equations (16.22) to (16.24) are equivalent to the original equations (16.6) to (16.9), provided

$$\Phi(\xi;1) = F(\xi), \quad \Theta(\xi;1) = S(\xi). \tag{16.26}$$

Thus, as q increases from 0 to 1, $\Phi(\xi;q)$ and $\Theta(\xi;q)$ vary (or deform) from the initial guesses $F_0(\xi), S_0(\xi)$ to the solutions $F(\xi), S(\xi)$ of Equations (16.6) to (16.9), respectively.

By Taylor's theorem and using (16.25), we obtain the power series

$$\Phi(\xi;q) = F_0(\xi) + \sum_{n=1}^{+\infty} F_n(\xi)\, q^n, \tag{16.27}$$

$$\Theta(\xi;q) = S_0(\xi) + \sum_{n=1}^{+\infty} S_n(\xi)\, q^n, \tag{16.28}$$

where

$$F_n(\xi) = \frac{1}{n!} \left.\frac{\partial^n \Phi(\xi;q)}{\partial q^n}\right|_{q=0}, \quad S_n(\xi) = \frac{1}{n!} \left.\frac{\partial^n \Theta(\xi;q)}{\partial q^n}\right|_{q=0}. \tag{16.29}$$

Assuming that the spatial-scale parameter λ, the auxiliary parameter γ in (16.15), the auxiliary parameters $\hbar_F, \hbar_S$ and the auxiliary functions $H_F(\xi)$, $H_S(\xi)$ are properly chosen so that the above series converge at $q = 1$, we have, using (16.26),

$$F(\xi) = F_0(\xi) + \sum_{n=1}^{+\infty} F_n(\xi), \tag{16.30}$$

$$S(\xi) = S_0(\xi) + \sum_{n=1}^{+\infty} S_n(\xi). \tag{16.31}$$

The corresponding mth-order approximations are given by

$$F(\xi) \approx F_0(\xi) + \sum_{n=1}^{m} F_n(\xi), \tag{16.32}$$

$$S(\xi) \approx S_0(\xi) + \sum_{n=1}^{m} S_n(\xi). \tag{16.33}$$

16.1.3 High-order deformation equation

For conciseness, define the vectors

$$\vec{F}_m = \{F_0(\xi), F_1(\xi), F_2(\xi), \cdots, F_m(\xi)\}$$

and

$$\vec{S}_m = \{S_0(\xi), S_1(\xi), S_2(\xi), \cdots, S_m(\xi)\}.$$

Boundary-layer Flows with Algebraic Property

Differentiating the zero-order deformation equations (16.22) to (16.24) n times with respective to q, then dividing by $n!$, and finally setting $q = 0$, we have the high-order deformation equations

$$\mathcal{L}_F [F_n(\xi) - \chi_n F_{n-1}(\xi)] = \hbar_F \, H_F(\xi) \, R_n^F(\vec{F}_{n-1}, \vec{S}_{n-1}), \qquad (16.34)$$

$$\mathcal{L}_S [S_n(\xi) - \chi_n S_{n-1}(\xi)] = \hbar_S \, H_S(\xi) \, R_n^S(\vec{F}_{n-1}, \vec{S}_{n-1}), \qquad (16.35)$$

subject to the boundary conditions

$$F_n(1) = S_n(1) = F_n(+\infty) = S_n(+\infty) = 0, \qquad (16.36)$$

where χ_n is defined by (2.42) and

$$R_n^F(\vec{F}_{n-1}, \vec{S}_{n-1}) = \lambda^2 \, F_{n-1}''(\xi) + S_{n-1}(\xi)$$
$$- \sum_{j=0}^{n-1} F_j(\xi) \, F_{n-1-j}(\xi), \qquad (16.37)$$

$$R_n^S(\vec{F}_{n-1}, \vec{S}_{n-1}) = \lambda^2 \, S_{n-1}''(\xi) - 3 \, \sigma \sum_{j=0}^{n-1} F_j(\xi) \, S_{n-1-j}(\xi). \qquad (16.38)$$

Note that the high-order deformation equations (16.34) to (16.36) are uncoupled and linear. It is therefore easy to solve them, using symbolic computation software.

Under the *rules of the solution expression* denoted by (16.13) and (16.14) and from Equations (16.34) and (16.35), the auxiliary functions $H_F(\xi)$ and $H_S(\xi)$ may be written as

$$H_F(\xi) = \xi^{\kappa_1}, \quad H_S(\xi) = \xi^{\kappa_2}, \qquad (16.39)$$

where κ_1 and κ_2 are integers. It is found that, when $\kappa_1 \geq 1$ and/or $\kappa_2 \geq 1$, the term $\ln \xi$ appears in the solution expressions, which does not conform to the *rules of solution expression* denoted by (16.13) and (16.14). When $\kappa_1 \leq -1$ and/or $\kappa_2 \leq -1$, $F(\xi)$ and $S(\xi)$ do not contain the terms ξ^{-2} and ξ^{-4}, respectively. This is not in accordance with the *rule of coefficient ergodicity*. To adhere to both the *rules of solution expression* and the *rule of coefficient ergodicity*, we must choose

$$\kappa_1 = \kappa_2 = 0,$$

corresponding to

$$H_F(\xi) = H_S(\xi) = 1. \qquad (16.40)$$

Now, the inhomogeneous terms of Equations (16.34) and (16.35) are completely known.

16.1.4 Recursive formulations

By solving the first several high-order deformation equations (16.34) to (16.36), $F_n(\xi)$ and $S_n(\xi)$ can be expressed by

$$F_n(\xi) = \xi^{-2} \sum_{j=0}^{2n+1} a_{n,j}\, \xi^{-j}, \qquad S_n(\xi) = \xi^{-4} \sum_{j=0}^{2n} b_{n,j}\, \xi^{-j}, \qquad (16.41)$$

where $a_{n,j}$ and $b_{n,j}$ are coefficients. Substituting them into Equations (16.34) to (16.36), we have the recursive formulae ($j \geq 1$)

$$a_{n,j} = \chi_n\, \chi_{2n+1-j}\, a_{n-1,j}$$
$$+ \frac{3\,\hbar_F \left[\chi_{2n+2-j}\, \lambda^2\, (j+1)(j+2) a_{n-1,j-1} + \chi_{2n+1-j} b_{n-1,j-1} - A_{n,j-1}\right]}{j(j+2)},$$
$$(16.42)$$

$$b_{n,j} = \chi_n\, \chi_{2n-j}\, b_{n-1,j}$$
$$+ \frac{5\,\hbar_S \left[\chi_{2n+1-j}\, \lambda^2\, (j+3)(j+4) b_{n-1,j-1} - 3\sigma\, B_{n,j-1}\right]}{j(j+4)}, \qquad (16.43)$$

and

$$a_{n,0} = -\sum_{j=1}^{2n+1} a_{n,j}, \qquad b_{n,0} = -\sum_{j=1}^{2n} b_{n,j}, \qquad (16.44)$$

where

$$A_{n,i} = \sum_{j=0}^{n-1} \sum_{r=\max\{0, i+2j-2n+1\}}^{\min\{2j+1, i\}} a_{j,r}\, a_{n-j-1, i-r}, \qquad (16.45)$$

$$B_{n,i} = \sum_{j=0}^{n-1} \sum_{r=\max\{0, i+2j-2n+2\}}^{\min\{2j+1, i\}} a_{j,r}\, b_{n-j-1, i-r}. \qquad (16.46)$$

Using (16.15), we have the first three coefficients

$$a_{0,0} = \gamma, \ a_{0,1} = -\gamma, \ b_{0,0} = 1. \qquad (16.47)$$

From these and using the above recursive formulae, we can calculate all other coefficients $a_{n,j}$ and $b_{n,j}$, successively. Thus, we have the explicit analytic solutions

$$F(\xi) = \sum_{n=0}^{+\infty} \sum_{j=0}^{2n+1} \frac{a_{n,j}}{\xi^{j+2}}, \qquad S(\xi) = \sum_{n=0}^{+\infty} \sum_{j=0}^{2n} \frac{b_{n,j}}{\xi^{j+4}}. \qquad (16.48)$$

Boundary-layer Flows with Algebraic Property

Using the transformation (16.5) we obtain

$$f'(\eta) = \sum_{n=0}^{+\infty} \sum_{j=0}^{2n+1} \frac{a_{n,j}}{(1+\lambda\,\eta)^{j+2}}, \quad \theta(\eta) = \sum_{n=0}^{+\infty} \sum_{j=0}^{2n} \frac{b_{n,j}}{(1+\lambda\,\eta)^{j+4}}. \qquad (16.49)$$

The mth-order approximations are

$$f'(\eta) \approx \sum_{n=0}^{m} \sum_{j=0}^{2n+1} \frac{a_{n,j}}{(1+\lambda\,\eta)^{j+2}}, \quad \theta(\eta) \approx \sum_{n=0}^{m} \sum_{j=0}^{2n} \frac{b_{n,j}}{(1+\lambda\,\eta)^{j+4}}, \qquad (16.50)$$

which give

$$f(\eta) \approx \sum_{n=0}^{m} \sum_{j=0}^{2n+1} \frac{a_{n,j}}{\lambda\,(j+1)} \left[1 - \frac{1}{(1+\lambda\,\eta)^{j+1}}\right] \qquad (16.51)$$

and

$$f''(0) \approx -\lambda \sum_{n=0}^{m} \sum_{j=0}^{2n+1} (j+2) a_{n,j}, \qquad (16.52)$$

$$f(+\infty) \approx \sum_{n=0}^{m} \sum_{j=0}^{2n+1} \frac{a_{n,j}}{\lambda\,(j+1)}, \qquad (16.53)$$

$$\theta'(0) \approx -\lambda \sum_{n=0}^{m} \sum_{j=0}^{2n} (j+4) b_{n,j}. \qquad (16.54)$$

16.1.5 Convergence theorem

THEOREM 16.1
If the solution series (16.30) and (16.31) are convergent, where $F_n(\xi)$ and $S_n(\xi)$ are governed by Equations (16.34) to (16.36) under the definitions (16.37) to (16.38), and (2.42), they must be the solution of Equations (16.6) to (16.9)

Proof: If the solution series (16.30) and (16.31) are convergent, it is necessary that
$$\lim_{m \to +\infty} F_m(\xi) = 0, \quad \lim_{m \to +\infty} S_m(\xi) = 0.$$
Then, from (16.16) and (16.17),
$$\mathcal{L}_F \left[\lim_{m \to +\infty} F_m(\xi)\right] = 0, \quad \mathcal{L}_S \left[\lim_{m \to +\infty} S_m(\xi)\right] = 0.$$
From Equations (16.34) and (16.35), we have, using (2.42),
$$\hbar_F \, H_F(\xi) \sum_{n=1}^{m} R_n^F(\vec{F}_{n-1}, \vec{S}_{n-1}) = \mathcal{L}_F [F_m(\xi)]$$

and
$$\hbar_S \, H_S(\xi) \sum_{n=1}^{m} R_n^S(\vec{F}_{n-1}, \vec{S}_{n-1}) = \mathcal{L}_S \left[S_m(\xi) \right].$$

Therefore,
$$\hbar_F \, H_F(\xi) \sum_{n=1}^{+\infty} R_n^F(\vec{F}_{n-1}, \vec{S}_{n-1}) = \mathcal{L}_F \left[\lim_{m \to +\infty} F_m(\xi) \right] = 0$$

and
$$\hbar_S \, H_S(\xi) \sum_{n=1}^{+\infty} R_n^S(\vec{F}_{n-1}, \vec{S}_{n-1}) = \mathcal{L}_S \left[\lim_{m \to +\infty} S_m(\xi) \right] = 0,$$

which give, since $\hbar_F \neq 0, \hbar_S \neq 0, H_F(\xi) \neq 0$ and $H_S(\xi) \neq 0$,
$$\sum_{n=1}^{+\infty} R_n^F(\vec{F}_{n-1}, \vec{S}_{n-1}) = 0$$

and
$$\sum_{n=1}^{+\infty} R_n^S(\vec{F}_{n-1}, \vec{S}_{n-1}) = 0.$$

Substituting (16.37) and (16.38) into above two expressions and simplifying them, we obtain

$$\lambda^2 \frac{\partial^2}{\partial \xi^2} \left[\sum_{n=0}^{+\infty} F_n(\xi) \right] + \sum_{n=0}^{+\infty} S_n(\xi) - \left[\sum_{n=0}^{+\infty} F_n(\xi) \right]^2 = 0 \qquad (16.55)$$

and

$$\lambda^2 \frac{\partial^2}{\partial \xi^2} \left[\sum_{n=0}^{+\infty} S_n(\xi) \right] - 3\,\sigma \left[\sum_{n=0}^{+\infty} F_n(\xi) \right] \left[\sum_{n=0}^{+\infty} S_n(\xi) \right] = 0. \qquad (16.56)$$

From (16.15) and (16.36), we obviously have

$$\sum_{n=0}^{+\infty} F_n(1) = \sum_{n=0}^{+\infty} F_n(+\infty) = \sum_{n=0}^{+\infty} S_n(+\infty) = 0, \quad \sum_{n=0}^{+\infty} S_n(1) = 1. \qquad (16.57)$$

Comparing Equations (16.55) to (16.57) with Equations (16.6) to (16.9), it is clear that the convergent series (16.30) and (16.31) are the solutions of the two coupled nonlinear differential equations. This ends the proof.

16.2 Result analysis

There exist four auxiliary parameters: $\hbar_F, \hbar_S, \lambda$, and γ. We have therefore a four-parameter family of solution expressions. According to Theorem 16.1, we need only to focus on the choice of the four auxiliary parameters to ensure that the solution series (16.30) and (16.31) converge.

Note that the initial guess $F_0(\xi)$ defined by (16.15) contains the auxiliary parameter γ. Substituting the initial approximations (16.15) into Equations (16.6) and (16.7) and balancing the main terms, we obtain a set of two algebraic equations of λ and γ, which give

$$\lambda = \sqrt{\frac{3\sigma}{20}}\left(1 - \frac{9\sigma}{10}\right)^{-1/4}, \qquad \gamma = \left(1 - \frac{9\sigma}{10}\right)^{-1/2}. \qquad (16.58)$$

Although the above expressions are valid only when $\sigma < 10/9$, it provides valuable information for the choice of λ and γ. From (16.58), it is obvious that

$$\gamma \sim 1, \qquad \lambda \sim \sqrt{\sigma}$$

for $\sigma \ll 1$.

In general, for any given Prandtl number σ, we can investigate the influence of the four auxiliary parameters $\lambda, \gamma, \hbar_F$, and $\hbar_S$ on the convergence of solution series and then choose a set of proper values. For example, let us consider the case $\sigma = 1$. Note that

$$f''(0) = \lambda\, F'(0), \qquad \theta'(0) = \lambda\, S'(0)$$

are related with the skin friction and thermal flux and therefore have important physical meaning. For simplicity, we first study the influence of the four auxiliary parameters on the convergence of $f''(0)$ and $\theta'(0)$. When $\hbar_F = \hbar_S = -1/2$ and $\gamma = 1, 2, 3$, both $\theta'(0)$ and $f''(0)$ are convergent in a region of λ that becomes the largest when $\gamma = 3$, as shown in Figures 16.1 and 16.2. From these two figures, it is clear that the series of $f''(0)$ and $\theta'(0)$ converge if $\lambda = 1/3, \gamma = 3$, and $\hbar_F = \hbar_S = -1/2$. To ensure this, we can further investigate the influence of the auxiliary parameters $\hbar_F$ and $\hbar_S$ on the convergence of solution series when $\lambda = 1/3, \gamma = 3$, and $\hbar_F = \hbar_S = \hbar$ by plotting the so-called $\hbar$-curves (see page 26 and §3.5.1) of $f''(0)$ and $\theta'(0)$, as shown in Figure 16.3. When $\sigma = 1$, the series of $f''(0)$ and $\theta'(0)$ converge by means of $\lambda = 1/3, \gamma = 3$, and $\hbar_F = \hbar_S = -1/2$, as shown in Table 16.1. Generally, for given Prandtl number σ, we can choose the auxiliary parameters $\lambda, \gamma, \hbar_F$, and $\hbar_S$ in the similar way to ensure that the series of $f''(0)$ and $\theta'(0)$ converge. For example, the series of $f''(0)$ and $\theta'(0)$ converge when $\sigma = 1/10$ by means of $\lambda = 1/5, \gamma = 1$, and $\hbar_F = \hbar_S = -1/2$, and when $\sigma = 10$ by $\lambda = 1/3, \gamma = 1, \hbar_F = -1/4$, and $\hbar_S = -1/10$, respectively. The solution series converge when $\sigma > 10$ by means of $\lambda = 1, \gamma = 1, \hbar_F = -1/4$, and $\hbar_S = -1/\sigma$, and when $\sigma < 1/10$ by means of $\gamma = 1, \hbar_F = \hbar_S = -1/2$, and $\lambda = 1/5$ or even smaller. Furthermore, as long as the series of $f''(0)$ and $\theta'(0)$ are convergent, the corresponding solution series of $f(\xi)$ and $\theta(\xi)$ also converge in the whole region $0 \leq \xi < +\infty$, as shown in Figures 16.4 and 16.5.

The convergence of the series of $f''(0)$ and $\theta'(0)$ can be accelerated using the homotopy-Padé technique (see page 38 and §3.5.2), as shown in Tables 16.2 and 16.3. When $\hbar_F = \hbar_S = \hbar$, the $[m, m]$ homotopy-Padé approximants do not depend upon $\hbar$.

Note that we obtain an explicit, purely analytic solution of the two coupled nonlinear equations (16.1) and (16.2) by means of the recursive formulae (16.42) to (16.47). The analytic solutions behave algebraically at infinity. In Chapter 15 the homotopy analysis method is successfully applied to solve the Falkner-Skan boundary layer flows that behave exponentially at infinity. The homotopy analysis method is therefore valid for these two different types of boundary layer flows.

TABLE 16.1
The mth-order approximations of $f''(0)$ and $\theta'(0)$ when $\sigma = 1$ by means of $\lambda = 1/3, \gamma = 3$, and $\hbar_F = \hbar_S = -1/2$ compared with Kuiken's result [111].

m	$f''(0)$	$\theta'(0)$
5	0.713814	-0.831716
10	0.706453	-0.765271
15	0.702547	-0.769478
20	0.697170	-0.771491
25	0.694380	-0.770640
30	0.693538	-0.770001
35	0.693342	-0.769872
40	0.693268	-0.769879
45	0.693227	-0.769876
50	0.693213	-0.769866
Kuiken's result	0.693212	-0.769861

TABLE 16.2
The $[m, m]$ homotopy-Padé approximations of $f''(0)$ compared with Kuiken's result [111].

$[m, m]$	$\sigma = 1/10$ $\lambda = 1/5, \gamma = 1$	$\sigma = 1$ $\lambda = 1/3, \gamma = 3$	$\sigma = 10$ $\lambda = 1/3, \gamma = 1$
$[5, 5]$	0.952170	0.705940	0.433555
$[10, 10]$	0.921936	0.693438	0.452229
$[15, 15]$	0.924108	0.693214	0.447038
$[20, 20]$	0.924087	0.693212	0.447107
$[25, 25]$	0.924088	0.693212	0.447117
$[30, 30]$	0.924086	0.693212	0.447117
$[35, 35]$	0.924084	0.693212	0.447117
$[40, 40]$	0.924083	0.693212	0.447117
$[45, 45]$	0.924083	0.693212	0.447117
$[50, 50]$	0.924083	0.693212	0.447117
Kuiken's result	0.924083	0.693212	0.447117

TABLE 16.3
The $[m, m]$ homotopy-Padé approximations of $\theta'(0)$ compared with Kuiken's result [111].

$[m, m]$	$\sigma = 1/10$ $\lambda = 1/5, \gamma = 1$	$\sigma = 1$ $\lambda = 1/3, \gamma = 3$	$\sigma = 10$ $\lambda = 1/3, \gamma = 1$
$[5, 5]$	-0.347058	-0.774151	-1.61583
$[10, 10]$	-0.350119	-0.770018	-1.49263
$[15, 15]$	-0.350027	-0.769866	-1.49733
$[20, 20]$	-0.350058	-0.769861	-1.49708
$[25, 25]$	-0.350058	-0.769861	-1.49710
$[30, 30]$	-0.350059	-0.769861	-1.49710
$[35, 35]$	-0.350059	-0.769861	-1.49710
$[40, 40]$	-0.350059	-0.769861	-1.49710
$[45, 45]$	-0.350059	-0.769861	-1.49710
$[50, 50]$	-0.350059	-0.769861	-1.49710
Kuiken's result	-0.350059	-0.769861	-1.49710

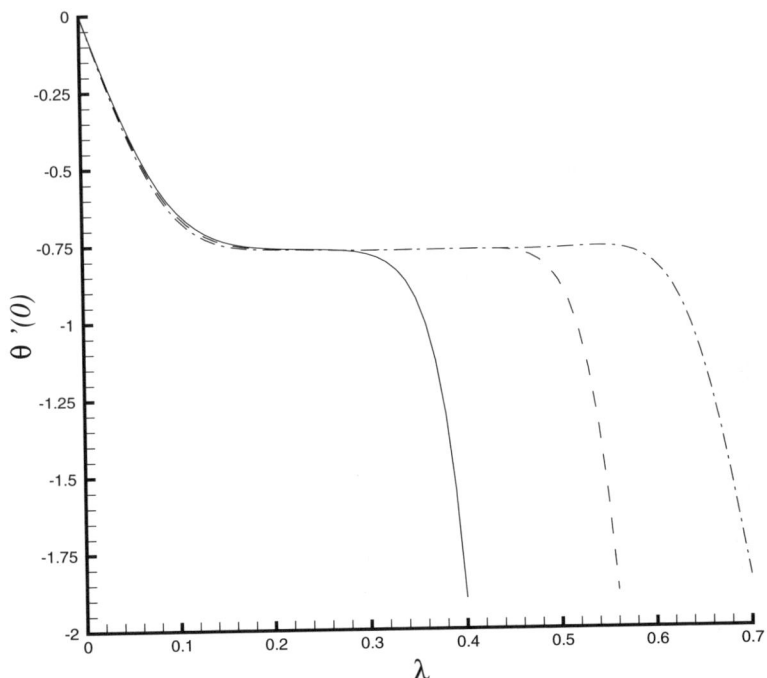

FIGURE 16.1
$\theta'(0)$ versus λ at the 24th order of approximation when $\sigma = 1$ and $\hbar_F = \hbar_S = -1/2$. Solid line: $\gamma = 1$; dashed line: $\gamma = 2$; dash-dotted line: $\gamma = 3$.

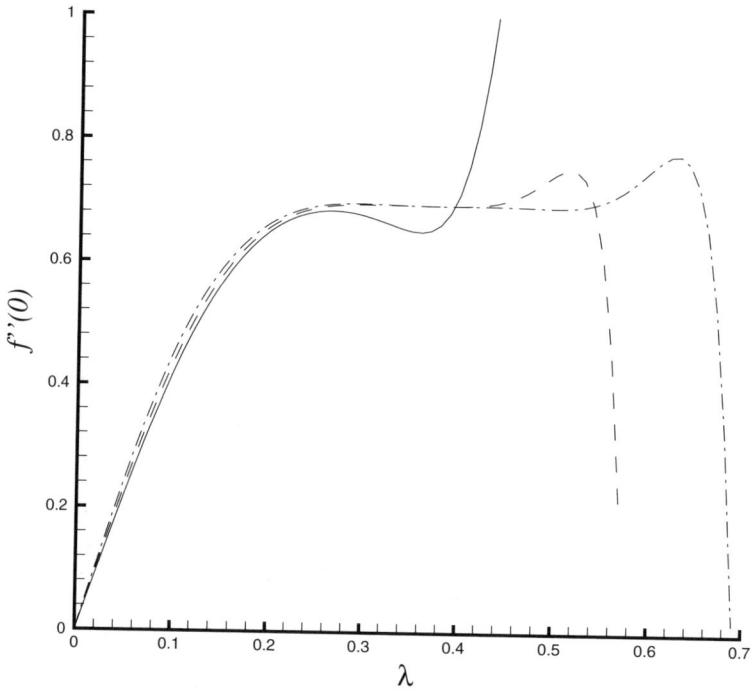

FIGURE 16.2
$f''(0)$ versus λ at the 24th order of approximation when $\sigma = 1$ and $\hbar_F = \hbar_S = -1/2$. Solid line: $\gamma = 1$; dashed line: $\gamma = 2$; dash-dotted line: $\gamma = 3$.

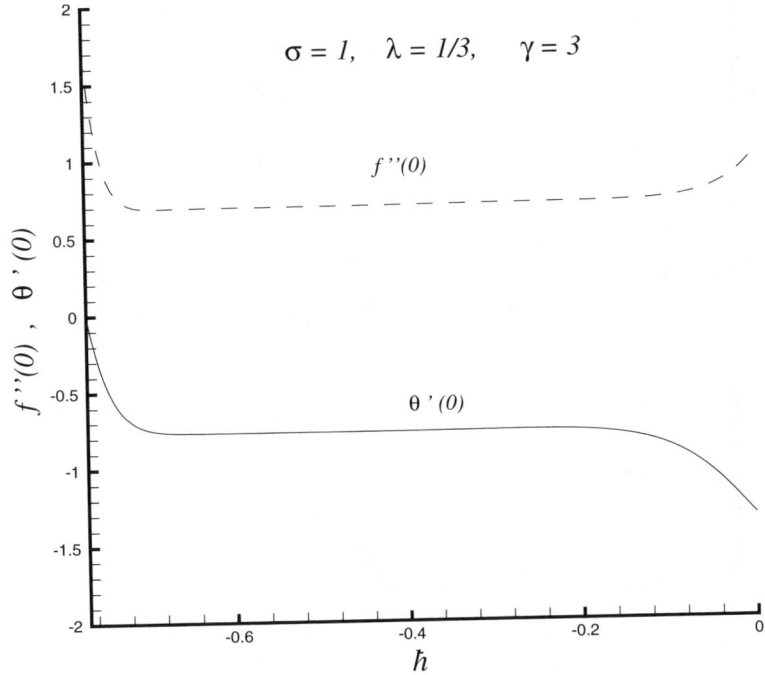

FIGURE 16.3
The $\hbar$-curves of $f''(0)$ and $\theta'(0)$ at the 24th order of approximations when $\sigma = 1$ by means of $\gamma = 3, \lambda = 1/3$, and $\hbar_F = \hbar_S = -1/2$. Solid line: $\theta'(0)$; dashed line: $f''(0)$.

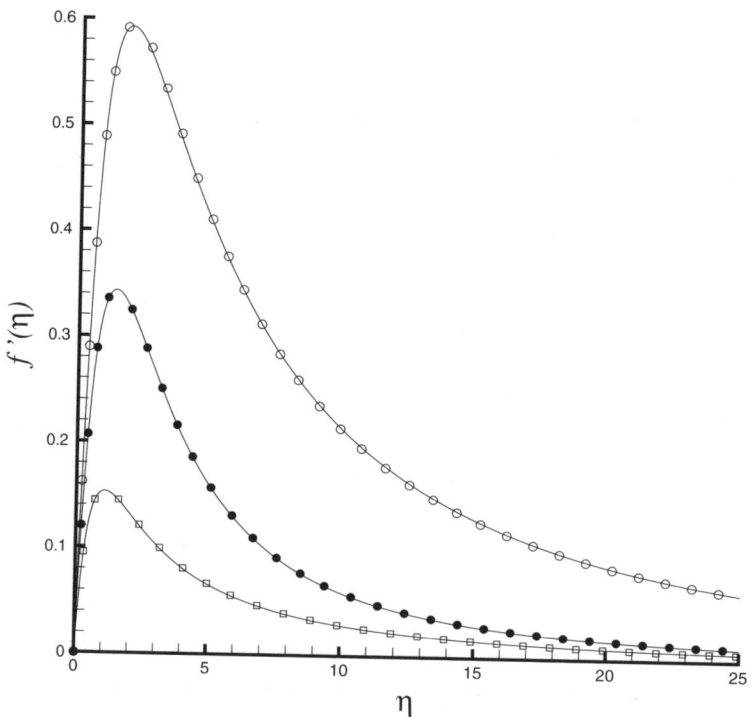

FIGURE 16.4
Comparison of $f'(\eta)$ with numerical results. Open circle: 30th-order approximation when $\sigma = 1/10$ by means of $\hbar_F = \hbar_S = -1/2, \lambda = 1/5$, and $\gamma = 1$; filled-circle: 20th-order approximation when $\sigma = 1$ by means of $\hbar_F = \hbar_S = -1/2, \lambda = 1/3$, and $\gamma = 3$; square: 40th-order approximation when $\sigma = 10$ by means of $\hbar_F = -1/4, \hbar_S = -1/10, \lambda = 1/3$, and $\gamma = 1$; solid lines: numerical results.

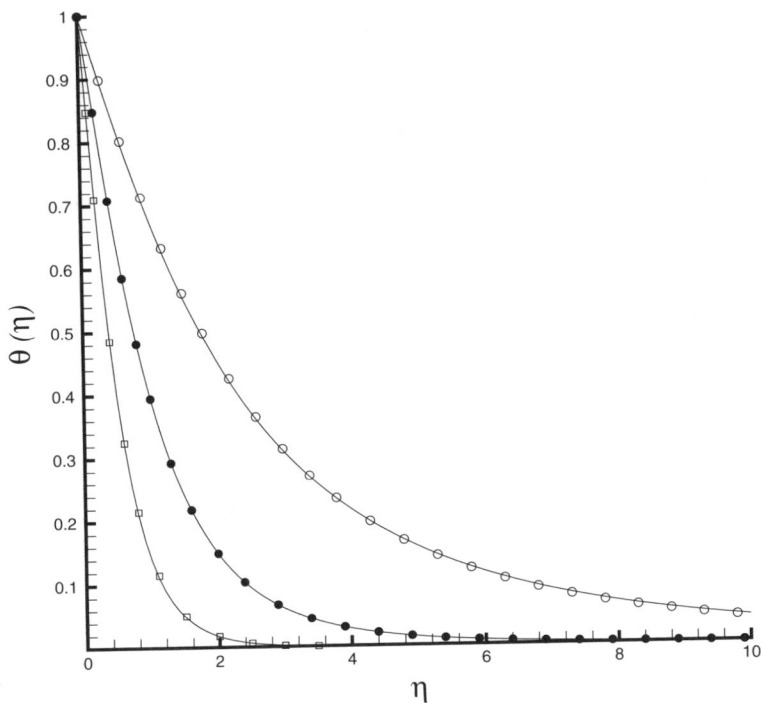

FIGURE 16.5
Comparison of $\theta(\eta)$ with numerical results. Open circle: 20th-order approximation when $\sigma = 1/10$ by means of $\hbar_F = \hbar_S = -1/2, \lambda = 1/5$, and $\gamma = 1$; filled-circle: 20th-order approximation when $\sigma = 1$ by means of $\hbar_F = \hbar_S = -1/2, \lambda = 1/3$, and $\gamma = 3$; square: 20th-order approximation when $\sigma = 10$ by means of $\hbar_F = -1/4, \hbar_S = -1/10, \lambda = 1/3$, and $\gamma = 1$; solid lines: numerical results.

17

Von Kármán swirling viscous flow

Consider the steady, laminar, axially symmetric viscous flow of an incompressible fluid induced by an infinite disk rotating steadily with angular velocity Ω about the z-axis in a cylindrical coordinate system (r, θ, z). The motion of the fluid is governed by the continuity equation

$$\frac{1}{r}\frac{\partial(rV_r)}{\partial r} + \frac{1}{r}\frac{\partial V_\theta}{\partial \theta} + \frac{\partial V_z}{\partial z} = 0 \qquad (17.1)$$

and the Navier-Stokes equations

$$V_r\frac{\partial V_r}{\partial r} + V_z\frac{\partial V_r}{\partial z} - \frac{V_\theta^2}{r} = \nu\left[\frac{\partial^2 V_r}{\partial r^2} + \frac{1}{r}\frac{\partial V_r}{\partial r} + \frac{\partial^2 V_r}{\partial z^2} - \frac{V_r}{r^2}\right] - \frac{1}{\rho}\frac{\partial p}{\partial r}, \qquad (17.2)$$

$$V_r\frac{\partial V_\theta}{\partial r} + V_z\frac{\partial V_\theta}{\partial z} + \frac{V_r V_\theta}{r} = \nu\left[\frac{\partial^2 V_\theta}{\partial r^2} + \frac{1}{r}\frac{\partial V_\theta}{\partial r} + \frac{\partial^2 V_\theta}{\partial z^2} - \frac{V_\theta}{r^2}\right], \qquad (17.3)$$

$$V_r\frac{\partial V_z}{\partial r} + V_z\frac{\partial V_z}{\partial z} = \nu\left[\frac{\partial^2 V_z}{\partial r^2} + \frac{1}{r}\frac{\partial V_z}{\partial r} + \frac{\partial^2 V_z}{\partial z^2}\right] - \frac{1}{\rho}\frac{\partial p}{\partial z}, \qquad (17.4)$$

subject to the nonslip boundary conditions

$$V_\theta = r\Omega, \quad V_r = V_z = 0, \qquad \text{when } z = 0, \qquad (17.5)$$

and the conditions at infinity

$$V_r = V_\theta = 0, \qquad \text{when } z = +\infty, \qquad (17.6)$$

where ρ denotes the fluid density, ν the kinematic viscosity coefficient, p the pressure, and V_r, V_θ, V_z the velocity components in the radial, azimuthal, and axial directions, respectively. Defining the similarity variable

$$\eta = z\sqrt{\frac{\Omega}{\nu}} \qquad (17.7)$$

and using the similarity transformation

$$V_r = (r\Omega)\, f(\eta), \qquad (17.8)$$
$$V_\theta = (r\Omega)\, g(\eta), \qquad (17.9)$$
$$V_z = \sqrt{\nu\Omega}\, w(\eta), \qquad (17.10)$$
$$p = -\rho\nu\Omega\, P(\eta), \qquad (17.11)$$

Von Kármán [112] devised the governing partial differential equations (17.1) to (17.6) to a set of ordinary differential equations

$$f'' = f^2 - g^2 + f' w, \qquad (17.12)$$
$$g'' = g' w + 2f g, \qquad (17.13)$$
$$w w' = P' + w'', \qquad (17.14)$$
$$2f + w' = 0, \qquad (17.15)$$

subject to the boundary conditions

$$f(0) = f(+\infty) = 0,\ g(0) = 1,\ g(+\infty) = 0,\ w(0) = 0, \qquad (17.16)$$

where the prime denotes the derivative with respect to η. From (17.15),

$$f = -\frac{w'}{2}. \qquad (17.17)$$

Substituting it into Equations (17.12) and (17.13), we have

$$w''' - w'' w + \frac{1}{2} w' w' - 2g^2 = 0, \qquad (17.18)$$
$$g'' - w g' + w' g = 0, \qquad (17.19)$$

subject to the boundary conditions

$$w(0) = w'(0) = w'(+\infty) = 0,\ g(0) = 1,\ g(+\infty) = 0. \qquad (17.20)$$

For details the reader is referred to Von Kármán [112] and Zandbergen and Dijkstra [113].

The above equations are coupled and strongly nonlinear. They were investigated by many researchers including Von Kármán [112], Cochran [114], Fettis [115], Rogers and Lance [116], Benton [117], McLeod [118], Zandbergen and Dijkstra [119], Ackroyd [120], and Hulzen [121]. These solutions are either numerical or numerical-analytic. In this chapter the homotopy analysis method is employed to give a purely analytic solution.

17.1 Homotopy analysis solution

Note that the velocity component at infinity in the axial direction is unknown, which has significant physical meaning. So, as Benton [117] did, we define

$$\gamma = -w(+\infty) \qquad (17.21)$$

and use the transformation

$$w(\eta) = -\gamma \left[1 - s(\eta)\right]. \qquad (17.22)$$

Von Kármán Swirling Viscous Flow

However, unlike Benton [117], we introduce another transformation

$$\xi = \lambda \, \eta, \tag{17.23}$$

where λ is the so-called spatial-scale parameter. Using (17.22) and (17.23), Equations (17.18) and (17.19) become

$$\gamma \lambda^3 s''' + \gamma^2 \lambda^2 (1-s) s'' + \frac{1}{2} \gamma^2 \lambda^2 s' s' - 2 g^2 = 0, \tag{17.24}$$

$$\lambda g'' + \gamma (1-s) g' + \gamma s' g = 0, \tag{17.25}$$

subject to the boundary conditions

$$s(0) = g(0) = 1, \ s(+\infty) = g(+\infty) = 0, \ s'(0) = s'(+\infty) = 0, \tag{17.26}$$

where the prime denotes differentiation with respect to ξ. Note that we have the freedom to choose the value of the spatial-scale parameter λ, but γ defined by (17.21) is unknown.

17.1.1 Zero-order deformation equation

According to Rogers and Lance's investigation [116], Von Kármán's swirling flow has exponential property at infinity. In 1978, Dijkstra gave a full asymptotic expansion that contains only exponentials. Hulzen [121] computed the series that consists of exponentials multiplied by polynomials. Recently, Yang and Liao [43] applied the homotopy analysis method to give, for the first time, an explicit, purely analytic solution expressed by exponentials multiplied by polynomials. Here, we give the solutions $s(\xi)$ and $g(\xi)$ expressed by the set of base functions

$$\{\exp(-n\,\xi) \mid n \geq 1\} \tag{17.27}$$

in the forms:

$$s(\xi) = \sum_{n=1}^{+\infty} a_n \exp(-n\xi), \qquad g(\xi) = \sum_{n=1}^{+\infty} b_n \exp(-n\xi), \tag{17.28}$$

where a_n and b_n are coefficients. The above expressions provide the so-called *rules of solution expression* for $s(\xi)$ and $g(\xi)$, respectively.

Let ϵ denote an auxiliary parameter. Under the *rules of solution expression* denoted by (17.28) and using (17.26), it is easy to choose the initial guesses

$$s_0(\xi) = 2\exp(-\xi) - \exp(-2\xi), \tag{17.29}$$
$$g_0(\xi) = \exp(-\xi) + \epsilon \left[\exp(-2\xi) - \exp(-\xi)\right] \tag{17.30}$$

of $s(\xi)$ and $g(\xi)$, respectively. Under the *rules of solution expression* denoted by (17.28) and from Equations (17.24) and (17.25), we choose the auxiliary linear operators

$$\mathcal{L}_s f = \frac{\partial^3 f}{\partial \xi^3} + 2\frac{\partial^2 f}{\partial \xi^2} - \frac{\partial f}{\partial \xi} - 2f, \tag{17.31}$$

$$\mathcal{L}_g f = \frac{\partial^2 f}{\partial \xi^2} - f \tag{17.32}$$

with the properties

$$\mathcal{L}_s \left[C_1 \exp(\xi) + C_2 \exp(-\xi) + C_3 \exp(-2\xi) \right] = 0 \tag{17.33}$$

and

$$\mathcal{L}_g \left[C_1 \exp(\xi) + C_2 \exp(-\xi) \right] = 0, \tag{17.34}$$

where C_1, C_2, and C_3 are coefficients. For conciseness, from Equations (17.24) and (17.25) we define the two nonlinear operators

$$\mathcal{N}_s \left[S(\xi;q), G(\xi;q), \Lambda(q), \Gamma(q) \right]$$
$$= \Gamma(q)\Lambda^3(q) \frac{\partial^3 S(\xi;q)}{\partial \xi^3} + \Gamma^2(q)\Lambda^2(q) \left[1 - S(\xi;q) \right] \frac{\partial^2 S(\xi;q)}{\partial \xi^2}$$
$$+ \left(\frac{1}{2}\right) \Gamma^2(q)\Lambda^2(q) \left[\frac{\partial S(\xi;q)}{\partial \xi} \right]^2 - 2G(\xi;q)^2 \tag{17.35}$$

and

$$\mathcal{N}_g \left[S(\xi;q), G(\xi;q), \Lambda(q), \Gamma(q) \right]$$
$$= \Lambda(q) \frac{\partial^2 G(\xi;q)}{\partial \xi^2} + \Gamma(q) \left[1 - S(\xi;q) \right] \frac{\partial G(\xi;q)}{\partial \xi}$$
$$+ \Gamma(q)G(\xi;q) \frac{\partial S(\xi;q)}{\partial \xi}, \tag{17.36}$$

where $q \in [0,1]$ is an embedding parameter, $S(\xi;q)$ and $G(\xi;q)$ are real functions of ξ and q, $\Lambda(q)$ and $\Gamma(q)$ are real functions of q, respectively. Let $\hbar_g$ and $\hbar_s$ denote two nonzero auxiliary parameters, $H_s(\xi)$ and $H_g(\xi)$ two nonzero auxiliary functions, $q \in [0,1]$ the embedding parameter, respectively. We construct the zero-order deformation equations

$$(1-q)\ \mathcal{L}_s \left[S(\xi;q) - s_0(\xi) \right]$$
$$= q\ \hbar_s\ H_s(\xi)\ \mathcal{N}_s[S(\xi;q), G(\xi;q), \Lambda(q), \Gamma(q)], \tag{17.37}$$

$$(1-q)\ \mathcal{L}_g \left[G(\xi;q) - g_0(\xi) \right]$$
$$= q\ \hbar_g\ H_g(\xi)\ \mathcal{N}_g[S(\xi;q), G(\xi;q), \Lambda(q), \Gamma(q)], \tag{17.38}$$

subject to the boundary conditions

$$S(0;q) = 1,\ S(+\infty;q) = 0,\ \left.\frac{\partial S(\xi;q)}{\partial \xi}\right|_{\xi=0} = \left.\frac{\partial S(\xi;q)}{\partial \xi}\right|_{\xi=+\infty} = 0 \tag{17.39}$$

and
$$G(0;q) = 1, \quad G(+\infty;q) = 0. \tag{17.40}$$

When $q = 0$, it is clear from (17.29), (17.30), and Equations (17.37) to (17.40) that
$$S(\xi;0) = s_0(\xi), \quad G(\xi;0) = g_0(\xi). \tag{17.41}$$

When $q = 1$, since $\hbar_s \ne 0, \hbar_g \ne 0, H_s(\xi) \ne 0$ and $H_g(\xi) \ne 0$, Equations (17.37) to (17.40) are equivalent to the original equations (17.24) to (17.26), provided
$$S(\xi;1) = s(\xi), \quad G(\xi;1) = g(\xi), \quad \Lambda(1) = \lambda, \quad \Gamma(1) = \gamma. \tag{17.42}$$

By Taylor's theorem and using (17.41), we have the power series in the expansion of q as follows:

$$S(\xi;q) = s_0(\xi) + \sum_{n=1}^{+\infty} s_n(\xi)\, q^n, \tag{17.43}$$

$$G(\xi;q) = g_0(\xi) + \sum_{n=1}^{+\infty} g_n(\xi)\, q^n, \tag{17.44}$$

$$\Lambda(q) = \lambda_0 + \sum_{n=1}^{+\infty} \lambda_n\, q^n, \tag{17.45}$$

$$\Gamma(q) = \gamma_0 + \sum_{n=1}^{+\infty} \gamma_n\, q^n, \tag{17.46}$$

where λ_0 and γ_0 are initial guesses of λ and γ, and

$$s_n(\xi) = \frac{1}{n!} \left.\frac{\partial^n S(\xi;q)}{\partial q^n}\right|_{q=0}, \tag{17.47}$$

$$g_n(\xi) = \frac{1}{n!} \left.\frac{\partial^n G(\xi;q)}{\partial q^n}\right|_{q=0}, \tag{17.48}$$

$$\lambda_n = \frac{1}{n!} \left.\frac{\partial^n \Lambda(q)}{\partial q^n}\right|_{q=0}, \tag{17.49}$$

$$\gamma_n = \frac{1}{n!} \left.\frac{\partial^n \Gamma(q)}{\partial q^n}\right|_{q=0}. \tag{17.50}$$

Note that Equations (17.37) and (17.38) contain two auxiliary parameters $\hbar_s$ and $\hbar_g$, and two auxiliary functions $H_s(\xi)$ and $H_g(\xi)$. Assuming that all of them are correctly chosen so that the above series are convergent at $q = 1$, we have, using (17.42), the solution series

$$s(\xi) = s_0(\xi) + \sum_{n=1}^{+\infty} s_n(\xi), \tag{17.51}$$

$$g(\xi) = g_0(\xi) + \sum_{n=1}^{+\infty} g_n(\xi), \qquad (17.52)$$

$$\lambda = \lambda_0 + \sum_{n=1}^{+\infty} \lambda_n, \qquad (17.53)$$

$$\gamma = \gamma_0 + \sum_{n=1}^{+\infty} \gamma_n. \qquad (17.54)$$

17.1.2 High-order deformation equation

For conciseness, define the vectors

$$\vec{s}_k = \{s_0(\xi), s_1(\xi), s_2(\xi), \cdots, s_k(\xi)\},$$

$$\vec{g}_k = \{g_0(\xi), g_1(\xi), g_2(\xi), \cdots, g_k(\xi)\},$$

and

$$\vec{\lambda}_k = \{\lambda_0, \lambda_1, \lambda_2, \cdots, \lambda_k\}, \qquad \vec{\gamma}_k = \{\gamma_0, \gamma_1, \gamma_2, \cdots, \gamma_k\}.$$

Differentiating the zero-order deformation equations (17.37) to (17.40) n times with respect to q, then dividing by $n!$, and finally setting $q = 0$, we have the high-order deformation equations

$$\mathcal{L}_s[s_n(\xi) - \chi_n s_{n-1}(\xi)] = \hbar_s H_s(\xi) R_n^s(\vec{s}_{n-1}, \vec{g}_{n-1}, \vec{\lambda}_{n-1}, \vec{\gamma}_{n-1}), \qquad (17.55)$$

$$\mathcal{L}_g[g_n(\xi) - \chi_n g_{n-1}(\xi)] = \hbar_g H_g(\xi) R_n^g(\vec{s}_{n-1}, \vec{g}_{n-1}, \vec{\lambda}_{n-1}, \vec{\gamma}_{n-1}), \qquad (17.56)$$

subject to the boundary conditions

$$s_n(0) = g_n(0) = s_n(+\infty) = g_n(+\infty) = 0, \quad s'_n(0) = s'_n(+\infty) = 0, (17.57)$$

where χ_n is defined by (2.42),

$$R_n^s(\vec{s}_{n-1}, \vec{g}_{n-1}, \vec{\lambda}_{n-1}, \vec{\gamma}_{n-1})$$
$$= \frac{1}{(n-1)!} \left. \frac{\partial^{n-1} \mathcal{N}_s[S(\xi;q), G(\xi;q), \Lambda(q), \Gamma(q)]}{\partial q^{n-1}} \right|_{q=0}$$
$$= \sum_{k=0}^{n-1} [\alpha_{n-1-k} \, s'''_k(\xi) + \beta_{n-1-k} \, s''_k(\xi)]$$
$$- \sum_{k=0}^{n-1} \beta_{n-1-k} \left[\sum_{j=0}^{k} s_j(\xi) \, s''_{k-j}(\xi) \right]$$
$$+ \frac{1}{2} \sum_{k=0}^{n-1} \beta_{n-1-k} \left[\sum_{j=0}^{k} s'_j(\xi) \, s'_{k-j}(\xi) \right]$$
$$- 2 \sum_{k=0}^{n-1} g_{n-1-k}(\xi) \, g_k(\xi), \qquad (17.58)$$

and

$$R_n^g(\vec{s}_{n-1}, \vec{g}_{n-1}, \vec{\lambda}_{n-1}, \vec{\gamma}_{n-1})$$
$$= \frac{1}{(n-1)!} \left. \frac{\partial^{n-1} \mathcal{N}_g [S(\xi;q), G(\xi;q), \Lambda(q), \Gamma(q)]}{\partial q^{n-1}} \right|_{q=0}$$
$$= \sum_{k=0}^{n-1} [\lambda_{n-1-k}\, g_k''(\xi) + \gamma_{n-1-k}\, g_k'(\xi)]$$
$$+ \sum_{k=0}^{n-1} \gamma_{n-1-k} \sum_{j=0}^{k} [s_j'(\xi)\, g_{k-j}(\xi) - s_j(\xi)\, g_{k-j}'(\xi)] \qquad (17.59)$$

under the definitions

$$\alpha_n = \sum_{k=0}^{n} \lambda_{n-k}\, \delta_k, \qquad (17.60)$$

$$\beta_n = \sum_{k=0}^{n} \gamma_{n-k}\, \delta_k, \qquad (17.61)$$

$$\delta_n = \sum_{k=0}^{n} \gamma_{n-k} \sum_{j=0}^{k} \lambda_j\, \lambda_{k-j}. \qquad (17.62)$$

Note that the high-order deformation equations (17.55) and (17.56) are linear and uncoupled, subject to the linear boundary conditions (17.57). It is easy to successively solve them, using symbolic computation software.

Note that there exist four unknowns: $s_n(\xi)$, $g_n(\xi)$, λ_{n-1}, and γ_{n-1}. However, we have only two differential equations (17.55) and (17.56) for $s_n(\xi)$ and $g_n(\xi)$. Thus, the problem is not closed and two additional algebraic equations are needed to determine λ_{n-1} and γ_{n-1}. Under the *rules of solution expression* denoted by (17.28) and from Equations (17.55) and (17.56), the auxiliary functions should be

$$H_s(\xi) = \exp(\kappa_s\, \xi), \qquad H_g(\xi) = \exp(\kappa_g\, \xi), \qquad (17.63)$$

where κ_s and κ_g are integers. Using (17.29) and (17.30), we have

$$R_1^s(\vec{s}_0, \vec{g}_0, \vec{\lambda}_0, \vec{\gamma}_0) = \sum_{k=1}^{4} c_{1,k}(\lambda_0, \gamma_0)\, \exp(-k\xi), \qquad (17.64)$$

$$R_1^g(\vec{s}_0, \vec{g}_0, \vec{\lambda}_0, \vec{\gamma}_0) = \sum_{k=1}^{3} d_{1,k}(\lambda_0, \gamma_0)\, \exp(-k\xi), \qquad (17.65)$$

where $c_{1,k}(\lambda_0, \gamma_0)$ and $d_{1,k}(\lambda_0, \gamma_0)$ are coefficients independent of ξ. There are two ways to solve the problem. One is to enforce the coefficients $c_{1,1}(\lambda_0, \gamma_0)$ and $d_{1,1}(\lambda_0, \gamma_0)$ to be zero, which gives

$$2\gamma_0(\gamma_0 - \lambda_0)\lambda_0^2 = 0, \qquad \gamma_0 - \lambda_0 = 0. \qquad (17.66)$$

The above set of algebraic equations has an infinite number of solutions

$$\gamma_0 = \lambda_0. \qquad (17.67)$$

It implies that $d_{1,1}(\lambda_0, \gamma_0) = 0$ is true as long as $c_{1,1}(\lambda_0, \gamma_0) = 0$. Thus, it does not work. The other is to make the coefficients $c_{1,1}(\lambda_0, \gamma_0)$ and $c_{1,2}(\lambda_0, \gamma_0)$ zero, which gives

$$2\gamma_0(\gamma_0 - \lambda_0)\lambda_0^2 = 0, \qquad (1-\epsilon)^2 + 3\gamma_0^2\lambda_0^2 - 4\gamma_0\lambda_0^3 = 0. \qquad (17.68)$$

This set of algebraic equations has the unique nonzero solution

$$\gamma_0 = \sqrt{|1-\epsilon|}, \qquad \lambda_0 = \sqrt{|1-\epsilon|}. \qquad (17.69)$$

In this way, the problem is closed and the *rule of solution existence* is satisfied. In this case, if $\kappa_s > 0$, the right-hand side of Equation (17.55) contains the term $\exp(-2\xi)$ and/or $\exp(-\xi)$. Thus, according to the property (17.33), $s_1(\xi)$ contains the term $\xi \exp(-2\xi)$ and/or $\xi \exp(-\xi)$, which however does not conform to the *rules of solution expression* denoted by (17.28). So, we have

$$\kappa_s \leq 0.$$

On the other hand, when $\kappa_s \leq -1$, the coefficient of the term $\exp(-3\xi)$ of $s(\xi)$ cannot be modified; and this however is not in accordance with the *rule of coefficient ergodicity*. So, to ensure that both of the *rules of solution expression* denoted by (17.28) and the *rule of coefficient ergodicity* hold, we must choose

$$\kappa_s = 0, \qquad (17.70)$$

corresponding to $H_s(\xi) = 1$. Similarly, we have

$$\kappa_g = 0, \qquad (17.71)$$

which gives $H_g(\xi) = 1$.

In summary, under the *rules of solution expression* denoted by (17.28) and the *rule of coefficient ergodicity*, and in order to work out the high-order deformation equations, we choose the auxiliary functions

$$H_s(\xi) = H_g(\xi) = 1 \qquad (17.72)$$

and solve the set of two algebraic equations

$$c_{n,1}(\vec{\lambda}_{n-1}, \vec{\gamma}_{n-1}) = 0, \qquad c_{n,2}(\vec{\lambda}_{n-1}, \vec{\gamma}_{n-1}) = 0 \qquad (17.73)$$

to obtain λ_{n-1} and γ_{n-1}, where $c_{n,1}(\vec{\lambda}_{n-1}, \vec{\gamma}_{n-1})$ and $c_{n,2}(\vec{\lambda}_{n-1}, \vec{\gamma}_{n-1})$ are coefficients of the terms

$$R_n^s(\vec{s}_{n-1}, \vec{g}_{n-1}, \vec{\lambda}_{n-1}, \vec{\gamma}_{n-1}) = \sum_{k=1}^{2n+2} c_{n,k}(\vec{\lambda}_{n-1}, \vec{\gamma}_{n-1}) \exp(-k\xi), \qquad (17.74)$$

Von Kármán Swirling Viscous Flow   283

$$R_n^g(\vec{s}_{n-1}, \vec{g}_{n-1}, \vec{\lambda}_{n-1}, \vec{\gamma}_{n-1}) = \sum_{k=1}^{2n+2} d_{n,k}(\vec{\lambda}_{n-1}, \vec{\gamma}_{n-1}) \exp(-k\xi). \qquad (17.75)$$

Thus, we can successively solve the linear differential equations (17.55) and (17.56) under the linear boundary conditions (17.57), together with the set of two algebraic equations (17.73) that are linear when $n \geq 2$. Then,

$$s_n(\xi) = \sum_{k=1}^{2n+2} a_{n,k} \exp(-k\xi), \qquad (17.76)$$

$$g_n(\xi) = \sum_{k=1}^{2n+2} b_{n,k} \exp(-k\xi), \qquad (17.77)$$

where $a_{n,k}$ and $b_{n,k}$ are coefficients. The recursive formulae for the coefficients $a_{n,k}$ and $b_{n,k}$ may be obtained if the above expressions are substituted into Equations (17.55) to (17.57).

17.1.3   Convergence theorem

THEOREM 17.1
If the solution series (17.51), (17.52), (17.53), and (17.54) are convergent, where $s_n(\xi)$ and $g_n(\xi)$ are governed by (17.55), (17.56), and (17.57) under the definitions (17.31), (17.32), (17.58), (17.59), and (2.42), they must be the solution of Equations (17.24) and (17.25) under the boundary conditions (17.26) .

Proof: If the series (17.51) and (17.52) are convergent, it is necessary that

$$\lim_{m \to +\infty} s_m(\xi) = 0, \quad \lim_{m \to +\infty} g_m(\xi) = 0. \qquad (17.78)$$

Then, using (17.31), (17.32), (2.42) and from Equations (17.55) and (17.56), we have

$$\hbar_s H_s(\xi) \sum_{n=1}^{+\infty} R_n^s(\vec{s}_{n-1}, \vec{g}_{n-1}, \vec{\lambda}_{n-1}, \vec{\gamma}_{n-1})$$

$$= \lim_{m \to +\infty} \mathcal{L}_s [s_m(\xi)] = \mathcal{L}_s \left[\lim_{m \to +\infty} s_m(\xi) \right] = 0 \qquad (17.79)$$

and

$$\hbar_g H_g(\xi) \sum_{n=1}^{+\infty} R_n^g(\vec{s}_{n-1}, \vec{g}_{n-1}, \vec{\lambda}_{n-1}, \vec{\gamma}_{n-1})$$

$$= \lim_{m \to +\infty} \mathcal{L}_g [g_m(\xi)] = \mathcal{L}_g \left[\lim_{m \to +\infty} g_m(\xi) \right] = 0. \qquad (17.80)$$

Since $\hbar_s \neq 0, \hbar_g \neq 0, H_s(\xi) \neq 0$ and $H_g(\xi) \neq 0$, the above equations yield

$$\sum_{n=1}^{+\infty} R_n^s(\vec{s}_{n-1}, \vec{g}_{n-1}, \vec{\lambda}_{n-1}, \vec{\gamma}_{n-1}) = 0 \tag{17.81}$$

and

$$\sum_{n=1}^{+\infty} R_n^g(\vec{s}_{n-1}, \vec{g}_{n-1}, \vec{\lambda}_{n-1}, \vec{\gamma}_{n-1}) = 0. \tag{17.82}$$

Substituting the definitions (17.58) and (17.59) into the above expressions and then simplifying them, due to the convergence of the series (17.51) to (17.54), we have

$$\left(\sum_{i=0}^{+\infty} \gamma_i\right)\left(\sum_{j=0}^{+\infty} \lambda_j\right)^3 \frac{d^3}{d\xi^3}\left[\sum_{k=0}^{+\infty} s_k(\xi)\right]$$
$$+ \left(\sum_{i=0}^{+\infty} \gamma_i\right)^2 \left(\sum_{j=0}^{+\infty} \lambda_j\right)^2 \left[1 - \sum_{k=0}^{+\infty} s_k(\xi)\right] \frac{d^2}{d\xi^2}\left[\sum_{k=0}^{+\infty} s_k(\xi)\right]$$
$$+ \frac{1}{2}\left(\sum_{i=0}^{+\infty} \gamma_i\right)^2 \left(\sum_{j=0}^{+\infty} \lambda_j\right)^2 \frac{d}{d\xi}\left[\sum_{k=0}^{+\infty} s_k(\xi)\right] \frac{d}{d\xi}\left[\sum_{k=0}^{+\infty} s_k(\xi)\right]$$
$$- 2\left[\sum_{k=0}^{+\infty} g_k(\xi)\right]^2 = 0 \tag{17.83}$$

and

$$\left(\sum_{j=0}^{+\infty} \lambda_j\right) \frac{d^2}{d\xi^2}\left[\sum_{k=0}^{+\infty} g_k(\xi)\right] + \left(\sum_{i=0}^{+\infty} \gamma_i\right)\left(1 - \sum_{j=0}^{+\infty} s_j(\xi)\right) \frac{d}{d\xi}\left[\sum_{k=0}^{+\infty} g_k(\xi)\right]$$
$$+ \left(\sum_{i=0}^{+\infty} \gamma_i\right)\left[\sum_{j=0}^{+\infty} g_j(\xi)\right] \frac{d}{d\xi}\left[\sum_{k=0}^{+\infty} s_k(\xi)\right] = 0. \tag{17.84}$$

Furthermore, using (17.29), (17.30), and (17.57), we have

$$\sum_{n=0}^{+\infty} s_n(0) = \sum_{n=0}^{+\infty} g_n(0) = 1, \quad \sum_{n=0}^{+\infty} s_n(+\infty) = \sum_{n=0}^{+\infty} g_n(+\infty) = 0, \tag{17.85}$$

and

$$\sum_{n=0}^{+\infty} s'_n(0) = \sum_{n=0}^{+\infty} s'_n(+\infty) = 0. \tag{17.86}$$

Comparing the above four expressions with Equations (17.24), (17.25), and (17.26), the convergent series (17.51), (17.52), (17.53), and (17.54) must be the solution of Von Kármán's swirling flow. This ends the proof.

17.2 Result analysis

According to Theorem 17.1, we need only to focus on ensuring that the solution series (17.51) to (17.54) are convergent. There exist three auxiliary parameters: ϵ, $\hbar_s$, and $\hbar_g$. We have therefore a three-parameter family of solution expressions. For simplicity, consider the case of

$$\hbar_s = \hbar_g = \hbar$$

and investigate first the influence of ϵ and $\hbar$ on the convergence of $\gamma = -w(+\infty)$ that has a clear physical meaning.

Obviously, for any chosen value of ϵ, γ is a power series of $\hbar$; thus we can investigate the influence of $\hbar$ on the convergence of γ by plotting the so-called $\hbar$-curves (see page 26 and §3.5.1) of γ, as shown in Figure 17.1. From these $\hbar$-curves, it is clear that the series of γ converges when $\epsilon = 0$ and $-1/5 \leq \hbar < 0$, or $\epsilon = 1/4$ and $-3/5 \leq \hbar < 0$. For example, when $\epsilon = 0$ and $\hbar_s = \hbar_g = -1/5$ or $\epsilon = 1/4$ and $\hbar_s = \hbar_g = -1/2$, the corresponding series of γ converges to Benton's [117] numerical result, as shown in Table 17.1. When $\epsilon = 1/4$ and $-3/5 \leq \hbar < 0$ the series converges faster than when $\epsilon = 0$ and $-1/5 \leq \hbar < 0$, indicating that the auxiliary parameters ϵ and $\hbar = \hbar_s = \hbar_g$ may control the convergence rate of the solution series. The homotopy-Padé technique (see page 38 and §3.5.2) can be employed to accelerate the convergence, as shown in Table 17.2.

From Figure 17.1, as ϵ increases from 0 to 0.5, the valid region of $\hbar$ enlarges but then reduces, indicating that there should exist a value of ϵ that corresponds to the largest valid region of $\hbar$. To obtain this value, we choose ϵ so that $\gamma_1 = \lambda_1 = 0$, corresponding to

$$\sqrt{1-\epsilon}\left(119 - 328\epsilon + 193\epsilon^2\right) - 5(3 + 19\epsilon) = 0,$$

whose solution is

$$\epsilon \approx 0.26167. \tag{17.87}$$

The valid region of $\hbar$ corresponding to $\epsilon = 0.26167$ seems to be longest, as shown in Figure 17.1. Moreover, when $\epsilon = 0.26167$ and $\hbar_s = \hbar_g = -1/2$, the series of γ converges even faster, as shown in Tables 17.1 and 17.2.

It is found that

$$\lambda_n = \gamma_n$$

for any integers $n \geq 0$. Thus, we have the relationship

$$\lambda = \gamma,$$

which agrees with the asymptotic expansion for large ξ given by Cochran [114]. Indeed, Von Kármán's swirling flow contains an elegant mathematical structure.

As long as the series of γ is convergent, the corresponding series of $s(\xi)$ and $g(\xi)$ also converge in the whole region $0 \leq \xi < +\infty$. For example, when $\epsilon = 0$ and $\hbar_s = \hbar_g = -1/5$, the analytic approximations of $w(\eta)$ and $g(\eta)$ converge to Benton's [117] numerical result, as shown in Figures 17.2 and 17.3. Moreover, when $\epsilon = 1/4$ and $\hbar_g = \hbar_s = -1/2$, even the [1,1] homotopy-Páde approximants of $g(\eta)$ and $w(\eta)$, i.e.

$$g(\eta) \approx \frac{\Delta_1(\eta)}{\Pi_1(\eta)}, \qquad w(\eta) \approx -\gamma \left[\frac{\Delta_2(\eta)}{\Pi_2(\eta)}\right], \qquad (17.88)$$

agree with Benton's numerical result [117], as shown in Figures 17.4 and 17.5, where

$$\begin{aligned}
\Delta_1(\eta) = &\left(935649 + 3881640\sqrt{3}\right) \exp(-\gamma\eta) \\
&+ \left(456252 + 2097200\sqrt{3}\right) \exp(-2\gamma\eta) \\
&+ \left(785007 - 311640\sqrt{3}\right) \exp(-3\gamma\eta) \\
&+ \left(220464 - 317520\sqrt{3}\right) \exp(-4\gamma\eta) \\
&- \left(22212 + 25200\sqrt{3}\right) \exp(-5\gamma\eta) - 4608 \exp(-6\gamma\eta),
\end{aligned}$$

$$\begin{aligned}
\Pi_1(\eta) = &\left(1247532 + 3022880\sqrt{3}\right) \\
&+ \left(192492 + 1858080\sqrt{3}\right) \exp(-\gamma\eta) \\
&+ \left(982512 + 544320\sqrt{3}\right) \exp(-2\gamma\eta) \\
&- \left(33552 + 100800\sqrt{3}\right) \exp(-3\gamma\eta) - 18432 \exp(-4\gamma\eta),
\end{aligned}$$

$$\begin{aligned}
\Delta_2(\eta) = &\left(1364904 - 477008\sqrt{3}\right) \\
&- \left(2855992 - 962752\sqrt{3}\right) \exp(-\gamma\eta) \\
&+ \left(1612135 - 497280\sqrt{3}\right) \exp(-2\gamma\eta) \\
&- \left(115206 - 14336\sqrt{3}\right) \exp(-3\gamma\eta) \\
&- \left(6545 + 2800\sqrt{3}\right) \exp(-4\gamma\eta) + 704 \exp(-5\gamma\eta),
\end{aligned}$$

$$\begin{aligned}
\Pi_2(\eta) = &\left(1364904 - 477008\sqrt{3}\right) \\
&+ \left(28446 + 8736\sqrt{3}\right) \exp(-\gamma\eta) \\
&- \left(57777 + 2800\sqrt{3}\right) \exp(-2\gamma\eta) + 5184 \exp(-3\gamma\eta),
\end{aligned}$$

in which $\gamma = 0.884474$. This provides us with a simple but accurate analytic expression of Von Kármán's swirling flow.

Von Kármán Swirling Viscous Flow

In this chapter we illustrate that the homotopy analysis method is valid for some three-dimensional viscous flows governed by the exact Navier-Stokes equations. The reader is referred to Liao [41] for viscous flows past a sphere in a uniform stream, a well known classical problem in fluid mechanics.

TABLE 17.1
The mth-order homotopy analysis approximations of $\gamma = -w(+\infty)$.

m	$\epsilon = 0$ $\hbar_s = \hbar_g = -1/5$	$\epsilon = 1/4$ $\hbar_s = \hbar_g = -1/2$	$\epsilon = 0.26167$ $\hbar_s = \hbar_g = -1/2$
10	0.879446	0.882352	0.882977
20	0.881898	0.884437	0.884454
30	0.883607	0.884477	0.884477
40	0.884173	0.884474	0.884474
50	0.884337	0.884474	0.884474

TABLE 17.2
The $[m, m]$ homotopy-Páde approximations of $\gamma = -w(+\infty)$.

$[m, m]$	$\epsilon = 0$ $\hbar_s = \hbar_g = -1/5$	$\epsilon = 1/4$ $\hbar_s = \hbar_g = -1/2$	$\epsilon = 0.26167$ $\hbar_s = \hbar_g = -1/2$
[5, 5]	0.879337	0.883856	0.885038
[10, 10]	0.884502	0.884482	0.884475
[15, 15]	0.884436	0.884474	0.884474
[20, 20]	0.884474	0.884474	0.884474
[25, 25]	0.884474	0.884474	0.884474

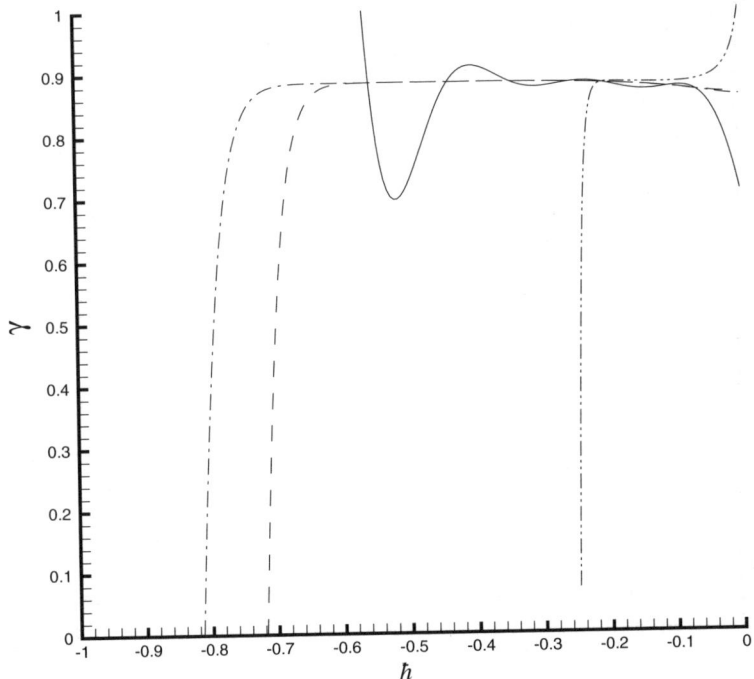

FIGURE 17.1
$\hbar$-curves of $\gamma = -w(+\infty)$ at the 19th order of approximation. Dash-dot-dotted line: $\epsilon = 0$; dashed line: $\epsilon = 1/4$; dash-dotted line: $\epsilon = 0.26167$; solid line: $\epsilon = 1/2$.

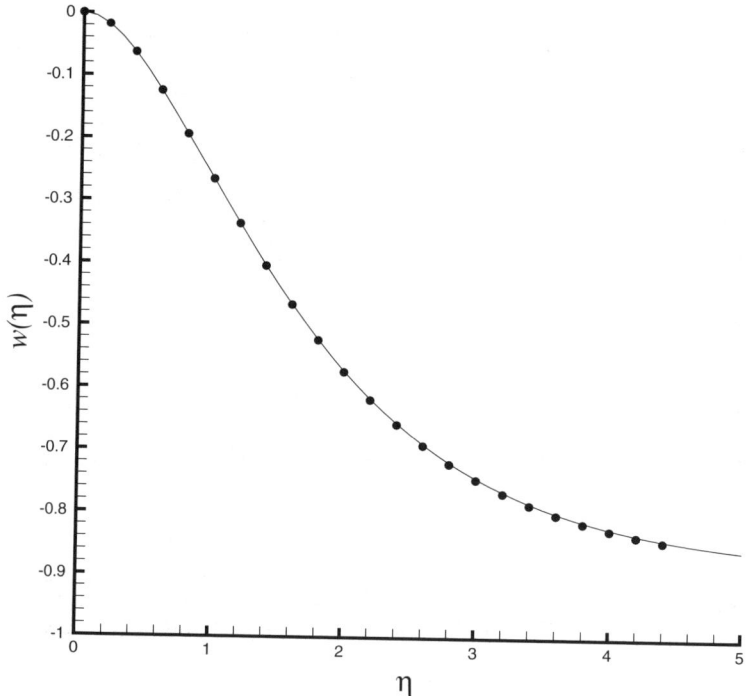

FIGURE 17.2
Comparison of the analytic approximation of $w(\eta)$ with the numerical result given by Benton [117]. Symbol: numerical result; solid line: 20th-order approximation by means of $\epsilon = 0$ and $\hbar_s = \hbar_g = -1/5$.

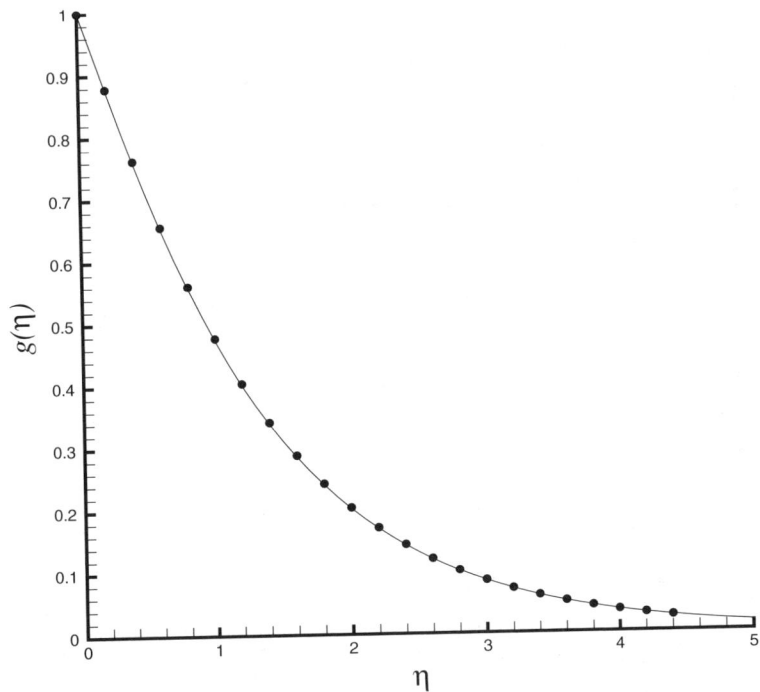

FIGURE 17.3
Comparison of the analytical approximation of $g(\eta)$ with the numerical result given by Benton [117]. Symbol: numerical result; solid line: 20th-order approximation by means of $\epsilon = 0$ and $\hbar_s = \hbar_g = -1/5$.

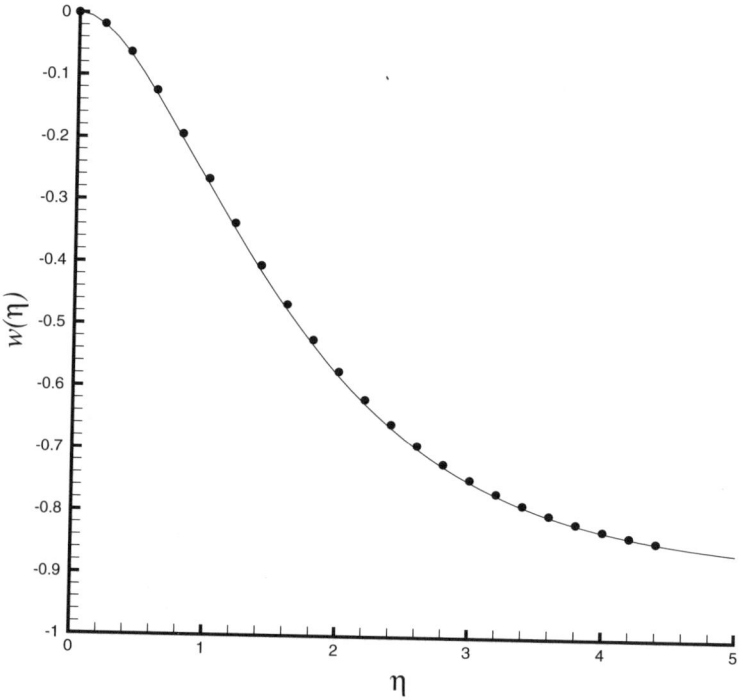

FIGURE 17.4
Comparison of the [1,1] homotopy-Páde approximation of $w(\eta)$ with the numerical result given by Benton [117]. Symbol: numerical result; solid line: [1,1] homotopy-Páde approximation (17.88).

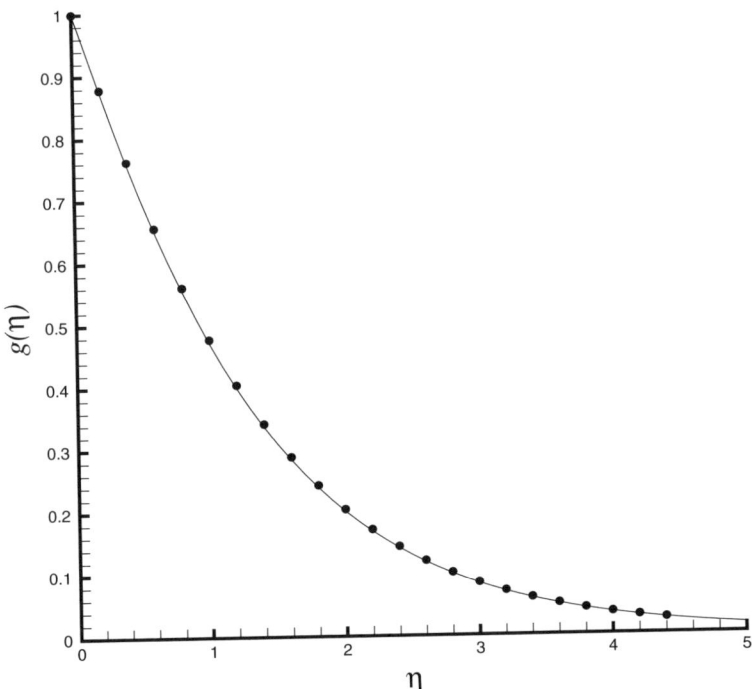

FIGURE 17.5
Comparison of the [1,1] homotopy-Páde approximation of $g(\eta)$ with the numerical result given by Benton [117]. Symbol: numerical result; solid line: [1,1] homotopy-Páde approximation (17.88).

18
Nonlinear progressive waves in deep water

Consider two-dimensional progressive gravity waves moving at a phase speed C on the surface of infinitely deep water. The problem is defined in a coordinate system (x, y) fixed to the waves, with the x-axis positive in the direction of wave propagation and the y-axis pointing vertically upward from the still-water level. Assume that the fluid is inviscid, incompressible, and without surface tension. Let $\phi(x, y)$ denote the velocity potential and $\zeta(x)$ the wave elevation, respectively. The fluid motion can be described by the Laplace equation

$$\nabla^2 \phi(x, y) = 0 \quad \text{for } (x, y) \in \Omega, \tag{18.1}$$

where

$$\Omega = \{(x, y) \mid -\infty < x < +\infty, -\infty < y < \zeta(x)\}.$$

The velocity potential $\phi(x, y)$ is subject to the free surface boundary conditions

$$C^2 \phi_{xx} + g\phi_y + \frac{1}{2}\nabla\phi\nabla(\nabla\phi\nabla\phi) - 2C\nabla\phi\nabla\phi_x = 0 \quad \text{at } y = \zeta(x), \tag{18.2}$$

$$\zeta(x) = \frac{1}{g}\left(C\phi_x - \frac{1}{2}\nabla\phi\nabla\phi\right) \quad \text{at } y = \zeta(x), \tag{18.3}$$

and the bottom condition

$$\lim_{y \to -\infty} \frac{\partial \phi}{\partial y} = 0, \tag{18.4}$$

where g is the acceleration of gravity and the subscripts x and y denote partial derivatives in the respective directions.

Although the governing equation (18.1) is linear, the free surface boundary conditions (18.2) and (18.3) are nonlinear and are defined on a surface that is unknown *a priori*. This classic water-wave problem does not have a simple solution and has attracted attention from many researchers since the mid-19th century. Stokes [122] first proposed a perturbation technique for this classical problem and later obtained an analytic solution to the fifth order in wave amplitude [123, 124]. Thereafter, researchers have applied Stokes' perturbation approach and derived higher-order solutions [125, 126, 127]. Using computer, Schwartz [128] extended Stokes' perturbation expansion to obtain a solution to the 58th order. The solution is obtained in the complex plane through a mapping function. His perturbation expansion has limited convergence and the Padé technique is employed to derive the solution at the limiting wave condition $(H/L)_{max} = 0.14118$, where H is the wave height and L denotes the wavelength.

Following Schwartz [128], Longuet-Higgins [129] took the Stokes-type expansion in wave amplitude to high orders and obtained stable solutions up to the wave steepness $H/L = 0.1411$. The results show that for a given wavelength the energy and phase speed are not monotonic functions of wave steepness. Besides, Longuet-Higgins [130, 131] investigated the stability of steady gravity waves to infinitesimal disturbances and found that subharmonic modes that become unstable when the wave height reaches a certain value, may become stable and then unstable again as the wave height continues to increase. Chen and Saffman [132] found by numerical techniques that symmetrical steady gravity waves of large amplitudes have bifurcations at $H/L \approx 0.13$. Additional high-order solutions based on Stokes' perturbation approach further illustrate the nonlinear characteristics of steep gravity waves [133, 134, 135, 136].

In this chapter we apply the homotopy analysis method to solve this boundary-value problem with nonlinear conditions on an unknown surface.

18.1 Homotopy analysis solution

18.1.1 Zero-order deformation equation

The velocity potential ϕ satisfies the Laplace equation (18.1) and the bottom boundary condition (18.4). So, it is easily understood that ϕ can be expressed by the set of base functions

$$\{\exp(mky) \sin(nkx) \mid m \geq 1, n \geq 1\} \tag{18.5}$$

in the form:

$$\phi(x,y) = \sum_{m=1}^{+\infty} \sum_{n=1}^{+\infty} \alpha_{m,n} \exp(mky) \sin(nkx), \tag{18.6}$$

where $k = 2\pi/L$ is the wave number, $\alpha_{m,n}$ is a coefficient. This provides us with the *rule of solution expression* for the velocity potential $\phi(x,y)$. Accordingly, the wave elevation $\zeta(x)$ can be expressed by the set of base functions

$$\{\cos(m\,k\,x) \mid m \geq 0\} \tag{18.7}$$

in the form:

$$\zeta(x) = \sum_{m=0}^{+\infty} \beta_m \cos(mkx), \tag{18.8}$$

Nonlinear Progressive Waves in Deep Water

which provides us with the *rule of solution expression* for the wave elevation $\zeta(x)$.

Under the *rule of solution expression* denoted by (18.6), it is expedient to select the solution of the linear Airy wave theory

$$\phi_0(x,y) = A\, C_0\, \exp(ky)\sin(kx), \tag{18.9}$$

$$C_0 = \sqrt{\frac{g}{k}}, \tag{18.10}$$

as the initial guesses of the velocity potential $\phi(x,y)$ and the phase speed C, where A is a constant to be determined later. In spite of the more obvious choice from the linear solution, we choose

$$\zeta_0(x) = 0 \tag{18.11}$$

as the initial guess of the surface elevation $\zeta(x)$ to simplify the subsequent formulation and the solution procedure. Based on the two linear terms of the free surface boundary condition (18.2), we choose an auxiliary linear operator

$$\mathcal{L}\left[\Phi(x,y;q), \Lambda(q)\right] = \Lambda^2(q)\frac{\partial^2 \Phi(x,y;q)}{\partial x^2} + g\frac{\partial \Phi(x,y;q)}{\partial y}, \tag{18.12}$$

where $q \in [0,1]$ is an embedding parameter, $\Lambda(q)$ is a real function of q, $\Phi(x,y;q)$ is a real function of x, y, and q. From the two free surface boundary conditions (18.2) and (18.3), we define two nonlinear operators

$$\begin{aligned}
&\mathcal{N}\left[\Phi(x,y;q), \Lambda(q)\right] \\
&= \Lambda^2(q)\Phi_{xx}(x,y;q) + g\Phi_y(x,y;q) \\
&\quad + \frac{1}{2}\nabla\Phi(x,y;q)\nabla\left[\nabla\Phi(x,y;q)\nabla\Phi(x,y;q)\right] \\
&\quad - 2\Lambda(q)\nabla\Phi(x,y;q)\nabla\Phi_x(x,y;q)
\end{aligned} \tag{18.13}$$

and

$$\begin{aligned}
&\mathcal{Z}\left[\Phi(x,y;q), \Lambda(q)\right] \\
&= \frac{1}{g}\left[\Lambda(q)\,\Phi_x(x,y;q) - \frac{1}{2}\nabla\Phi(x,y;q)\nabla\Phi(x,y;q)\right].
\end{aligned} \tag{18.14}$$

The homotopy analysis method is based on a continuous variation from an initial trial to the exact solution. In the water-wave problem, we construct the mappings $\phi(x,y) \to \Phi(x,y;q)$, $\zeta(x) \to \eta(x;q)$, and $C \to \Lambda(q)$ so that, as the embedding parameter q increases from 0 to 1, $\Phi(x,y;q)$, $\eta(x;q)$, and $\Lambda(q)$ vary from the initial guesses to the exact solution $\phi(x,y)$, $\zeta(x)$, and C respectively. To ensure this, based on Equations (18.1) to (18.4), we construct the zero-order deformation equation

$$\nabla^2 \Phi(x,y;q) = 0 \qquad \text{for } (x,y) \in \overline{\Omega}(q), \tag{18.15}$$

subject to the boundary conditions on the unknown free surface $y = \eta(x;q)$,

$$(1-q)\,\mathcal{L}\,[\Phi(x,y;q) - \phi_0(x,y), \Lambda(q)]$$
$$= q\,\hbar_1\,H_1(x)\,\mathcal{N}[\Phi(x,y;q), \Lambda(q)], \qquad (18.16)$$

$$(1-q)\,[\eta(x;q) - \zeta_0(x)]$$
$$= q\,\hbar_2\,H_2(x)\,\{\eta(x;q) - \mathcal{Z}[\Phi(x,y;q), \Lambda(q)]\}, \qquad (18.17)$$

and the boundary condition on the bottom

$$\lim_{y \to -\infty} \frac{\partial \Phi(x,y;q)}{\partial y} = 0, \qquad (18.18)$$

where $q \in [0,1]$ is the embedding parameter, $\hbar_1, \hbar_2$ are two nonzero auxiliary parameters, $H_1(x), H_2(x)$ are two nonzero auxiliary functions, the domain

$$\overline{\Omega}(q) = \{(x,y) \mid -\infty < x < +\infty, -\infty < y < \eta(x;q)\}$$

should preserve the connectedness as q spans the interval $[0,1]$.

When $q = 0$, the governing equation (18.15) and the boundary conditions (18.16) to (18.18) yield the initial approximation

$$\Phi(x,y;0) = \phi_0(x,y), \quad \eta(x,0) = \zeta_0(x), \quad \Lambda(0) = C_0, \qquad (18.19)$$

where C_0 is the initial guess of the phase speed. When $q = 1$, since

$$\hbar_1 \neq 0, \hbar_2 \neq 0, H_1(x) \neq 0, H_2(x) \neq 0,$$

Equations (18.15) to (18.18) are equivalent to Equations (18.1) to (18.4), provided

$$\Phi(x,y;1) = \phi(x,y), \quad \eta(x,1) = \zeta(x), \quad \Lambda(1) = C. \qquad (18.20)$$

As q increases from 0 to 1, the boundary-value problem defined by Equations (18.15) to (18.18) thus provides a continuous variation to transform the initial trial into the exact solution.

Using Taylor's theorem and Equation (18.19), we expand $\Phi(x,y;q), \eta(x;q)$, and $\Lambda(q)$ in the power series of q as follows:

$$\Phi(x,y;q) = \phi_0(x,y) + \sum_{m=1}^{+\infty} \frac{\phi_0^{[m]}(x,y)}{m!} q^m, \qquad (18.21)$$

$$\eta(x;q) = \zeta_0(x) + \sum_{m=1}^{+\infty} \frac{\zeta_0^{[m]}(x)}{m!} q^m, \qquad (18.22)$$

$$\Lambda(q) = C_0 + \sum_{m=1}^{+\infty} \frac{C_0^{[m]}}{m!} q^m, \qquad (18.23)$$

Nonlinear Progressive Waves in Deep Water

where

$$\phi_0^{[m]}(x,y) = \left.\frac{\partial^m \Phi(x,y;q)}{\partial q^m}\right|_{q=0}, \qquad (18.24)$$

$$\zeta_0^{[m]}(x) = \left.\frac{\partial^m \eta(x;q)}{\partial q^m}\right|_{q=0}, \qquad (18.25)$$

$$C_0^{[m]} = \left.\frac{d^m \Lambda(q)}{dq^m}\right|_{q=0}. \qquad (18.26)$$

Note that Equations (18.16) and (18.17) contain two auxiliary parameters $\hbar_1, \hbar_2$, and two auxiliary functions $H_1(x), H_2(x)$. Assuming that all of them are correctly chosen so that the above series are convergent at $q = 1$, from (18.20) we have

$$\phi(x,y) = \phi_0(x,y) + \sum_{m=1}^{+\infty} \frac{\phi_0^{[m]}(x,y)}{m!}, \qquad (18.27)$$

$$\zeta(x) = \zeta_0(x) + \sum_{m=1}^{+\infty} \frac{\zeta_0^{[m]}(x)}{m!}, \qquad (18.28)$$

$$C = C_0 + \sum_{m=1}^{+\infty} \frac{C_0^{[m]}}{m!}. \qquad (18.29)$$

18.1.2 High-order deformation equation

For brevity, define the vectors

$$\vec{\phi}_n = \left\{\phi_0(x,y), \phi_0^{[1]}(x,y), \phi_0^{[2]}(x,y), \cdots, \phi_0^{[n]}(x,y)\right\},$$

$$\vec{\zeta}_n = \left\{\zeta_0(x), \zeta_0^{[1]}(x), \zeta_0^{[2]}(x), \cdots, \zeta_0^{[n]}(x)\right\},$$

and

$$\vec{C}_n = \left\{C_0, C_0^{[1]}, C_0^{[2]}, \cdots, C_0^{[n]}\right\}.$$

Besides, define the so-called deformation derivatives

$$\Phi^{[m]}(x,y;q) = \frac{\partial^m \Phi(x,y;q)}{\partial q^m}, \qquad (18.30)$$

$$\eta^{[m]}(x;q) = \frac{\partial^m \eta(x;q)}{\partial q^m}, \qquad (18.31)$$

$$\Lambda^{[m]} = \frac{d^m \Lambda(q)}{dq^m}. \qquad (18.32)$$

Differentiating Equations (18.15) and (18.18) m times with respect to q and setting $q = 0$, we have the high-order deformation equation

$$\nabla^2 \phi_0^{[m]}(x,y) = 0 \quad \text{in } (x,y) \in \Omega_0 \qquad (18.33)$$

and the condition on the bottom

$$\lim_{y \to -\infty} \frac{\partial \phi_0^{[m]}(x,y)}{\partial y} = 0, \tag{18.34}$$

where

$$\Omega_0 = \{(x,y) \mid -\infty < x < +\infty, -\infty < y \leq \zeta_0(x)\}.$$

It should be emphasized that the free surface boundary conditions (18.16) and (18.17) are satisfied at $y = \eta(x;q)$, which is dependent on q. Thus, it holds for $\Phi(x,y;q)$ at $y = \eta(x;q)$ that

$$\frac{D^m \Phi(x,y;q)}{Dq^m} = \left[\frac{\partial}{\partial p} + \eta^{[1]}(x;q)\frac{\partial}{\partial y}\right]^m \Phi(x,y;q), \tag{18.35}$$

where $\eta^{[1]}(x;q)$ is defined by (18.31). The differential operator D^m/Dq^m, which contains the linear term $\partial^m/\partial q^m$, is determined from a simple procedure described later in this chapter. We simply write

$$\frac{D^m \Phi(x,y;q)}{Dq^m} = \Phi^{[m]}(x,y;q) + \mathcal{R}_m[\Phi(x,y;q), \Lambda(q)], \tag{18.36}$$

where $\mathcal{R}_m$ is a nonlinear operator and $\Phi^{[m]}(x,y;q)$ is defined by (18.30). Note that for functions independent of $y = \eta(x;q)$, such as $\Lambda(q)$ and $\eta(x;q)$, we have

$$\frac{D^m \eta(x;q)}{Dq^m} = \frac{\partial^m \eta(x;q)}{\partial q^m} = \eta^{[m]}(x;q), \tag{18.37}$$

$$\frac{D^m \Lambda(q)}{Dq^m} = \frac{d^m \Lambda(q)}{dq^m} = \Lambda^{[m]}(q), \tag{18.38}$$

which are consistent with (18.31) and (18.32), respectively.

Thereafter, differentiating Equations (18.16) and (18.17) m times with respect to q and setting $q = 0$, we have the respective free surface boundary conditions defined at $y = \zeta_0(x)$ as

$$\sum_{i=0}^{m} \binom{m}{i} \frac{D^i [\Lambda^2(q)]}{Dq^i}\bigg|_{q=0} \frac{D^{m-i} \Phi_{xx}(x,y;q)}{Dq^{m-i}}\bigg|_{q=0}$$
$$+ g \frac{D^m \Phi_y(x,y;q)}{Dq^m}\bigg|_{q=0}$$
$$= m \; \chi_m \frac{D^{m-1} \mathcal{L}[\Phi(x,y;q), \Lambda(q)]}{Dq^{m-1}}\bigg|_{q=0}$$
$$+ m \; \hbar_1 \; H_1(x) \frac{D^{m-1} \mathcal{N}[\Phi(x,y;q), \Lambda(q)]}{Dq^{m-1}}\bigg|_{q=0} \tag{18.39}$$

Nonlinear Progressive Waves in Deep Water

and
$$\zeta_0^{[m]}(x) = m\, W_m(x, \vec{\zeta}_{m-1}, \vec{C}_{m-1}), \tag{18.40}$$

where χ_m is defined by (2.42) and

$$W_m(x, \vec{\zeta}_{m-1}, \vec{C}_{m-1}) = \chi_m\, \zeta_0^{[m-1]}(x)$$
$$+ \hbar_2\, H_2(x) \left[\zeta_0^{[m-1]}(x) - \left. \frac{D^{m-1} \mathcal{Z}[\Phi(x,y;q), \Lambda(q)]}{Dq^{m-1}} \right|_{q=0} \right]. \tag{18.41}$$

Substituting Equation (18.36) into (18.39), at $y = \zeta_0(x)$ we have

$$C_0^2 \frac{\partial^2 \phi_0^{[m]}(x,y)}{\partial x^2} + g \frac{\partial \phi_0^{[m]}(x,y)}{\partial y} = S_m(x, \vec{\phi}_{m-1}, \vec{\zeta}_m, \vec{C}_m), \tag{18.42}$$

where

$$S_m(x, \vec{\phi}_{m-1}, \vec{\zeta}_m, \vec{C}_m)$$
$$= \left\{ m\, \chi_m\, \frac{D^{m-1} \mathcal{L}[\Phi(x,y;q), \Lambda(q)]}{Dq^{m-1}} \right.$$
$$+ m\, \hbar_1\, H_1(x)\, \frac{D^{m-1} \mathcal{N}[\Phi(x,y;q), \Lambda(q)]}{Dq^{m-1}}$$
$$- C_0^2\, \mathcal{R}_m[\Phi_{xx}(x,y;q), \Lambda(q)] - g\, \mathcal{R}_m[\Phi_y(x,y;q), \Lambda(q)]$$
$$\left. - \sum_{i=1}^{m} \binom{m}{i} \frac{D^i[\Lambda^2(q)]}{Dq^i} \frac{D^{m-i}[\Phi_{xx}(x,y;q)]}{Dq^{m-i}} \right\}\Bigg|_{q=0}. \tag{18.43}$$

Note that the resulting boundary conditions (18.40) and (18.42) are satisfied on the initial approximation of the surface elevation $\zeta_0(x)$ and the reason for choosing $\zeta_0(x) = 0$ is now evident.

The boundary-value problem at the mth-order approximation is defined by the governing equation (18.33) and the boundary conditions (18.34), (18.40), and (18.42). It is clear that the term

$$W_m(x, \vec{\zeta}_{m-1}, \vec{C}_{m-1})$$

is only dependent upon results up to the $(m-1)$th approximation. Thus, $\zeta_0^{[m]}(x)$ can be directly calculated from Equation (18.40). Thereafter, there exist two unknowns: $\phi_0^{[m]}(x,y)$ and $C_0^{[m]}$. However, we have only one governing equation (18.33) with the boundary conditions (18.34) and (18.42) for $\phi_0^{[m]}(x,y)$. So, the problem is not closed and an additional algebraic equation is needed to determine $C_0^{[m]}$.

Under the *rules of solution expression* denoted by (18.6) and (18.8) and from Equations (18.40) and (18.42), the auxiliary functions $H_1(x)$ and $H_2(x)$ may appear as

$$H_1(x) = \cos(n_1 kx), \quad H_2(x) = \cos(n_2 kx),$$

where n_1, n_2 are integers. For simplicity, we choose

$$n_1 = n_2 = 0,$$

corresponding to

$$H_1(x) = H_2(x) = 1. \tag{18.44}$$

Then, under the *rules of solution expression* denoted by (18.6) and (18.8), the term $S_m(x, \vec{\phi}_{m-1}, \vec{\zeta}_m, \vec{C}_m)$ can be expressed by

$$S_m(x, \vec{\phi}_{m-1}, \vec{\zeta}_m, \vec{C}_m) = \sum_{n=1}^{m} b_{m,n}(\vec{C}_m) \sin(nkx) \quad \text{for } m \geq 1, \tag{18.45}$$

where $b_{m,n}(\vec{C}_m)$ is a coefficient dependent of the vector $\vec{C}_m$. Obviously, when $b_{m,1}(\vec{C}_m) \neq 0$, due to Equation (18.42), the solution $\phi_0^{[m]}(x,y)$ of the high-order deformation equations contains the secular terms, which do not conform to the *rule of solution expression* denoted by (18.6). To avoid this, we must enforce

$$b_{m,1}(\vec{C}_m) = 0 \quad \text{for } m \geq 1, \tag{18.46}$$

which provides us with one additional algebraic equation in the form

$$\alpha_m(\vec{C}_{m-1}) C_0^{[m]} + \beta_m(\vec{C}_{m-1}) = 0,$$

where $\alpha_m(\vec{C}_{m-1})$ and $\beta_m(\vec{C}_{m-1})$ are coefficients. Using this equation, $C_0^{[m]}$ is obtained. In this way, the problem is closed and the *rule of solution existence* is satisfied.

Thereafter, it is easy to obtain the solution

$$\phi_0^{[m]}(x,y) = \sum_{n=1}^{m} a_{m,n} \exp(nky) \sin(nkx), \tag{18.47}$$

where

$$a_{m,n} = \frac{b_{m,n}(\vec{C}_m)}{(kn)g - C_0^2(kn)^2} \quad \text{for } 2 \leq n \leq m. \tag{18.48}$$

Note that the coefficient $a_{m,1}$ is still unknown. To relate the solution and the wave height H, we use

$$\zeta_0^{[m]}(0) - \zeta_0^{[m]}(L/2) = \begin{cases} H & \text{for } m = 1 \\ 0 & \text{for } m \geq 2. \end{cases} \tag{18.49}$$

Nonlinear Progressive Waves in Deep Water

This relationship provides a linear algebraic equation in the form

$$\gamma_m \, a_{m,1} + \delta_m = 0,$$

where γ_m and δ_m are coefficients, from which the solution of $a_{m,1}$ can be evaluated. The value of A in the initial approximation $\phi_0(x,y)$ given by (18.9) is determined from (18.40) and (18.49) as

$$A = -\left(\frac{g\,H}{2\,\hbar_2\,k\,C_0^2}\right). \quad (18.50)$$

The nonlinear water-wave problem is now reduced to the two linear algebraic equations for $C_0^{[m]}$ and $a_{m,1}$. The solutions of the two equations complete the expression for $\phi_0^{[m]}(x,y)$ as well as the mth-order approximation of the solution. The formulation can be easily adapted for symbolic computation. In this way we obtain the high-order approximations $\zeta_0^{[m]}(x), C_0^{[m]}$, and $\phi_0^{[m]}(x,y)$, successively, in the order $m = 1, 2, 3, \cdots$.

The operator D^m/Dq^m for $m \geq 1$ can be determined by following the procedure outlined here. The potential $\Phi(x,y;q)$ on the free surface at $y = \eta(x;q)$ can be expanded about $q = 0$ by a Taylor series to give

$$\Phi(x,y;q) = \sum_{m=0}^{+\infty} \left.\frac{D^m\Phi(x,y;q)}{Dq^m}\right|_{q=0} \left(\frac{q^m}{m!}\right). \quad (18.51)$$

Similarly, this can be expanded by a Taylor series about the free surface at $y = \eta(x;0)$ as

$$\Phi(x,y;q) = \sum_{n=0}^{+\infty}\sum_{r=0}^{+\infty} \left.\frac{\partial^n \Phi^{[r]}(x,y;q)}{\partial y^n}\right|_{q=0} \left(\frac{q^r}{n!\,r!}\right) [\eta(x;q) - \eta(x,0)]^n. \quad (18.52)$$

Equating the two expressions for $\Phi(x,y;q)$ and invoking (18.19) and (18.22), we obtain

$$\sum_{m=0}^{+\infty} \left.\frac{D^m \Phi(x,y;q)}{Dq^m}\right|_{q=0} \left(\frac{q^m}{m!}\right)$$
$$= \sum_{n=0}^{+\infty}\sum_{r=0}^{+\infty} \left.\frac{\partial^n \Phi^{[r]}(x,y;q)}{\partial y^n}\right|_{q=0} \left(\frac{q^r}{n!\,r!}\right) \left[\sum_{s=1}^{+\infty}\left(\frac{q^s}{s!}\right)\zeta_0^{[s]}(x)\right]^n. \quad (18.53)$$

Expanding the right-hand side of the above equation and comparing the coefficients of the same power of q give the definition of the operator D^m/Dq^m for $m \geq 1$. This can be accomplished by symbolic computation. For details, the reader is referred to Liao and Cheung [50].

18.2 Result analysis

There exist two auxiliary parameters: $\hbar_1$ and $\hbar_2$. We have therefore a two-parameter family of solution expressions. For simplicity, we set

$$\hbar_1 = \hbar_2 = \hbar.$$

Physically, the phase speed C, velocity potential $\phi(x,y)$, and surface elevation $\zeta(x)$ are dependent upon the wave steepness. All of them are mathematically dependent on $\hbar$, which influences the convergence rate and region of the solution series (18.27), (18.28), and (18.29). In practice, a finite number of terms are used in the solution series. The Mth-order approximation of (18.27), (18.28), and (18.29) becomes

$$\phi(x,y) \approx \phi_0(x,y) + \sum_{m=1}^{M} \frac{\phi_0^{[m]}(x,y)}{m!}, \qquad (18.54)$$

$$\zeta(x) \approx \zeta_0(x) + \sum_{m=1}^{M} \frac{\zeta_0^{[m]}(x)}{m!}, \qquad (18.55)$$

$$C \approx C_0 + \sum_{m=1}^{M} \frac{C_0^{[m]}}{m!}. \qquad (18.56)$$

Most researchers focus their attention on the dispersion relationship between the phase speed C and the wave height H. Schwartz [128] formulated the Mth-order approximation of the phase speed as

$$\left(\frac{C}{C_0}\right)^2 \approx \sum_{j=0}^{M} a_j (kH)^{2j}, \qquad (18.57)$$

where a_j is a coefficient. Schwartz applied the Padé technique to improve the convergence and obtained the solution with the maximum wave steepness $(H/L)_{max} = 0.14118$. In our approach, the Mth-order approximation of the phase speed is

$$\frac{C}{C_0} \approx \sum_{j=0}^{M} b_j (kH)^{2j}, \qquad (18.58)$$

where b_j is a coefficient. In general, for a given value of kH, the influence of $\hbar$ on the convergence of the above series can be investigated by plotting the so-called $\hbar$-curves (see page 26 and §3.5.1) of C/C_0. As long as the series of the phase speed is convergent, the corresponding series of the velocity potential $\phi(x,y)$ and wave elevation $\zeta(x)$ also converge.

The accuracy and convergency of the phase speed can be enhanced by the homotopy-Padé technique (see page 38 and §3.5.2). It is found that the $[\kappa, \kappa]$ homotopy-Padé approximant of the phase speed is expressed by

$$\frac{C}{C_0} \approx \frac{1 + \sum_{n=1}^{\kappa(\kappa+1)/2} \Gamma_{2\kappa,n} (kH)^{2n}}{1 + \sum_{n=1}^{\kappa(\kappa+1)/2} \Delta_{2\kappa,n} (kH)^{2n}}, \qquad (18.59)$$

where $\Gamma_{2\kappa,j}$ and $\Delta_{2\kappa,j}$ are coefficients independent of $\hbar$. Note that the $[\kappa, \kappa]$ homotopy-Padé expression (18.59) is to $O(H^{2\kappa^2+2\kappa})$, which is considerably higher than $O(H^{2\kappa})$ achieved by the $[\kappa, \kappa]$ Padé expansion used by Schwartz [128].

Table 1 lists the dimensionless phase speed, C^2/C_0^2, computed at various levels of the homotopy-Padé approximation (18.59) and from Schwartz's perturbation solution to $O(H^{116})$ [128]. For wave steepness up to $H/L = 0.10$, the homotopy-Padé approximation converges at [6,6] and yields results identical to Schwartz's for the number of decimals considered. The computed dimensionless phase speed at this level of approximation is to $O(H^{82})$, which is lower than that considered by Schwartz. At the 20th-order approximation of the solution series, C^2 given by the [10,10] homotopy-Padé approximation is to $O(H^{220})$ and converges to slightly different results in comparison to Schwartz's for wave steepness $H/L > 0.12$. The homotopy-Padé approximation converges rapidly with the number of terms and the 20th- and 22nd-order approximations of the solution give identical or similar results over the range of wave steepness considered, indicating reasonable convergence at the 20th order and the validity of the proposed homotopy-Padé technique.

The [10,10] and [11,11] homotopy-Padé approximations of C/C_0 are compared with Longuet-Higgins' perturbation solution [129] in Table 2. The two homotopy-Padé approximations and Longuet-Higgins' results are identical for wave steepness up to $H/L = 0.121921$, whereas the [10,10] Homotopy-Padé approximation remains convergent up to $H/L = 0.137249$ for the number of decimals considered. The phase speeds computed by the various methods as the wave steepness approaches the limiting condition are compared in Figure 18.1. Both the present and Longuet-Higgins' approaches gives the maximum phase speed at the same wave steepness $H/L = 0.138712$ and demonstrate that phase speed is not a monotonic function of wave steepness. The homotopy-Padé approximations agree with Longuet-Higgins' results up to $H/L = 0.14$, but show more rapid decrease of the phase speed toward the limiting wave condition beyond that. The phase speed given by Schwartz [128] at $H/L = 0.14$ is slightly lower in comparison to the other predictions.

As observed in the previous and present studies, the physics of steep gravity waves is complicated and different approaches produce different solutions toward the limiting wave condition. The limiting wave is physically unstable and might be mathematically as well. It would be interesting to employ our approach to investigate the bifurcations of gravity waves for $H/L \approx 0.13$, found numerically by Chen and Saffman [132].

TABLE 18.1
Comparison of the $[\kappa, \kappa]$ homotopy-Padé approximation of C^2/C_0^2 with results given by Schwartz [128].

H/L	Schwartz's result	$\kappa = 6$	$\kappa = 8$	$\kappa = 10$	$\kappa = 11$
0.040	1.01592	1.01592	1.01592	1.01592	1.01592
0.070	1.04955	1.04955	1.04955	1.04955	1.04955
0.100	1.10367	1.10367	1.10367	1.10367	1.10367
0.120	1.15182	1.15190	1.15184	1.15182	1.15181
0.130	1.17820	1.17865	1.17834	1.17821	1.17821
0.135	1.18996	1.19148	1.19061	1.19003	1.19003
0.140	1.1930	1.20150	1.19833	1.19369	1.19385

Source: Kluwer Academic Publishers, *Journal of Engineering Mathematics*, vol. 45, No. 2, 2003, pp. 105-116, "Homotopy analysis of nonlinear progressive waves in deep water", Liao and Cheung, Table 1, Kluwer Academic Publishers Copyright ©2003 Kluwer Academic Publishers, with kind permission of Kluwer Academic Publishers.

TABLE 18.2
Comparison of the $[\kappa,\kappa]$ homotopy-Padé approximation of C/C_0 with result given by Longuet-Higgins [129].

H/L	Longuet-Higgins' result	$\kappa = 10$	$\kappa = 11$
0	1.00000	1.00000	1.00000
0.045266	1.01016	1.01016	1.01016
0.064351	1.02065	1.02065	1.02065
0.079187	1.03143	1.03143	1.03143
0.091809	1.04247	1.04247	1.04247
0.102959	1.05366	1.05366	1.05366
0.108093	1.05926	1.05926	1.05926
0.112962	1.06482	1.06482	1.06482
0.117572	1.07029	1.07029	1.07029
0.121921	1.07558	1.07558	1.07558
0.125993	1.08059	1.08060	1.08060
0.129760	1.08516	1.08517	1.08517
0.133178	1.08904	1.08906	1.08906
0.136178	1.09184	1.09188	1.09188
0.136723	1.09222	1.09228	1.09228
0.137249	1.09255	1.09260	1.09260
0.137755	1.09275	1.09284	1.09285
0.138242	1.09290	1.09300	1.09301
0.138712	1.09295	1.09306	1.09308
0.139170	1.09291	1.09302	1.09305
0.139610	1.09279	1.09285	1.09290
0.140060	1.09258	1.09250	1.09258
0.140530	1.09240	1.09189	1.09202
0.141100	1.09230	1.09066	1.09089

Source: Kluwer Academic Publishers, *Journal of Engineering Mathematics*, vol. 45, No. 2, 2003, pp. 105-116, "Homotopy analysis of nonlinear progressive waves in deep water", Liao and Cheung, Table 2, Kluwer Academic Publishers Copyright ©2003 Kluwer Academic Publishers, with kind permission of Kluwer Academic Publishers.

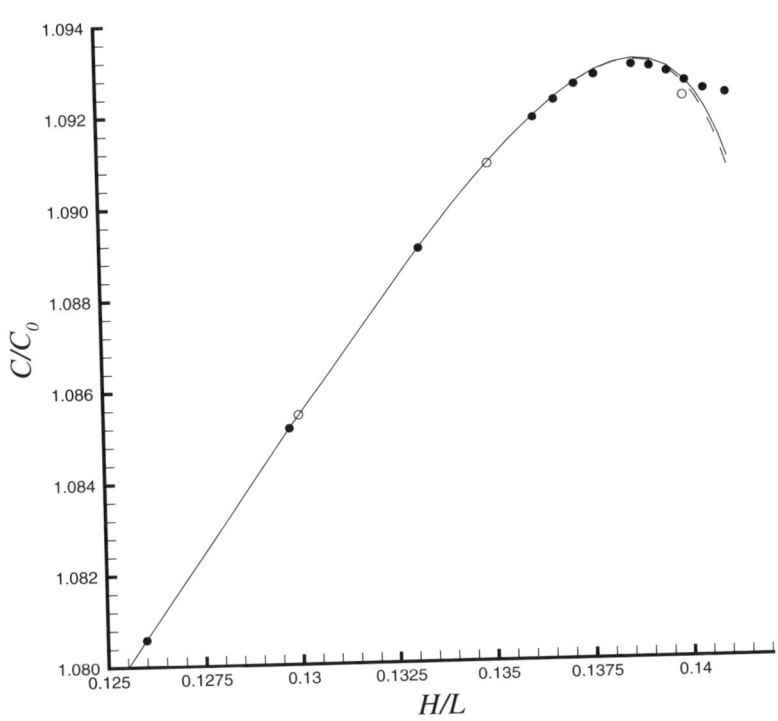

FIGURE 18.1
Phase speed C/C_0 versus wave steepness H/L for nonlinear progressive waves in deep water. Dashed line: [10,10] homotopy-Padé approximation; solid line: [11,11] homotopy-Padé approximation; open circle: Schwartz' results [128]; filled circle: Longuet-Higgins' results [129]. (From Kluwer Academic Publishers, *Journal of Engineering Mathematics*, vol. 45, No. 2, 2003, pp. 105-116, "Homotopy analysis of nonlinear progressive waves in deep water", Liao and Cheung, Figure 1, Kluwer Academic Publishers Copyright ©2003 Kluwer Academic Publishers, with kind permission of Kluwer Academic Publishers.)

Bibliography

[1] Cole, J.D. *Perturbation Methods in Applied Mathematics.* Blaisdell Publishing Company, Waltham, Massachusetts, 1968.

[2] Von Dyke, M. *Perturbation Methods in Fluid Mechanics.* The Parabolic Press, Stanford, California, 1975.

[3] Nayfeh, A.H. *Introduction to Perturbation Techniques.* John Wiley & Sons, New York, 1981.

[4] Nayfeh, A.H. *Problems in Perturbation.* John Wiley & Sons, New York, 1985.

[5] Grasman, J. *Asymptotic Methods for Relaxation Oscillations and Applications*, volume 63 of *Applied Mathematical Sciences.* Springer-Verlag, New York, 1987.

[6] Lagerstrom, P.A. *Matched Asymptotic Expansions: Ideas and Techniques*, volume 76 of *Applied Mathematical Sciences.* Springer-Verlag, New York, 1988.

[7] Hinch, E.J. *Perturbation Methods.* Cambridge Texts in Applied Mathematics. Cambridge University Press, Cambridge, 1991.

[8] Murdock, J.A. *Perturbations: Theory and Methods.* John Wiley & Sons, New York, 1991.

[9] Bush, A.W. *Perturbation Methods For Engineers and Scientists.* CRC Press Library of Engineering Mathematics. CRC Press, Boca Raton, Florida, 1992.

[10] Kevorkian, J. and Cole, J.D. *Multiple Scales and Singular Perturbation Methods*, volume 114 of *Applied Mathematical Sciences.* Springer-Verlag, New York, 1995.

[11] Kahn, P.B. and Zarmi, Y. *Nonlinear Dynamics: Exploration through Normal Forms.* John Wiley & Sons., Inc., New York, 1998.

[12] Nayfeh, A.H. *Perturbation Methods.* John Wiley & Sons, New York, 2000.

[13] Dai, S.Q. et al. Top 10 progress of theoretical and applied mechanics in the 20th century. *Advances in Mechanics*, 31(3):322–326, 2001. (in Chinese).

[14] Stokes, G.G. *Cambridge Philos.*, 9:8, 1851.

[15] Oseen, C.W. Über die Stokessche Formel und die verwandte Aufgabe in der Hydrodynamik. *Arkiv. Mat. Astron. Phsik.*, 6, 1910.

[16] Goldstein, S. Concerning some solutions of the boundary layer equations in hydrodynamics. *Proc. Cambridge Philos. Soc.*, 26, 1929.

[17] Whitehead, N. *Quart. J. Math.*, 23:143, 1889.

[18] Proudman and Pearson, J.R. Expansion at small Reynolds number for the flow past a sphere and a circular cylinder. *J. of Fluid Mechanics*, 2:237–262, 1957.

[19] Chester, W. and Breach, D.R. On the flow past a sphere at low Reynolds number. *J. Fluid Mechanics*, 37:751–760, 1969.

[20] White, F.M. *Viscous Fluid Flow.* McGraw-Hill, New York, 1991.

[21] Lyapunov, A.M. (1892). *General problem on stability of motion.* Taylor & Francis, London, 1992. (English translation).

[22] Karmishin, A.V., Zhukov, A.T., and Kolosov, V.G. *Methods of dynamics calculation and testing for thin-walled structures.* Mashinostroyenie, Moscow, 1990. (in Russian).

[23] Adomian, G. Nonlinear stochastic differential equations. *J. Math. Anal. and Applic.*, 55:441–452, 1976.

[24] Adomian, G. and Adomian, G.E. A global method for solution of complex systems. *Math. Model.*, 5:521–568, 1984.

[25] Adomian, G. *Solving Frontier Problems of Physics: The Decomposition Method.* Kluwer Academic Publishers, Boston and London, 1994.

[26] Liao, S.J. *The proposed homotopy analysis technique for the solution of nonlinear problems.* PhD thesis, Shanghai Jiao Tong University, 1992.

[27] Liao, S.J. A kind of linearity-invariance under homotopy and some simple applications of it in mechanics. Technical Report 520, Institute of Shipbuilding, University of Hamburg, Jan. 1992.

[28] Liao, S.J. A kind of approximate solution technique which does not depend upon small parameters (II): an application in fluid mechanics. *Int. J. of Non-Linear Mech.*, 32:815–822, 1997. (reprinted with permission from Elsevier).

[29] Liao, S.J. An explicit, totally analytic approximation of Blasius viscous flow problems. *Int. J. of Non-Linear Mech.*, 34(4):759–778, 1999. (reprinted with permission from Elsevier).

[30] Liao, S.J. A new analytic algorithm of Lane-Emden equation. *Applied Mathematics and Computation*, 142(1):1–16, 2003.

BIBLIOGRAPHY

[31] Hilton, P.J. *An introduction to homotopy theory.* Cambridge University Press, 1953.

[32] Sen, S. *Topology and geometry for physicists.* Academic Press, Florida, 1983.

[33] Grigolyuk, E.I. and Shalashilin, V.I. *Problems of Nonlinear Deformation: The Continuation Method Applied to Nonlinear Problems in Solid Mechanics.* Kluwer Academic Publishers, Dordrecht, Hardbound, 1991.

[34] Alexander, J.C. and Yorke, J.A. The homotopy continuation method: numerically implementable topological procedures. *Trans Am Math Soc*, 242:271–284, 1978.

[35] Liao, S.J. A second-order approximate analytical solution of a simple pendulum by the process analysis method. *J. Appl. Mech.*, 59:970–975, 1992.

[36] Liao, S.J. A kind of approximate solution technique which does not depend upon small parameters: a special example. *Int. J. of Non-Linear Mech.*, 30:371–380, 1995.

[37] Liao, S.J. and Chwang, A.T. Application of homotopy analysis method in nonlinear oscillations. *ASME J. of Appl. Mech.*, 65:914–922, 1998.

[38] Liao, S.J. An analytic approximate technique for free oscillations of positively damped systems with algebraically decaying amplitude. *Int. J. Non-Linear Mech.*, 38(8):1173–1183, 2003.

[39] Liao, S.J. An analytic approximate approach for free oscillations of self-excited systems. *Int. J. Non-Linear Mech.*, 39(2):271–280, 2004.

[40] Liao, S.J. A uniformly valid analytic solution of 2D viscous flow past a semi-infinite flat plate. *J. of Fluid Mech.*, 385:101–128, 1999. Cambridge University Press Copyright ©1999 Cambridge University Press, reprinted with permission.

[41] Liao, S.J. An analytic approximation of the drag coefficient for the viscous flow past a sphere. *Int. J. of Non-Linear Mech.*, 37:1–18, 2002. (reprinted with permission from Elsevier).

[42] Liao, S.J. On the homotopy analysis method for nonlinear problems. *Applied Mathematics and Computation.* (in press).

[43] Yang, C. and Liao, S.J. On the explicit, purely analytic solution of Von Kármán swirling viscous flow. Submitted.

[44] Liao, S.J. and Campo, A. Analytic solutions of the temperature distribution in Blasius viscous flow problems. *J. of Fluid Mech.*, 453:411–425, 2002.

[45] Wang, C. et al. On the explicit analytic solution of cheng-chang equation. *Int. J. Heat and Mass Transfer*, 46(10):1855–1860, 2003.

[46] Ayub, M., Rasheed, A., and Hayat, T. Exact flow of a third grade fluid past a porous plate using homotopy analysis method. *Int. J. Engineering Science*, 41:2091–2109, 2003.

[47] Hayat, T., Masood Khan, and Ayub, M. On the explicit analytic solutions of an Oldroyd 6-constant fluid. *Int. J. Engineering Science.* (in press).

[48] Liao, S.J. On the analytic solution of magnetohydrodynamic flows of non-Newtonian fluids over a stretching sheet. *J. of Fluid Mech.*, 488:189–212, 2003.

[49] Liao, S.J. Application of Process Analysis Method to the solution of 2D nonlinear progressive gravity waves. *J. of Ship Res.*, 36(1):30–37, 1992.

[50] Liao, S.J. and Cheung, K.F. Homotopy analysis of nonlinear progressive waves in deep water. *Journal of Engineering Mathematics*, 45(2):105–116, 2003. (with kind permission of Kluwer Academic Publishers).

[51] Liao, S.J. An explicit analytic solution to the Thomas-Fermi equation. *Applied Mathematics and Computation*, 144:495–506, 2003.

[52] Lindstedt, A. Über die Integration einer für die Störungstheorie wichtigen Differentialgleichung. *Astron, Nach.*, 103:211–220, 1882.

[53] Bohlin, K.P. Über eine neue Annäherungsmethode in der Störungstheorie . *Akad Handl. Bihang.*, 14, 1889.

[54] Poincaré, H. (1892). *New Methods of Celestial Mechanics*, volume I-III. NASA TTF-450, 1967. (English translation).

[55] Glydén, H. Nouvelles recerches sur les séries employées dans les théories des planéts. *Acta Math.*, 9:1–168, 1893.

[56] Lighthill, M.J. A technique for rendering approximate solutions to physical problems unformly valid. *Phil. Mag.*, 40:1179–1201, 1949.

[57] Lighthill, H.J. A technique for rendering approximate solutions to physical problems uniformly valid. *Z. Flugwiss.*, 1:267–275, 1961.

[58] Malkin, T.G. *Methods of Poincaré and Lyapunov in theory of non-linear oscillations.* Moscow, 1949. (in Russian).

[59] Kuo, Y.H. On the flow of an incompressible viscous fluid past a flat plate at moderate Reynolds numbers. *J. Math and Phys.*, 32:83–101, 1953.

[60] Kuo, Y.H. Viscous flow along a flat plate moving at high supersonic speeds. *J. Aeron. Sci.*, 23:125–136, 1956.

BIBLIOGRAPHY

[61] Tsien, H.S. The Poincaré-Lighthill-Kuo method. *Advan. Appl. Mech.*, 4:281–349, 1956.

[62] Newton, I. *On the Binomial Theorem For Fractional and Negative Exponents (A Source Book in Mathematics, Edited by Walcott, G. D.)*. McGraw Hill Book Company, New York and London, 1929.

[63] Rach, R. On the Adomian method and comparisons with Picard's method. *J. Math. Anal. and Applic.*, 10:139–159, 1984.

[64] Rach, R. A convenient computational form for the A_n polynomials. *J. Math. Anal. and Applic.*, 102:415–419, 1984.

[65] Adomian, G. and Rach, R. On the solution of algebraic equations by the decomposition method. *Math. Anal. Appl.*, 105(1):141–166, 1985.

[66] Cherruault, Y. Convergence of Adomian's method. *Kyberneters*, 8(2):31–38, 1988.

[67] Adomian, G. A review of the decomposition method and some recent results for nonlinear equations. *Comp. and Math. with Applic.*, 21:101–127, 1991.

[68] Abbaoui, K. and Cherruault, Y. Convergence of Adomian's method applied to nonlinear equations. *Math. Compact. Model.*, 20(9):69–73, 1994.

[69] Khuri, S.A. A new approach to the cubic Schrodinger equation: An application of the decomposition technique. *Applied Mathematics and Computation*, 97:251–254, 1998.

[70] Michael, G. A note on the decomposition method for operator equations. *Applied Mathematics and Computation*, 106:215–220, 1999.

[71] Wazwaz, A.M. The decomposition method applied to systems of partial differential equations and to the reaction-diffusion Brusselator model. *Applied Mathematica and Computation*, 110:251–264, 2000.

[72] Wazwaz, A.M. A new algorithm for solving differential equations of Lane-Emden type. *Applied Mathematics and Computation*, 118:287–310, 2001.

[73] Babolian, E. and Biazar, J. Solving the problem of biological species living together by Adomian decomposition method. *Applied Mathematica and Computation*, 129:339–343, 2001.

[74] Ramos, J.I. and Soler, E. Domain decomposition techniques for reaction-diffusion equations in two-dimensional regions with re-entrant corners. *Applied Mathematics and Computation*, 118:189–221, 2001.

[75] Babolian, E. and Biazar, J. On the order of convergence of Adomian method. *Applied Mathematics and Compututation*, 130:383–387, 2002.

[76] Babolian, E. and Biazar, J. Solution of nonlinear equations by modified Adomian decomposition method. *Applied Mathematics and Computation*, 132:167–172, 2002.

[77] Shawagfeh, N.T. Analytical approximate solutions for nonlinear fractional differential equations. *Applied Mathematica and Computation*, 131:517–529, 2002.

[78] Casasús L. and Al-Hayani, W. The decomposition method for ordinary differential equations with discontinuities. *Applied Mathematics and Computation*, 131:245–251, 2002.

[79] Wazwaz, A.M. Exact solutions for variable coefficients fourth-order parabolic partial differential equations in higher-dimensional spaces. *Applied Mathematics and Computation*, 130:415–424, 2002.

[80] Huang, R.E. et al. The empirical mode decomposition and the Hilbert spectrum for nonlinear and non-stationary time series analysis. *Proc. R. Soc. Lond. A*, 454:903–995, 1998.

[81] Fermi, E. Un metodo statistico par la determinzione di alcune Proprietá dell'atome. *Rend. Accad. Naz. del Lincei, Cl. Sci. Fis., Mat. e. Nat.*, 6:602–607, 1927.

[82] Thomas, L.H. The calculation of atomic fields. *Proc. Cambridge Philos. Soc.*, 23:542–548, 1927.

[83] Bush, V. and Caldwell, S.H. Thomas-Fermi equation solution by the differential analyzer. *Phys. Rev.*, 38:1898–1901, 1931.

[84] Burrows, B.L. and Core, P.W. A variational iterative approximate solution of the Thomas-Fermi equation. *Quart. Appl. Math.*, 42:73–76, 1984.

[85] Milton, K.A., Bender, C.M., and Pinsky, S.S. A new perturbation approach to nonlinear problems. *J. Math. Phys.*, 30(7):1447–1450, 1989.

[86] Laurenzi, B.J. An analytic solution to the Thomas-Fermi equation. *J. Math. Phys.*, 31(10):2535–2537, 1990.

[87] Cedillo, A. A perturbative approach of the Thomas-Fermi equation in terms of the density. *J. Math. Phys.*, 34:2713, 1993.

[88] Chan, C.Y. and Hon, Y.C. A constructive solution for a generalized Thomas-Fermi theory of ionized atoms. *Quart. Appl. Math.*, 45:591–599, 1987.

[89] Hon, Y.C. A decomposition method for the Themos-Fermi equation. *SEA Bull. Math.*, 20(3):55–58, 1996.

[90] Venkatarangan, S.N. and Rajalashmi, K. Modification of Adomian's decomposition method to solve equation containing radicals. *Comput. Math. Appl.*, 29(6):75–80, 1995.

BIBLIOGRAPHY

[91] Wazwaz, A.M. The modified decomposition method and the Padé approximants for solving Thomas-Fermi equation. *Mathematics and Computation*, 105:11–19, 1999.

[92] Luning, C.D. and Perry, W.L. An iterative technique for solution of the Thomas-Fermi equation utilizing a non-linear eigenvalue problem. *Quarterly of Applied Mathematics*, 35:257–268, 1977.

[93] Wu, M.S. Modified variational solution of the Thomas-Fermi equation for atoms. *Phys. Rev. A*, 26(1):57–61, 1982.

[94] Civan, F. and Sliepcevich, C.M. On the solution of the Thomas-Fermi equation by differential quadrature. *J. Comput. Phys.*, 56:343–348, 1984.

[95] Chan, C.Y. and Du, S.W. A constructive method for the Thomas-Fermi equation. *Quarterly of Applied Mathematics*, 44:303–307, 1986.

[96] Allan, M. Chebyshev series solution of the Thomas-Fermi equation. *Comp. Phys. Comm.*, 67:389–391, 1992.

[97] Pert, G.J. Approximations for the rapid evaluation of the Thomas-Fermi equation. *J. Phys. B*, 32(6):5067–5082, 1999.

[98] Kobayashi et al. Some coefficients of the TFD function. *J. Phys. Soc. Japan*, 10:759–765, 1955.

[99] Scudo, F.M. Vito Volerra and theoretical ecology. *Theoret. Population Biol.*, 2:1–23, 1971.

[100] Small, R.D. *Mathematical Modelling: Classroom Notes in Applied Mathematics*. SIAM, Philadelphia, PA, 1989.

[101] TeBeest, K.G. Numerical and analytical solutions of Volterra's population model. *SIAM Rev.*, 39(3):484–493, 1997.

[102] Wazwaz, A.M. Analytical approximations and Padé approximations for Volterra's population model. *Applied Mathematics and Computation*, 100:13–25, 1999.

[103] Kahn, P.B. and Zarmi, Y. *Nonlinear Dynamics: Exploration through Normal Forms*. John Wiley & Sons, Int., New York, 1997.

[104] Blasius, H. Grenzschichten in Füssigkeiten mit kleiner Reibung. *Z. Math. Phys.*, 56:1–37, 1908.

[105] Howarth, L. On the solution of the laminar boundary layer equations. *Proc. Roy. Soc. London A*, 164:547–579, 1938.

[106] Falkner, V.M. and Skan, S.W. Some approximate solutions of the boundary layer equations. *Phil. Mag.*, 12:865–896, 1931.

[107] Hartree, D.R. On an equation occuring in Falker-Skan's approximate treatment of the equations of the boundary layer. *Proc. Cambr. Phil. Soc.*, 33:223–239, 1937.

[108] Stewartson,K. Further solutions of the Falkner-Skan equation. *Proc. Camb. Phil. Soc.*, 50:454–465, 1954.

[109] Libby, P.A. and Liu, T.M. Further solutions of the Falkner-Skan equation. *AIAA Journal*, 5:1040–1042, 1967.

[110] Kuiken, H.K. On boundary layers in fluid mechanics that decay algebraically along stretches of wall that are not vanishingly small. *Q. J. Mech. Appl. Math.*, 27:387–405, 1981.

[111] Kuiken, H.K. A backward free-convective boundary layer. *IMA Journal of Applied Mathematics*, 34:397–413, 1981.

[112] Von Kármán, T. Über läminare und turbulence Reibung. *ZAMM*, 1:233–252, 1921.

[113] Zandbergen, P.J. and Dijkstra, D. Von Kármán swirling flows. *Ann. Rev. Fluid Mech.*, 19:465–491, 1987.

[114] Cochran, W.G. The flow due to a rotating disk. *Proc. Camb. Phil. Soc.*, 30:365–375, 1934.

[115] Hettis, H.E. On the integration of a class of differential equations occuring in boundary layer and other hydrodynamic problems. In *Proc. 4th Midwestern Conf. on Fluid Mech.*, Purdue, 1955.

[116] Rogers,M.H. and Lance, G.N. The rotationally symmetric flow of a viscous fluid in the presence of an infinite rotating disk. *J. Fluid Mech.*, 7:617–631, 1960.

[117] Benton, E.R. On the flow due to a rotating disk. *J. Fluid Mech.*, 24(4):781–800, 1966.

[118] McLeod, J.B. Von Kármán's swirling flow problem. *Arch. Ration. Mech. Anal.*, 33:91–102, 1969.

[119] Zandbergen, P.J. and Dijkstra, D. Non-unique solutions of the Navier-Stokes equations for the Kármán swirling flow. *Journal of Engineering Mathematics*, 11:167–188, 1977.

[120] Ackroyd, J.A.D. On the steady flow produced by a rotating disc either surface suction or injection. *J. Eng. Math.*, 12:207–220, 1978.

[121] Hulzen, V. Computational problems in producing Taylor coefficients for the rotating disk problem. *Sigsam Bull. ACM*, 14:36–49, 1980.

[122] Stokes, G.G. On the theory of oscillation waves. *Trans. Cambridge Phil. Soc.*, 8:441–455, 1849.

[123] Stokes, G.G. Supplement to a paper on the theory of oscillation waves. *Mathematical and Physical papers*, 1:314–326, 1880.

[124] Stokes, G.G. On the highest waves of uniform propagation. *Proc. of Cambridge Phil. Soc.*, 4:361–365, 1883.

[125] Korteweg, D.J. and de Vries, G. On the change of form of long waves advancing in a rectangular canal and on a new type of long stationary waves. *Phil. Mag.*, 39:422–443, 1895.

[126] Wehausen, J.V. and Laitone, E.V. *Handbush der Physik*, volume 9. Springer-Verlag, Berlin, 1960.

[127] Chiang, C.M. The nonlinear evolution of stokes waves in deep water. *J. of Fluid Mech.*, 47:337–352, 1971.

[128] Schwartz, L.W. Computer extension and analytic continuation of Stokes' expansion for gravity waves. *J. of Fluid Mech.*, 62(3):553–578, 1974.

[129] Longuet-Higgins, M.S. Integral properties of periodic gravity waves of finite amplitudes. *Proc. R. Soc. Lond.*, 342:157–174, 1975.

[130] Longuet-Higgins, M.S. The instability of gravity of finite amplitude in deep water (1): superharmonics. *Proc. R. Soc. Lond. A*, 360:471–488, 1978.

[131] Longuet-Higgins, M.S. The instability of gravity of finite amplitude in deep water (2): subharmonics. *Proc. R. Soc. Lond. A*, 360:489–505, 1978.

[132] Chen, B. and Saffman, P.G. Numerical evidence for the existence of new types of gravity waves of permanent form on deep water. *Studies in Appl. Math.*, 62:1–21, 1980.

[133] Cokelet, E.D. Steep gravity waves in water of arbitrary uniform depth. *Phil. Trans. R. Soc. Lond. A*, 286:183–230, 1977.

[134] Longuet-Higgins, M.S. and Cokelet, E.D. The deformation of steep surface waves on water (I): a numerical method of computation. *Proc. R. Soc. Lond. A*, 350:1–26, 1976.

[135] Longuet-Higgins, M.S. and Cokelet, E.D. The deformation of steep surface waves on water (II): growth of normal mode instabilities. *Proc. R. Soc. Lond. A*, 364:1–28, 1978.

[136] Longuet-Higgins, M.S. and Fox, M.J.H. Theory of the almost-highest wave (Part 2): Matching and analytic extension. *J. Fluid Mech.*, 85:769–786, 1978.

Index

δ-expansion method, 4, 12, 75
$\hbar$-curve, 26, 28, 63, 79, 91, 105, 122, 140, 157, 206, 228, 248, 265, 285, 304

Adomian polynomial, 70, 74
Adomian's decomposition method, 5, 11, 69
approach function of the first kind, 37
approach function of the second kind, 37
approaching paths, 37
artificial small parameter, 4
asymptotic expression, 134
asymptotic property, 134, 141
auxiliary function, 14, 53, 87, 100, 116, 135, 150, 166, 181, 199, 218, 241, 259, 278, 298
auxiliary linear operator, 14, 53, 87, 100, 116, 134, 150, 166, 180, 198, 218, 241, 259, 278, 297
auxiliary parameter, 14, 53, 87, 100, 116, 135, 150, 166, 181, 199, 218, 241, 259, 278, 298

bifurcation, 85, 91, 104
bifurcation diagram, 105, 106
Blasius' viscous flow, 217
bottom condition, 295

chaos, 81
conservative system, 165, 179
continuation method, 6
continuity equation, 275

convergence theorem, 18, 57, 89, 103, 120, 138, 155, 204, 220, 246, 263, 283

deep water, 295
deformation, 6, 54
deformation derivative, 15, 54, 182, 299
dispersion relationship, 304
Duffing oscillator, 85, 99, 105

eigenfunction, 115
eigenvalue, 115
embedding function, 56
embedding parameter, 6, 14, 53, 116, 135, 150, 181, 218, 224, 242, 278
empirical mode decomposition method, 81

Falkner-Skan boundary layer flow, 240
Falkner-Skan viscous flow, 250
free oscillation, 165, 179
free surface boundary condition, 295

glass-fiber production process, 257

high-order deformation equation, 16, 55, 76, 88, 101, 118, 136, 152, 167, 183, 201, 219, 225, 242, 261, 280, 299
homotopy, 6, 53
homotopy continuation method, 6
homotopy-Padé technique, 40, 64, 79, 105, 122, 157, 229, 266, 285, 305

incompressible boundary layer, 239
initial guess, 14, 53, 87, 100, 116, 134, 166, 180, 198, 218, 224, 241, 258, 277, 297
instrinic mode functions, 82
integro-differential equation, 149

kinetic energy, 165, 179

laminar viscous flow, 217, 239
Laplace equation, 295
limit cycle, 197
Lyapunov's artificial small parameter method, 4, 11, 73

monotonic function, 296
multidimensional dynamical system, 197
multiple eigenfunction, 123
multiple solution, 99, 104, 250

Navier-Stokes equation, 275, 287
nonperturbation techniques, 4
　δ-expansion method, 12, 75
　Adomian's decomposition method, 11, 69
　Lyapunov's artificial small parameter method, 11, 73
normalized eigenfunction, 115

odd nonlinearity, 165

Padé technique, 39, 64, 304
perturbation techniques, 3
phase speed, 295
population growth, 149
Prandtl number, 257
progressive gravity wave, 295

quadratic nonlinearity, 179

rule of coefficient ergodicity, 20, 62, 90, 104
rule of solution existence, 21, 62
rule of solution expression, 20, 60, 86, 116, 134, 150, 166, 180, 198, 218, 224, 241, 258, 277, 296

secular term, 30, 61
similar solution, 239
similarity transformation, 257, 275
similarity variable, 217, 257, 275
skin friction, 265
spatial-scale parameter, 224, 227, 228, 241, 247, 258, 277
stream function, 217

Taylor's theorem, 16, 54, 70, 87, 101, 135, 151, 167, 182, 200, 219, 242, 260, 279, 298
temperature distribution, 257
thermal flux, 265
Thomas-Fermi atom model, 133
Thomas-Fermi equation, 133
three-dimensional viscous flow, 287

valid region of $\hbar$, 26, 28, 63, 79, 91, 105, 122, 157, 206, 228, 285
velocity potential, 295, 296
velocity profile, 257
Volterra model, 149

water-wave, 295
wave amplitude, 295
wave elevation, 295, 296
wave height, 296
wave number, 296
wave propagation, 295
wave steepness, 296, 304
wavelength, 296

zero-order deformation equation, 14, 54, 70, 76, 87, 100, 116, 135, 150, 166, 181, 199, 218, 224, 241, 259, 278, 297